Lecture Notes in Statistics 185

Edited by P. Bickel, P. Diggle, S. Fienberg, U. Gather,
I. Olkin, S. Zeger

T0138143

Adrian Baddeley
Pablo Gregori
Jorge Mateu
Radu Stoica
Dietrich Stoyan
(Editors)

Case Studies in Spatial Point Process Modeling

With 107 Figures

 Springer

Adrian Baddeley
Department of Mathematics
University of Western
    Australia
Nedlands 6907 Australia
adrian@maths.uwa.edu.au

Pablo Gregori
Department of
    Mathematics
Universitat Jaume 1 of
    Castellon
Castellon 12071 Spain
gregori@mat.uji.es

Jorge Mateu
Department of
    Mathematics
Universitat Jaume 1 of
    Castellon
Castellon 12071 Spain
mateu@mat.uji.es

Radu Stoica
INRA - Biometrie,
    Domaine St. Paul,
Site Agroparc
84914 Avignon, Cedex 9,
    France
Radu.Stoica@avignon.inra.fr

Dietrich Stoyan
Institut für Stochastik
Prüferstraße 9
TU Bergakademie Freiberg
D-09596 Freiberg
Germany
stoyan@orion.hrz.tu-freiberg.de

Library of Congress Control Number: 2005931123

ISBN-10: 0-387-28311-0
ISBN-13: 978-0387-28311-1

Printed on acid-free paper.

Camera ready copy provided by the editors.

Printed in the United States of America.   (SBA)

9 8 7 6 5 4 3 2 1

springeronline.com

Preface

The week before Easter 2004 a conference on spatial point process modelling and its applications was held in Benicàssim (Castellón, Spain). The organizers targeted two aims. The first goal was to bring together most of the known people to guarantee the high scientific quality of the meeting to foster the theoretical and practical use of spatial point processes. The second one consisted of enabling young researchers to present their work and to obtain a valuable feed-back coming from the reknown specialists in the domain. The contributions of all the participants were published in the proceedings book of the conference.

The majority of the contributions in this book represents the reviewed version of the papers presented during the conference. In order to offer the reader a larger spectrum of this domain, authors that could not attend the conference were also invited to contribute.

The book is constituted by 16 chapters divided in three parts and gathering 44 authors coming from 13 different countries.

The first part of the volume – represented by its two first contributions – is dedicated to basic notions and tools for understanding and manipulating spatial point processes.

In the first contribution, D. Stoyan presents a general overview of the theoretical foundations for spatial point process. The author defines a point process and a marked point process, and describes the construction of the first and second order moment measures, which leads to the nowadays well known summary statistics such as the K-function, L-function or the pair-correlation function. The Poisson point process plays an important role, since in practice it is often used as null model for hypothesis testing and as reference model for the construction of realistic models for point patterns.

The second contribution, written by A.J. Baddeley and R. Turner, enters directly in the "flesh" of the problem presenting the concrete use of spatial point processes for modelling spatial point patterns, via the spatstat package – a software library for the R language. Four main points can be tackled by this package: basic manipulation of point patterns, exploratory data analysis, parametric model-fitting and simulation of spatial point processes. The very important issue of model validation is also addressed. The contribution contains also the necessary mathematical details and/or literature references in order to avoid the use of this software as a "black box". Two complete case studies are presented at the end of the contribution.

There is no serious practical application without a rigourous theoretical development. Therefore the second part of the book is more oriented towards theoretical and methodological advances in spatial point processes theory. Topics of this part of the book contain *analytical properties of the Poisson process* (presented in the contribution by S. Zuyev), *Bayesian analysis of Markov point processes* (by K. K. Berthelsen and J. Møller), *statistics for locally scaled point processes* (by M. Prokěsová, U. Hahn and E. B. Vedel Jensen), *nonparametric testing of distribution functions in germ-grain models* (by Z. Pawlas and L. Heinrich), and *principal component analysis applied to point processes through a simulation study* (by J. Illian, E. Benson, J. Crawford and H. Staines). Remarkable is the fact, that almost all these contributions show direct applications of the presented development.

The third part of this volume is entirely dedicated to concrete, precise case studies, that are solved within the point processes theory. The presented applications are of big impact: *material science* (by F. Ballani), *human epidemiology* (by M. A. Martínez-Beneito *et al.*) , *social sciences* (by N.A.C. Cressie, O. Perrin and C. Thomas-Agnan), *animal epidemiology* (by Webster *et al.* and P.J. Diggle, S. J. Eglen and J. B. Troy), *biology* (by F. Fleischer *et al.* and by A. Stein and N. Georgiadis), and *seismology* (by J. Zhuang, Y. Ogata and D. Vere-Jones and by A. Veen and F.P. Schoenberg). In their contributions, the authors show skill and cleverness in using, combining and continuously evolving the point processes tools in order to answer the proposed questions.

We hope the reader will enjoy reading the book and will find it instructive and inspiring for going a step further in this very open research field.

The Editors are grateful to all the authors that made possible finishing the book within an acceptable time scheduling. A word of thanks is given to Springer-Verlag and, in particular, to John Kimmel for creating the opportunity of making this project real.

Castellón (Spain)   *Adrian Baddeley*
May 2005   *Pablo Gregori*
                                                                Jorge Mateu
                                                                Radu Stoica
                                                              Dietrich Stoyan
                                                                    Editors

Contents

Part I Basic Notions and Manipulation of Spatial Point Processes

Fundamentals of Point Process Statistics
Dietrich Stoyan 3

Modelling Spatial Point Patterns in R
Adrian Baddeley, Rolf Turner 23

Part II Theoretical and Methodological Advances in Spatial Point
Processes

Strong Markov Property of Poisson Processes and Slivnyak
Formula

Bayesian Analysis of Markov Point Processes

Part III Practical Applications of Spatial Point Processes

On Modelling of Refractory Castables by Marked Gibbs and Gibbsian-like Processes

Source Detection in an Outbreak of Legionnaire's Disease

Doctors' Prescribing Patterns in the Midi-Pyrénées rRegion of France: Point-process Aggregation

**Assessing Spatial Point Process Models Using Weighted
K-functions: Analysis of California Earthquakes**

List of Contributors

1. **Juan J. Abellán** Small Area Health Statistics Unit, Imperial College of London, UK
2. **Adrian Baddeley** School of Mathematics and Statistics, University of Western Australia, Crawley 6009 WA, Australia
3. **Felix Ballani** Institut für Stochastik, Technische Universität Bergakademie Freiberg, Prüferstr. 9, 09596 Freiberg, Germany
4. **Michael Beil** Department of Internal Medicine I, University Hospital Ulm, D-89070 Ulm, Germany
5. **Erica Benson** SIMBIOS, University of Abertay Dundee, Scotland, UK
6. **Kasper K. Berthelsen** Aalborg University, Department of Mathematical Sciences, F. Bajers Vej 7G, DK-9220 Aalborg, Denmark
7. **Helen E. Clough** Department of Veterinary Clinical Sciences, University of Liverpool, UK
8. **John Crawford** SIMBIOS, University of Abertay Dundee, Scotland, UK
9. **Nocl A.C. Cressie** Department of Statistics, The Ohio State University, Columbus, OH 43210, USA
10. **Peter J. Diggle** Department of Statistics, Lancaster University, UK
11. **Stephen J. Eglen** University of Cambridge, UK
12. **José Fenollar** Centro de Salud Pública de Alcoi, Generalitat Valenciana, Spain
13. **Frank Fleischer** Department of Applied Information Processing and Department of Stochastics, University of Ulm, D-89069 Ulm, Germany
14. **Nigel P. French** Institute of Veterinary, Animal and Biomedical Sciences, Massey University, New Zealand.
15. **Nick Georgiadis** Mpala Research Centre, PO Box 555, Nanyuki, Kenya
16. **Robert B. Green** Department of Neuropathology, Veterinary Laboratories Agency, UK
17. **Ute Hahn** University of Augsburg, Department of Applied Stochastics, 86135 Augsburg, Germany
18. **Lothar Heinrich** University of Augsburg, Institute of Mathematics. Universitätsstr. 14, D-86135 Augsburg, Germany

19. **Janine Illian** SIMBIOS, University of Abertay Dundee, Scotland, UK
20. **Eva B. Vedel Jensen** University of Aarhus, The T.N. Thiele Centre of Applied Mathematics in Natural Science, Department of Mathematical Sciences, Ny Munkegade, 8600 Aarhus C, Denmark
21. **Guillermo Jorques** Centro de Salud Pública de Alcoi, Generalitat Valenciana, Spain
22. **Marian Kazda** Department of Systematic Botany and Ecology, University of Ulm, D-89069 Ulm, Germany
23. **Antonio López-Quílez** Departament d'Estadística i Investigació Operativa, Universitat de València, Spain
24. **Miguel A. Martínez-Beneito** Dirección General de Salud Pública, Generalitat Valenciana, Spain
25. **Jesper Møller** Aalborg University, Department of Mathematical Sciences, F. Bajers Vej 7G, DK-9220 Aalborg, Denmark
26. **Yosihiko Ogata** Institute of Statistical Mathematics, Japan
27. **Zbyněk Pawlas** Charles University, Faculty of Mathematics and Physics, Department of Probability and Mathematical Statistics. Sokolovská 83, 186 75 Praha 8, Czech Republic
28. **Olivier Perrin** GREMAQ – Université Toulouse 1 (and LERNA-INRA), 21 allée de Brienne, 31000 Toulouse, France
29. **Michaela Prokešová** Charles University, Department of Probability, Sokolovská 83, 18675 Praha 8, Czech Republic
30. **Volker Schmidt** Department of Stochastics, University of Ulm, D-89069 Ulm, Germany
31. **Frederic Paik Schoenberg** UCLA Department of Statistics, 8125, Math Sciences Building, Box 951554, Los Angeles, CA 90095-1554, USA
32. **Harry Staines** SIMBIOS, University of Abertay Dundee, Scotland, UK
33. **Alfred Stein** International Institute for Geo-Information Science and Earth Observation (ITC), PO Box 6, 7500 AA Enschede, The Netherlands
34. **Dietrich Stoyan** Institut für Stochastik, Technische Universität Bergakademie Freiberg, Agricolastr. 1, 09596 Freiberg, Germany
35. **Christine Thomas-Agnan** GREMAQ – Université Toulouse 1 (and LSP – Université Toulouse 3), 21 allée de Brienne, 31000 Toulouse, France
36. **John B. Troy** Northwestern University, USA
37. **Rolf Turner** Department of Mathematics and Statistics, University of New Brunswick, Fredericton, N.B., E3B 5A3 Canada
38. **Hermelinda Vanaclocha** Dirección General de Salud Pública, Generalitat Valenciana, Spain
39. **Alejandro Veen** UCLA Department of Statistics, 8125 Math Sciences Building, Box 951554, Los Angeles, CA 90095-1554, USA
40. **David Vere-Jones** Victoria University of Wellington, New Zealand
41. **Simon Webster** Department of Veterinary Clinical Sciences, University of Liverpool, UK
42. **Jiancang Zhuang** Institute of Statistical Mathematics, Japan

43. **Óscar Zurriaga** Dirección General de Salud Pública, Generalitat Valenciana, Spain
44. **Sergei Zuyev** Department of Statistics and Modelling Science, University of Strathclyde, Glasgow, G1 1XH, UK

Part I

Basic Notions and Manipulation of Spatial Point Processes

Fundamentals of Point Process Statistics

Dietrich Stoyan

Institut für Stochastik, Technische Universität Bergakademie Freiberg,
Agricolastr. 1, 09596 Freiberg, Germany, stoyan@orion.hrz.tu-freiberg.de

Summary. Point processes are mathematical models for irregular or random point patterns. A short introduction to the theory of point processes and their statistics, emphasizing connections between the presented theory and the use done by several authors and contributions appearing in this book is presented.

Key words: Marked point processes, Second-order characteristics, Spatial point processes overview, Statistical inference

1 Basic Notions of the Theory of Spatial Point Processes

The following text is a short introduction to the theory of point processes and their statistics, written mainly in order to make this book self-contained. For more information the reader is referred to the text [2] and the books [7, 16, 20, 21].

Point processes are mathematical models for irregular or random point patterns. The mathematical definition of a point process on $\mathbb{R}^d$ is as a random variable N taking values in the measurable space $[\mathbb{N}, \mathcal{N}]$, where $\mathbb{N}$ is the family of all sequences $\{x_n\}$ of points of $\mathbb{R}^d$ satisfying the local finiteness condition, which means that each bounded subset of $\mathbb{R}^d$ contains only a finite number of points. In this book only simple point processes are considered, i.e. $x_i \neq x_j$ if $i \neq j$.

The order of the points x_n is without interest, only the set $\{x_n\}$ matters. Thus the x_n are dummy variables and have no particular interpretation; for example x_1 need not be the point closest to the origin o.

The σ-algebra $\mathcal{N}$ is defined as the smallest σ-algebra of subsets of $\mathbb{N}$ to make measurable all mappings $\varphi \mapsto \varphi(B)$, for B running through the bounded Borel sets.

The reader should note that the term "process" does not imply a dynamic evolution over time and therefore the phrase "random point field" would be a more exact term; it is used in [21]. *Spatio-temporal point processes* explicitly involving temporal as well as spatial dispersion of points constitute a separate theory, see the paper by Zhuang et al. in this volume.

The *distribution* of a point process N is determined by the probabilities

$$\mathbf{P}(N \in Y) \qquad \text{for } Y \in \mathcal{N}.$$

The *finite-dimensional distributions* are of particular importance. These are probabilities of the form

$$\mathbf{P}(N(B_1) = n_1, \ldots, N(B_k) = n_k)$$

where $B_1, \ldots, B_k$ are bounded Borel sets and $n_1, \ldots, n_k$ non-negative integers. Here $N(B_i)$ is the number of points of N in B_i. The distribution of N on $[\mathbb{N}, \mathcal{N}]$ is uniquely determined by the system of all these values for $k = 1, 2, \ldots$. A still smaller subsystem is that of the *void probabilities*

$$v_B = \mathbf{P}(N(B) = 0) = \mathbf{P}(N \cap B = \emptyset)$$

for Borel sets B. Here N denotes the set of all points of the point process, the so-called support. If the point process is simple, as assumed here, then the distribution of N is already determined by the system of values of v_K as K ranges through the compact sets.

Let B be a convex compact Borel set in $\mathbb{R}^d$ with o being an inner point of B. The *contact distribution function* H_B with respect to the *test set B* is defined by

$$H_B(r) = 1 - \mathbf{P}(N(rB)) = 0) \qquad \text{for } r \geq 0. \tag{1}$$

In the special case of $rB = b(o, r) =$ sphere of radius r centred at o the contact distribution function is denoted as $F(r)$ or $H_s(r)$ and called the *spherical contact distribution function* or *empty space distribution function*. It can be interpreted as the distribution function of the random distance from the origin of $\mathbb{R}^d$ to the closest point of N. The function $H_B(r)$ is of a similar nature, but the metric is given by B.

A point process N is said to be *stationary* if its characteristics are invariant under translation: the processes $N = \{x_n\}$ and $N_x = \{x_n + x\}$ have the same distribution for all x in $\mathbb{R}^d$. So

$$\mathbf{P}(N \in Y) = \mathbf{P}(N_x \in Y) \tag{2}$$

for all Y in $\mathcal{N}$ and all x in $\mathbb{R}^d$. If we put $Y_x = \{\varphi \in \mathbb{N} : \varphi_{-x} \in Y\}$ for $Y \in \mathcal{N}$ then equation (2) can be rewritten as

$$\mathbf{P}(N \in Y) = \mathbf{P}(N \in Y_{-x}).$$

The notion of *isotropy* is entirely analogous: N is isotropic if its characteristics are invariant under rotation. Stationarity and isotropy together yield *motion-invariance*. The assumption of stationarity simplifies drastically the statistics of point patterns and therefore many papers in this book assume at least stationarity.

The *intensity measure* Λ of N is a characteristic analogous to the mean of a real-valued random variable. Its definition is

$$\Lambda(B) = \mathbf{E}\,(N(B)) \qquad \text{for Borel } B\,. \tag{3}$$

So $\Lambda(B)$ is the mean number of points in B. If N is stationary then the intensity measure simplifies; it is a multiple of Lebesgue measure ν_d, i.e.

$$\Lambda(B) = \lambda\nu_d(B) \tag{4}$$

for some (possibly infinite) non-negative constant λ, which is called the *intensity* of N it can be interpreted as the mean number of points of N per unit volume.

The point-related counterpart to $F(r)$ or $H_s(r)$ in the stationary case is the *nearest neighbour distance distribution function* $G(r)$ or $D(r)$, i.e. the d.f. of the distance from the typical point of N to its nearest neighbour.

Note that the application of $F(r)$ and $G(r)$ in the characterization of point processes is different. This is particularly important for cluster processes. In such cases $G(r)$ mainly describes distributional aspects in the clusters, while $F(r)$ characterizes particularly the empty space between the clusters. This different behaviour also explains the success of the *J-function* introduced by [12] defined as

$$J(r) = \frac{1 - G(r)}{1 - F(r)} \quad \text{for} \quad r \geq 0. \tag{5}$$

2 Marked Point Processes

A point process is made into a *marked point process* by attaching a characteristic (the *mark*) to each point of the process. Thus a marked point process on $\mathbb{R}^d$ is a random sequence $M = \{[x_n; m_n]\}$ from which the points x_n together constitute a point process (not marked) in $\mathbb{R}^d$ and the m_n are the marks corresponding to the x_n. The marks m_n may have a complicated structure. They belong to a given *space of marks* $\mathbb{M}$ which is assumed to be a Polish space. The Borel σ-algebra of $\mathbb{M}$ is denoted by $\mathcal{M}$. Specific examples or marked points are:

- For x the centre of a particle, m the volume of the particle;
- For x the position of a tree, m the stem diameter of the tree;
- For x the centre of an atom, m the type of the atom;
- For x the location (suitably defined) of a convex compact set, m the centred (shifted to origin) set itself.

Point process statistics often uses constructed marks. Examples are:

- m = distance to the nearest neighbour of x;
- m = number of points within distance r from x.

The marks can be continuous variables, as in the first two examples, indicators of types as in the third example (in which case the terms "multivariate point process" or "multitype point process" are often used, in the case of two marks the term "bivariate point processes") or actually very complicated indeed, as in the last example which occurs in the marked point process interpretation of germ-grain models (see [20]).

There is a particular feature of marked point processes: Euclidean motions of marked point processes are defined as transforms which move the points but leave the marks unchanged. So M_x, the translate of M by x, is given by

$$M_x = \{[x_1 + x; m_1], [x_2 + x; m_2], \ldots\}.$$

Rotations act on marked point processes by rotating the points but *not* altering the marks.

A marked point process M is said to be *stationary* if for all x the translated process M_x has the same distribution as M. It is *motion-invariant* if for all Euclidean motions m the process mM has the same distribution as M.

The definition of the *intensity measure* Λ of a marked point process M is analogous to that of the intensity measure of M when M is interpreted as a non-marked point process:

$$\Lambda(B \times L) = \mathbf{E}\left(M(B \times L)\right).$$

When M is stationary

$$\Lambda = \lambda \times \nu_d \times P_M, \tag{6}$$

where P_M denotes the mark distribution.

3 The Second-order Moment Measure

In the classical theory of random variables the moments (particularly mean and variance) are important tools of statistics. Point process theory has analogues to these. However, numerical means and variances must be replaced by the more complicated moment measures.

The second-order factorial moment measure of the point process N is the measure $\alpha^{(2)}$ on $\mathbb{R}^{2d}$ defined by

$$\int_{\mathbb{R}^{2d}} f(x_1, x_2) \, \alpha^{(2)}(\mathrm{d}(x_1, x_2)) = \mathbf{E}\left(\sum_{x_1, x_2 \in N}^{\neq} f(x_1, x_2)\right) \tag{7}$$

where f is any non-negative measurable function on $\mathbb{R}^{2d}$. The sum in (7) is extended over all pairs of different points; this is indicated by the symbol $\sum^{\neq}$.

It is

$$\mathbf{E}\left(N(B_1)N(B_2)\right) = \alpha^{(2)}(B_1 \times B_2) + \Lambda(B_1 \cap B_2)$$

and

$$\mathbf{var}\,(N(B)) \; = \; \alpha^{(2)}(B \times B) + \Lambda(B) - (\Lambda(B))^2 \,.$$

If N is stationary then $\alpha^{(2)}$ is translation invariant in an extended sense:

$$\alpha^{(2)}(B_1 \times B_2) = \alpha^{(2)}((B_1 + x) \times (B_2 + x))$$

for all x in $\mathbb{R}^d$.

Suppose that $\alpha^{(2)}$ is locally finite and absolutely continuous with respect to Lebesgue measure ν_{2d}. Then $\alpha^{(2)}$ has a density $\varrho^{(2)}$, the *second-order product density*:

$$\alpha^{(2)}(B_1 \times B_2) = \int\limits_{B_1} \int\limits_{B_2} \varrho^{(2)}(x_1, x_2) \mathrm{d}x_1 \mathrm{d}x_2 \,. \tag{8}$$

Moreover, for any non-negative bounded measurable function f

$$\mathbf{E}\left(\sum_{x_1, x_2 \in N}^{\neq} f(x_1, x_2)\right) = \int \int f(x_1, x_2) \varrho^{(2)}(x_1, x_2) \mathrm{d}x_1 \mathrm{d}x_2 \,.$$

The product density has an intuitive interpretation, which probably accounts for its historical precedence over the product measure and the K-function introduced below. (Note that there are also n^{th} order product densities and moment measures.) Suppose that C_1 and C_2 are disjoint spheres with centres x_1 and x_2 and infinitesimal volumes $\mathrm{d}V_1$ and $\mathrm{d}V_2$. Then $\varrho^{(2)}(x_1, x_2)\mathrm{d}V_1\mathrm{d}V_2$ is the probability that there is each a point of N in C_1 and C_2. If N is stationary then $\varrho^{(2)}$ depends only on the difference of its arguments and if furthermore N is motion-invariant then it depends only on the distance r between x_1 and x_2 and it is simply written as $\varrho^{(2)}(r)$. The *pair correlation function* $g(r)$ results by normalization:

$$g(r) = \varrho^{(2)}(r)/\lambda^2. \tag{9}$$

Without using the product density $\varrho^{(2)}$, the second factorial moment measure can be expressed by the *second reduced moment measure* $\mathcal{K}$ as

$$\alpha^{(2)}(B_1 \times B_2) = \lambda^2 \int\limits_{B_1} \mathcal{K}(B_2 - x)\mathrm{d}x$$

$$= \lambda^2 \int\limits_{\mathbb{R}^d} \int\limits_{\mathbb{R}^d} \mathbf{1}_{B_1}(x)\mathbf{1}_{B_2}(x + h)\mathcal{K}(\mathrm{d}h)\mathrm{d}x \,. \tag{10}$$

The term $\lambda \mathcal{K}(B)$ can be interpreted as the mean number of points in $B \backslash \{o\}$ under the condition that there is a point of N in o. The exact definition of $\mathcal{K}$ uses the theory of Palm distributions, see [20].

If a second-order product density $\varrho^{(2)}$ exists, then there is the following relationship between $\varrho^{(2)}$ and $\mathcal{K}$:

$$\lambda^2 \mathcal{K}(B) = \int_B \varrho^{(2)}(x) \mathrm{d}x \qquad \text{for Borel } B . \tag{11}$$

The description of the second moment measure simplifies still further in the motion-invariant case (when isotropy is added to stationarity). It then suffices to consider the *second reduced moment function K* or *Ripley's K-function* defined by

$$K(r) = \mathcal{K}(b(o, r)) \qquad \text{for } r \geq 0 .$$

The quantity $\lambda K(r)$ is the mean number of points of N within a sphere of radius r centred at the typical point, which is not itself counted.

The K-function is very popular in point process statistics and the present book contains many interesting applications. Also $\mathcal{K}$ plays some role in point process statistics, namely in the context of directional analysis, see [20, Sect. 4.3].

Other functions than K are often used to describe the second-order behaviour of a point process. Which function is to be preferred depends mainly on convenience, but also on statistical considerations. Some functions originate from the physical literature in which they have been used for a long time. The most important for the present volume are: The *pair-correlation function g*:

$$g(r) = \varrho^{(2)}(r) / \lambda^2$$

and the *L-function*:

$$L(r) = ((K(r)/b_d))^{1/d} , \tag{12}$$

where b_d is the volume of unit sphere of $\mathbb{R}^d$. The pair correlation function satisfies

$$g(r) = \frac{\mathrm{d}K(r)}{\mathrm{d}r} \bigg/ \left(d b_d r^{d-1} \right) . \tag{13}$$

The forms of these functions correspond to various properties of the underlying point process. Maxima of $g(r)$, or values of $K(r)$ larger than $b_d r^d$ for r in specific intervals, indicate frequent occurrences of interpoint distances at such r; equally minima of $g(r)$ or low values of $K(r)$ indicate inhibition at these r's. Model identification may be suggested by comparison of empirical pair-correlation functions, or reduced second moment functions, with model-related theoretical counterparts. The book [21] contains a big collection of pair correlation functions which may help the reader in statistical applications to find a good model for his/her data.

In the case of multivariate point processes it is useful to consider the function K_{ij} defined by

$$\lambda_j \cdot K_{ij}(r) = \text{mean number of } j\text{-points in } b(o, r)$$
$$\text{given that there is an } i\text{-point at } o \, .$$

In this case refined pair correlation functions $g_{ij}(r)$ can be defined. [21, Sect. 4.6] suggest further characteristics for such processes, namely *mark connection functions* $p_{ij}(r)$. Similar characteristics were also defined for point processes with real-valued marks, called *mark correlation function*.

4 Introduction to Statistics for Planar Point Processes

General remarks

The theory of statistics for point processes comprises a part of spatial statistics as described by [6] and [18]. Point process statistics as such is particularly powerful. For the theory the reader is referred to [2, 7, 16, 20, 21]. The following text describes briefly the estimation of some of the important non-parametric *summary statistics* such as intensity λ, empty function F or spherical contact distribution H_s, nearest neighbour distance distribution function G or D and the second-order characteristics g, K and L. These characteristics play also an important role in the parametric case, which uses point process models. For such statistical analysis they are used for constructing minimum contrast estimators, i.e. for parameter estimation via least squares (and related) methods, see [21]. In analogy to classical statistics, functional summary characteristics such as L, D or H_s (the latter two are distribution functions) are of particular interest for tests. In contrast, characteristics which are density functions, such as g or the probability density function for D, are more useful and simpler for exploratory analysis. However, the contribution by Diggle et al. shows that for clever statisticians also the K-function is an excellent tool in exploratory analysis.

Often statistical analysis of a point process depends on observation of *one sample* only, and that via a bounded sampling window W. Patterns arising in forestry, geology, geography, and ecology are often truly unique samples of a stochastic phenomenon. In other cases data collection is so complicated that only one sample is collected. Typically, in such cases it is assumed that the observed patterns are samples of *stationary ergodic* point processes, an assumption not susceptible to statistical analysis if there is only one sample, but one that is necessary if any statistical analysis is to be possible. Throughout this section ergodicity is assumed. In practice, either it is plausible from the very nature of the data, or else one must proceed on an *ad hoc* basis as if the assumption were true, and subject one's conclusions to the proviso that while the summary characteristics may be of possible value in the non-stationary case they will not then have the same interpretation. For example, an empirical point density (a mean number of points per unit area) can still be calculated even if the observed point pattern is non-stationary. It has value as a description of the spatial average behaviour of the pattern but does not possess all the properties of an estimator of a stationary point process intensity.

An ever present problem of spatial statistics is that of *edge-effects*. The problem intervenes in point process statistics in the estimation of g, K, H_s, and D, where for every point information from its neighbourhood is required, which is often not completely available for points close to the edge of the window W of observation.

Two simple methods to correct edge-effects are plus and minus sampling. In *plus-sampling* all points of the pattern in the window W are considered and, if necessary, information from outside of the window is taken to calculate the summary characteristics. So plus-sampling may require more information than that contained in W. In contrast, *minus-sampling* uses only a subset of the points in W, namely those for which the contributions for the summary characteristics can be completely calculated. In this way biases are avoided but losses of information accepted.

For some of the characteristics considered more economical edge-corrections are available. It is important to note that different forms of edge-corrections are necessary; they are tailored to the summary characteristics of interest. All the edge-corrections appearing in this contribution can be considered to be Horvitz-Thompson estimators, see [2].

The aim of edge-corrections is to obtain unbiased estimators or, at least, *ratio-unbiased* estimators (quotients where numerator and denominator are unbiased).

Estimation of the intensity λ

The classical and best estimator of the intensity λ is $\hat{\lambda}$, where

$$\hat{\lambda} = N(W)/\nu_d(W) \tag{14}$$

is the number of points in W per area of W. Straightforward calculation shows it to be unbiased. If N is ergodic then $\hat{\lambda}$ is strongly consistent, in the sense that $\hat{\lambda} \to \lambda$ almost surely as the window size is increased. Below other intensity estimators will appear which are used for the construction of ratio-unbiased estimators of functional summary characteristics. They are particularly adapted to the characteristic of interest and ensure small mean squared errors.

Estimation of second-order characteristics

The reduced second-order moment measure, more precisely $\lambda^2 \mathcal{K}(B)$, can be estimated by $\kappa_s(B)$, where

$$\kappa_s(B) = \sum_{x,y \in N \cap W}^{\neq} \frac{1_B(y-x)}{\nu_d(W_x \cap W_y)}, \tag{15}$$

valid for bounded Borel B such that $\nu_d(W \cap W_z)$ is positive for all z in B. This is an unbiased estimator. The problem is the factor λ^2, see below. For $B = b(o, r)$, an unbiased estimator of $\lambda^2 K(r)$ is obtained.

In the isotropic case a better estimator of $\lambda^2 K(r)$ is available:

$$\kappa_i(r) = \sum_{x,y \in N \cap W} \frac{1(0 < \|x-y\| \leq r)k(x,y)}{\nu_d(W^{(\|x-y\|)})} \tag{16}$$

for $0 \leq r < r^*$ where

$$r^* = \sup\left\{ r : \nu_d\left(W^{(r)}\right) > 0 \right\}$$

and

$$W^{(r)} \quad = \quad \{x \in W : \partial(b(x,r)) \cap W \neq \emptyset\},$$

and, in the planar case, $k(x,y) = 2\pi/\alpha_{xy}$ where α_{xy} is the sum of all angles of the arcs in W of a circle centre x and radius $\|x-y\|$. If $\alpha_{xy} = 0$ then $k(x,y) = 0$. Generalization to the d-dimensional case is straightforward.

It is known that $\kappa_i(r)$ has a smaller mean squared error than $\kappa_s(r)$ if the point process analysed is really isotropic. But $\kappa_i(r)$ is sensitive to deviations from the isotropy assumption. Thus it may be preferable to use always $\kappa_s(r)$ instead of $\kappa_i(r)$.

Estimators of $\mathcal{K}(B)$ and $K(r)$ are obtained by division by estimators of λ^2. (Note that here is not written "squared estimators of λ".) Some statisticians use

$$\widehat{\lambda^2} = n(n-1)/\nu_d(W),$$

where n is the number of points in the window W. This estimator is unbiased in the case of a Poisson process. [22] recommend the use of $(\hat{\lambda}_V(r))^2$ as estimator of λ^2 with

$$\hat{\lambda}_V(r) = \sum_{x \in N} 1(x)\nu_d(W \cap b(x,r)) \left/ \left(db_d \int_0^r u^{d-1}\overline{\gamma}_W(u)du \right) \right.,$$

where $\overline{\gamma}_W(r)$ is the isotropic version of the set covariance of the window W; for parallelepipedal and spherical windows formulas are available. Note that

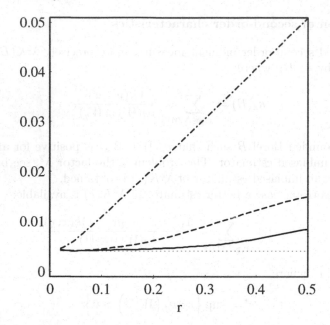

Fig. 1. Square roots of mean squared errors of L-estimators for a Poisson process of intensity $\lambda = 100$ in the unit square.
—— = Ripley's estimator with $\hat{\lambda}$,
– – – = Ohser-Stoyan's estimator with $\hat{\lambda}$,
– · – = Ripley's estimator with known λ,
· · · = Ohser-Stoyan's estimator with $\hat{\lambda}_V(r)$

the intensity estimator depends on the variable r. Its use reduces drastically the mean squared error of the $K(r)$-estimator.

For statistical purposes it is useful to stabilize variances. This can be done by using the L-function, estimating $L(r)$ by

$$\hat{L}(r) = (\hat{K}(r)/b_d)^{1/d} \qquad (17)$$

where $\hat{K}(r)$ is one of the estimators of $K(r)$. [22] verified for $d = 1, 2, 3$ the stabilization effect. The mean squared error of $\hat{L}(r)$ with $\hat{\lambda}_V(r)$ is very small and nearly independent of r, see Fig. 1. This figure also shows that Ripley's estimator $\kappa_i(r)$ is in this isotropic case better than $\kappa_s(r)$. Furthermore, it demonstrates that adaption of intensity estimators is really important: If λ is known (as it is possible in simulations), this does not help and leads to bad estimates of second-order characteristics. On the other hand, $\kappa_s(r)$ combined with $\hat{\lambda}_V(r)$ is better than $\kappa_i(r)$ with $\hat{\lambda}$. Figure 1 shows the mean squared error for a Poisson point process with $\lambda = 100$. The values for other λ can be obtained by multiplication with $10/\sqrt{\lambda}$. For other models the behaviour is similar; for regular processes the mean squared error is smaller and for cluster processes larger.

Since the K-function is so successful in point process statistics, also an analogue was defined for the inhomogeneous case, see [3] and the contribution by Webster et al. in this book.

The *product density* $\varrho^{(2)}$ can be estimated by an edge-corrected density estimator such as

$$\hat{\varrho}^{(2)}(r) = \sum_{x,y \in N \cap W} \frac{\mathbf{k}(\|x-y\|-r)}{db_d r^{d-1} \nu_d(W_x \cap W_y)} \tag{18}$$

where $\mathbf{k}$ is a kernel function, see [22]. The best kernel function is the box kernel

$$\mathbf{k}(x) = 1_{[-h,h]}(x)/2h,$$

where h is the bandwidth, and not the Epanechnikov kernel. If motion-invariance is sure, the edge-correction term $\nu_d(W_x \cap W_y)$ can be replaced by $\nu_d(W^{\|x-y\|})/k(x,y)$ (as in Ripley's estimator of the K-function).

Estimators of the *pair correlation function* are obtained by division of the product density estimators by estimators of λ^2. The best known estimator is $(\hat{\lambda}_S(r))^2$ with

$$\hat{\lambda}_S(r) = \sum_{x \in N} 1_W(x) \nu_{d-1}(W \cap \partial h(x,r)) / (db_d r^{d-1} \overline{\gamma}_W(r)) \tag{19}$$

found by the astronomers [8] and [11].

A very good approximation of the variance of the pair correlation function estimator obtained by combination of (18) and (19) is

$$\sigma^2(r) - \frac{\int\limits_{-h}^{h} (k(s))^2 ds \cdot g(r)}{\frac{1}{2} db_d r^{d-1} \overline{\gamma}_W(r) \lambda^2} \quad \text{for } r \geq 0. \tag{20}$$

Some statisticians estimate the pair correlation function using formula (13), starting from an estimate of the K-function and using numerical differentiation, see e. g. the contributions of Illian et al. and Stein and Georgiadis. This is probably a reaction on the fact that until now (March 2005) kernel estimators of $g(r)$ are only available if the window W is circular or rectangular, but that there is in R an estimator of $K(r)$ for arbitrary polygonal windows. But numerical differentiation is probably a method of minor quality (note the analogy to probability density estimation, where numerical differentiation of the empirical distribution function is not very popular) and the situation could be easily changed: In the source code of a K-estimation program only some lines must be changed, instead of $1(0 < \|x-y\| \leq r)$ the term $\mathbf{k}(\|x-y\|-r)/(db_d r^{d-1})$ must be inserted; the main work for the K-estimator, the determination of the Horvitz-Thompson weights, can be used also in the case of pair correlation estimation. The estimator can still be improved by using $\hat{\lambda}_S(r)$ instead of $\hat{\lambda}$.

The variability of all these estimators is difficult to analyse, so usually simulations are made, see e. g. [22]. In the Poisson process case some analytical calculations have been carried out, see [8, 11, 19].

Estimation of the spherical contact distribution H_s

It is well known that

$$H_s(r) = \mathbf{P}(o \in N \oplus b(o, r)) \qquad \text{for } r \geq 0.$$

Thus $H_s(r)$ is equal to the areal fraction of the random closed set

$$\Xi = N \oplus b(o, r) = \bigcup_{x \in N} b(x, r).$$

An unbiased estimator is given by

$$\hat{H}_s(r) = \frac{\nu_d\left((W \ominus b(o, r)) \cap \bigcup_{x \in N} b(x, r)\right)}{\nu_d(W \ominus b(o, r))} \tag{21}$$

for $0 \leq r \leq \frac{1}{2}\mathrm{diam}(W)$. In this estimator the principle of minus-sampling is used.

Many statisticians determine the measures in numerator and denominator by means of a grid. [5] produced an algorithm which determines the exact areas $(d = 2)$ and volumes $(d = 3)$.

By the way, K. Mecke had the idea to introduce summary characteristics which use instead of the Lebesgue measure, which plays an important role in the definition of $H_s(r)$, other Minkowski measures, in the planar case length and Euler characteristic. The paper [15] demonstrates the use of the corresponding characteristics in point process statistics for patterns of tree locations and shows that they give valuable information which $H_s(r)$ cannot offer.

Estimation of the nearest neighbour distance distribution function

Recall that the nearest-neighbour distance distribution function is the distribution function of the distance from the typical point to its nearest neighbour. It can be also defined based on a marking of the points x of the process with the distance $d(x)$ from x to its nearest neighbour. The resulting marked point process M inherits the stationarity property from $N = \{x_n\}$. The corresponding mark distribution function is precisely D or G. This formulation via M clarifies the logic behind the expression for estimating D.

The first estimator to be described is the minus-sampling or border estimator:

$$\hat{D}_b(r) = \sum_{[x;d] \in M} \mathbf{1}_{W \ominus b(o,r)}(x)\mathbf{1}(0 < d \le r)/N(W \ominus b(o,r)) \quad \text{for } r \ge 0 \,. \quad (22)$$

This is probably the most natural estimator. Only the points in the eroded window $W \ominus b(o,r)$ are used for the estimation. If N is ergodic then $\hat{D}_b$ is asymptotically unbiased. But additionally to the fact that this estimator ignores unnecessarily many of the points in the window (those which have the nearest neighbour in W), it has the disadvantage to be sometimes not monotonic increasing in r and to exceed the value 1.

Another unbiased estimator has been suggested in [9], which does not suffer from these disadvantages and uses an edge-correction which could be called "nearest neighbour correction":

$$\hat{D}_H(r) = \hat{\mathcal{D}}_H(r)/\hat{\lambda}_H \qquad \text{for } r \ge 0 \qquad (23)$$

where

$$\hat{\mathcal{D}}_H(r) = \sum_{[x;d] \in M} \mathbf{1}_{W \ominus b(o,d)}(x)\mathbf{1}(0 < d \le r)\Big/\nu_d(W \ominus b(o,d))$$

and

$$\hat{\lambda}_H = \sum_{[x;d] \in M} \mathbf{1}_{W \ominus b(o,d)}(x)\Big/\nu_d(W \ominus b(o,d)).$$

The principle underlying $\hat{\mathcal{D}}_H(r)$ is simple: To use precisely those points x for which it is known that the nearest neighbour is both within W and closer than r to x. The estimator $\hat{\mathcal{D}}_H(r)$ is unbiased for $\lambda D(r)$, while $\hat{\lambda}_H$ is an unbiased intensity estimator, which is adapted to $\hat{\mathcal{D}}_H(r)$. (The proof of unbiasedness uses one of the deeper results of the theory of point processes, the Campbell-Mecke theorem.)

[23] compared by simulation the estimation variances of the above estimators of $D(r)$ or $G(r)$ (and further estimators such as the Kaplan-Meier estimator) for Poisson, cluster and hard-core processes and found that Hanisch's estimator is the best one.

It is not difficult to construct also estimators of the distribution function of the distance to the second, third, ... neighbour or of probability density functions of neighbour distance functions with Hanisch's edge correction.

Application of the various summary characteristics

Second-order summary characteristics such as $K(r), L(r)$ and $g(r)$ and distance characteristics such as $G(r)$ and $F(r)$ belong to the toolbox of spatial statisticians. In particular physicists follow the "dogma of second-order" and believe that second-order characteristics (in particular the pair correlation function) yield the best information (and all information needed). Also prominent statisticians tend to believe this. For example, [14] and [17] compared the use of H_s and K in point process statistics and came to the conclusion that second-order characteristics such as K are more suitable than functions such as D or H_s for model testing (via simulation tests) or for parameter estimation (via minimum contrast method). However, the paper by [1] showed that there are point processes which are different in distribution but have identical K-function. So it is at least useful to use other characteristics as supplementary summary characteristics. Particularly useful are K. Mecke's morphological functions, the use of which is demonstrated in [15]. The family of these functions includes the spherical contact distribution function. Their advantage as alternative summary characteristics results also from the fact that these characteristics are of a stationary and not of a Palm nature.

Simulation methods in point process statistics

Modern point process statistics uses simulation methods in a large extent. They are used simply for generating point patterns in the context of model choice, but also for parameter estimation and model test. The reader is referred to [16].

Here only some short remarks on simulation tests are given since such tests appear in some contributions of this book. Such tests, introduced to point process statistics by [4] and [17], provide a flexible means of statistical investigation.

Suppose the hypothesis to be considered is that a given point pattern, observed through a window W, is a sample of a point process N. One chooses a test characteristic χ, such as

$$\hat{\lambda},$$

$$\sup_{r_1 \leq r \leq r_2} \left| \hat{L}(r) - r \right|,$$

$$\sup_{r_1 \leq r \leq r_2} \left| D(r) - \hat{H}_s(r) \right|$$

and calculates its value χ_o for the given sample. This value is compared with the ordered sample

$$\chi^{(1)} \leq \chi^{(2)} \leq \cdots \leq \chi^{(m)}$$

obtained by simulating N m times, observing it through the window W, and then calculating the statistic χ for each simulation by the same estimator as for the sample.

If χ_0 takes a very small or very large position in the series of ordered $\chi^{(k)}$, then the hypothesis may be cast in doubt. It is then possible to perform a test of the hypothesis at an *exact* significance level: for example a two-sided test of significance level $\alpha = 2k/(m+1)$ is obtained by the rejection of the hypothesis when

$$\chi_0 \leq \chi^{(k)} \qquad \text{or} \qquad \chi_0 \geq \chi^{(m-k+1)} .$$

When using functional summary characteristics such as $L(r)$ or $G(r)$ it is popular to generate by simulation confidence bands and to observe whether the empirical summary characteristic is completely within the band. In the negative case the model hypothesis is rejected. For this test the error probability is difficult to determine exactly. The contributions by Diggle et al. and Martínez-Beneito et al. in this book give instructive examples: Diggle et al. speak carefully about "envelopes from 99 simulations", while Martínez-Beneito et al. use the term "95% confidence band". (For every fixed r they are right, but for the function graph as a whole the band is too broad and a test based on such a band has an error probability greater than 0.05.)

It is recommended to use in a simulation test of goodness-of-fit another summary characteristic than that used for estimating model parameters. For example, if the parameter estimation is based on second-order characteristics, then distance distributions should be used in the model test.

5 The Homogeneous Poisson Point Process

The Poisson point process is the simplest and most important model for random point pattern. It plays a central role as null model and as a starting point for the construction of realistic models for point patterns; it is the model for a completely random point pattern (CSR = complete spatial randomness). Simulation procedures often include the construction of a Poisson point process, which is then modified into the form required.

A stationary Poisson point process N is characterised by two fundamental properties:

(i) *Poisson distribution of point counts*: the number of points of N in a bounded Borel set B has a Poisson distribution of mean $\lambda \nu_d(B)$ for some constant λ;

(ii) *Independent scattering*: the numbers of points of N in k disjoint Borel sets form k independent random variables, for arbitrary k.

Property (ii) is also known as the "completely random" or "purely random" property.

The positive number λ occurring in (i) is the intensity of the stationary Poisson point process. It gives the mean number of points to be found in a unit volume, and it is given by

$$\lambda \nu_d(B) = \mathbf{E}(N(B)) \qquad \text{for all bounded Borel sets } B .$$

Basic properties

Let N be a stationary Poisson point process of intensity λ. From properties (i) and (ii) the whole distribution of the stationary Poisson point process can be determined once the intensity λ is known.

(a) *Finite-dimensional distributions.* It can be shown directly from (i) and (ii) that if $B_1, \ldots, B_k$ are disjoint bounded Borel sets then the random variables $N(B_1), \ldots, N(B_k)$ are independent Poisson distributed with means $\lambda\nu_d(B_1), \ldots, \lambda\nu_d(B_k)$. Thus

$$\mathbf{P}(N(B_1) = n_1, \ldots, N(B_k) = n_k)$$
$$= \frac{\lambda^{n_1+\cdots+n_k}(\nu_d(B_1))^{n_1} \cdot \ldots \cdot (\nu_d(B_k))^{n_k}}{n_1! \cdot \ldots \cdot n_k!} \exp\left(-\sum_{i=1}^{k} \lambda\nu_d(B_i)\right) (24)$$

(b) *Stationarity and isotropy.* The stationary Poisson point process N is stationary and isotropic.

(c) *Void-probabilities.* The *void-probabilities* of the Poisson process are

$$v_B = \exp(-\lambda\nu_d(B)). \qquad (25)$$

The contact distribution functions are given by

$$H_B(r) = 1 - v_{rB} = 1 - \mathbf{P}(N(rB) = 0) \qquad \text{for } r \geq 0.$$

In the particular, the *spherical contact distribution function* is

$$H_s(r) = 1 - \exp(-\lambda b_d r^d) \qquad (26)$$

for $r \geq 0$.

(d) *Conditioning.* If N is a stationary Poisson point process then one can consider the restriction of N to a compact set W under the condition that $N(W) = n$. The point process formed by these n points in W has the same distribution as n independent and uniformly in W distributed points.

Summary characteristics

The nearest neighbour distance distribution function is given by

$$D(r) = 1 - \exp\left(-\lambda b_d r^d\right) \qquad (27)$$

for $r \geq 0$.

The right-hand sides of (26) and (27) are equal, i.e. the spherical contact distribution function and nearest neighbour distance distribution for the stationary Poisson point process are equal.

The *reduced second-order moment measure* function $\mathcal{K}$ is equal to the Lebesgue measure and it is

$$K(r) = b_d r^d \quad \text{for } r \geq 0.$$

Consequently, the L-function has the simple form

$$L(r) = r$$

and the pair correlation function is still simpler:

$$g(r) = 1.$$

The general Poisson point process

The stationary Poisson point process has an intensity measure which is proportional to Lebesgue measure. The mean number of points per unit area does not vary over space. Many point patterns arising in applications exhibit fluctuations that make such a lack of spatial variation implausible.

The general Poisson point process N provides a more general stochastic model, appropriate for such point patterns. It is characterised by a diffuse measure Λ on $\mathbb{R}^d$ which is called the *intensity measure* of N and which is the intensity measure of that point process. A *general Poisson point process N with intensity measure Λ* is a point process possessing the two following properties:

(i') *Poisson distribution of point counts*: the number of points in a bounded Borel set B has a Poisson distribution with mean $\Lambda(B)$

$$\mathbf{P}(N(B) = m) = (\Lambda(B))^m \cdot \exp(-\Lambda(B))/m! \quad \text{for } m = 0, 1, 2, \ldots \quad (28)$$

(ii') *Independent scattering*: the numbers of points in k disjoint Borel sets form k independent random variables.

It is clear from property (i') that such a process N is *not* stationary in general.

If the measure Λ has a density with respect to Lebesgue measure then it can be written as

$$\Lambda(B) = \int_B \lambda(x) \mathrm{d}x \quad \text{for Borel sets } B.$$

The density $\lambda(x)$ is called the *intensity function* of the general Poisson point process. It has an appealing and intuitive infinitesimal interpretation: $\lambda(x)\mathrm{d}V$ is the infinitesimal probability that there is a point of N in a region of infinitesimal volume $\mathrm{d}V$ situated at x.

Simulation of a stationary Poisson point process

The starting point for simulating the stationary Poisson point process is the uniform distribution property (d) above. Thus the simulation of a stationary Poisson point process in a compact region W falls naturally into two stages. First the number n of points in W is determined by simulating a Poisson random variable, and then the positions of the points n in W are determined by simulating independent uniform points in W.

A general Poisson point process can be simulated by means of a thinning procedure, see [20].

Statistics for the stationary Poisson point process

The discussion of the central role of the stationary Poisson point process establishes the importance of statistical methods for deciding whether or not a given point pattern is Poisson, while the estimation of the only model parameter, intensity λ, follows (14). This section is a brief survey of such methods. It confines itself to the case of *planar* point patterns; the methods presented carry over to point patterns on the line, in space, and on a sphere.

Probably all functional summary characteristics have been used for such tests. The until now most successful and most popular tests use the L-function. In that case the test statistics is

$$\tau = \max_{r \leq r_0} |\hat{L}(r) - r|$$

with $\hat{L}(r)$ being Ripley's estimator using (16) and (14), with $\hat{\lambda}^2$. Here r_0 is an upper bound on the interpoint distance r, perhaps 25 % of window side length.

If τ is large, then the Poisson hypothesis has to be rejected. (The alternative hypothesis is "no Poisson point process" without further specification.) The critical value of τ for the significance level $\alpha = 0.05$ is

$$\tau_{0.95} = 1.45\sqrt{a}/n, \qquad (29)$$

where a is the window area and n the number of points observed [19, p. 46]. This formula was obtained by simulations and can be used if nur_0^3/a^2 is small for a "wide range of" r_0. (Here u is the boundary length of W.) For $\alpha = 0.01$ the factor 1.63 can be used. [10] found that the choice of r_0 is more important than it seems when reading [19] and [21]. They also showed that the use of $\hat{\lambda}_V(r)$ improves the power of the test.

In the case of not rectangular windows simulation tests can be used.

6 Other Point Processes Models

There are many other, more complicated but for applications more realistic point process models. Examples are cluster processes (see the contribution by

Fleischer et al. in this book) and Markov point processes. The latter play a role in several contributions in this book; another name for them is Gibbsian point process, and there are relationships to Statistical Physics where the idea of such processes was originally developed.

A particular case are pairwise interaction processes, where the correlations or interactions of points are modelled in an additive form by contributions coming from point pairs. The contribution by Diggle et al. in this book gives an introduction to these processes which may be sufficient for this book. For more details the reader is referred to the nice book [13].

References

[1] A.J. Baddeley and B.W. Silverman. A cautionary example for the use of second order methods for analyzing point patterns. *Biometrics*, 40:1089–1094, 1984.

[2] A.J. Baddeley. Spatial sampling and censoring. In: Barndorff-Nielsen, O. E., Kendall, W.S., Lieshout, M. N. M. van (eds.) *Stochastic Geometry. Likelihood and Computation*. Chapman and Hall, London, New York, 1999.

[3] A.J. Baddeley, J. Møller and R.P. Waagepetersen. Non and semi parametric estimation of interaction in inhomogeneous point patterns. *Statistica Neerlandica*, 54:329–50, 2000.

[4] J.E. Besag and P.J. Diggle. Simple Monte Carlo tests for spatial data. *Bulletin of the International Statistical Institute*, 47:77–92, 1977.

[5] U. Brodatzki and K.R. Mecke. Simulating stochastic geometrics: morphology of overlapping grains. *Computer Physics Communications*, 147:218–221, 2002.

[6] N.A.C. Cressie. *Statistics for Spatial Data*, John Wiley & Sons, New York, 1991.

[7] P.J. Diggle. *Statistical Analysis of Spatial Point Patterns* (second edition). London: Arnold, 2003.

[8] A.J.S. Hamilton. Toward better ways to measure the galaxy correlation function. *Astrophysics Journal*, 417:19–35, 1993.

[9] K.-H. Hanisch. Some remarks on estimators of the distribution function of nearest-neighbor distance in stationary spatial point patterns. *Statistics*, 15, 409–412, 1984.

[10] L.P. Ho and S.N. Chiu. Testing the complete spatial randomness by Diggle's test without an arbitrary upper limits. *Journal of Statistical Computation and Simulation*. To appear 2005.

[11] S.D. Landy and A.S. Szalay. Bias and variance of angular correlation functions. *Astrophysics Journal*, 412:64–71, 1993.

[12] M.N.M. van Lieshout and A.J. Baddeley. A nonparametric measure of spatial interaction in point patterns. *Statistica Neerlandica*, 50:344–361, 1996.

[13] M.N.M. van Lieshout. *Markov Point Processes and their Applications*. London: Imperial College Press, 2000.

[14] H.W. Lotwick. Some models for multitype spatial point processes, with remarks on analysing multitype patterns. *Journal of Applied Probability*, 21:575–582, 1984.

[15] K.R. Mecke and D. Stoyan. Morphological characterization of point patterns. *Biometrical Journal*, 47, 2005.

[16] J. Møller and R.P. Waagepetersen. *Statistical Inference and Simulation for Spatial Point Processes*. London: Chapman and Hall, 2004.

[17] B.D. Ripley. Modelling spatial patterns (with discussion). *Journal of the Royal Statistical Society B*, 39:172–212, 1977.

[18] B.D. Ripley. *Spatial Statistics*, John Wiley & Sons, New York, 1981.

[19] B.D. Ripley. *Statistical Inference for Spatial Processes*, Cambridge University Press, Cambridge, 1988.

[20] D. Stoyan, W.S. Kendall and J. Mecke. *Stochastic Geometry and its Applications*. John Wiley & Sons, Chichester, 1995.

[21] D. Stoyan and H. Stoyan. *Fractals Random Shapes and Point Fields. Methods of geometrical statistics*. John Wiley & Sons, Chichester, 1994.

[22] D. Stoyan and H. Stoyan. Improving ratio estimators of second order point process characteristics. *Scandinavian Journal of Statistics*, 27:641–656, 2000.

[23] D. Stoyan. On estimators of the nearest neighbour distance distribution function for stationary point processes. *Metrika*, 54, 2005.

Modelling Spatial Point Patterns in R

Adrian Baddeley[1] and Rolf Turner[2]

[1] School of Mathematics and Statistics, University of Western Australia, Crawley
6009 WA, Australia, adrian@maths.uwa.edu.au
[2] Department of Mathematics and Statistics, University of New Brunswick,
Fredericton, N.B., E3B 5A3 Canada, rolf@math.unb.ca

Summary. We describe practical techniques for fitting stochastic models to spatial point pattern data in the statistical package R. The techniques have been implemented in our package spatstat in R. They are demonstrated on two example datasets.

Key words: EDA for spatial point processes, Point process model fitting and simulation, R, Spatstat package

1 Introduction

This paper describes practical techniques for fitting stochastic models to spatial point pattern data using the statistical language R. The techniques are demonstrated with a detailed analysis of two real datasets.

We have implemented the techniques as a package spatstat in the R language. Both spatstat and R are freely available from the R website [19].

Sections 2 and 3 introduce the spatstat package. Theory of point process models is covered in Sect. 4, while Sect. 5 describes how to fit models in spatstat, and Sect. 6 explains how to interpret the fitted models obtained from the package. Models involving external covariates are discussed in Sect. 7, and models for multitype point patterns in Sect. 8. Estimation of irregular parameters is discussed in Sect. 9. Section 10 discusses formal inference for models. Examples are analysed in Sects. 11–12.

2 The spatstat Package

We assume the reader is conversant with basic ideas of spatial point pattern analysis [28, 68] and with the R language [38, 40, 56].

Spatstat is a contributed R package for the analysis of spatial point pattern data [4]. It contains facilities for data manipulation, tools for exploratory data analysis, convenient graphical facilities, tools to simulate a wide range of point pattern models, versatile model-fitting capabilities, and model diagnostics. A detailed introduction to spatstat has been provided in [4]. Here we give a brief overview of the package.

2.1 Scope

Spatstat supports the following activities. Firstly **basic manipulation** of point patterns is supported; a point pattern dataset can easily be created, plotted, inspected, transformed and modified. **Exploratory data analysis** is possible using summary functions such as the K-function, pair correlation function, empty space function, kernel-smoothed intensity maps, etc. (see e.g. [28, 68]). A key feature of spatstat is its generic algorithm for **parametric model-fitting** of spatial point process models to point pattern data. Models may exhibit spatial inhomogeneity, interpoint interaction (of arbitrary order), dependence on covariates, and interdependence between marks. Finally, **simulation** of point process models, including models fitted to data, is supported.

Figure 1 shows an example of a point pattern dataset which can be handled by the package; it consists of points of two types (plotted as two different symbols) and is observed within an irregular sampling region which has a hole in it. The label or "mark" attached to each point may be a categorical variable, as in Fig. 1, or a continuous variable.

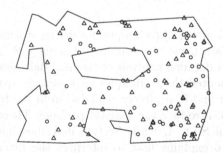

Fig. 1. Artificial example demonstrating the complexity of datasets which spatstat can handle

Point patterns analysed in spatstat may also be spatially inhomogeneous, and may exhibit dependence on covariates. The package can deal with a variety of covariate data structures. It will fit point process models which depend on the covariates in a general way, and can also simulate such models.

2.2 Data Types in spatstat

A point pattern dataset is stored as a single "object" X which may be plotted simply by typing plot(X). Here spatstat uses the object-oriented features of

R ("classes and methods") to make it easy to manipulate, analyse, and plot datasets.

The basic data types in spatstat are POINT PATTERNS, WINDOWS, and PIXEL IMAGES. See Fig. 2. A point pattern is a dataset recording the spatial locations of all "events" or "individuals" observed in a certain region. A window is a region in two-dimensional space. It usually represents the "study area". A pixel image is an array of "brightness" values for each grid point in a rectangular grid inside a certain region. It may contain covariate data (such as a satellite image) or it may be the result of calculations (such as kernel smoothing).

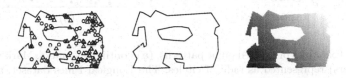

Fig. 2. A point pattern, a window, and a pixel image

A point pattern is represented in spatstat by an object of the class "ppp". A dataset in this format contains the coordinates of the points, optional "mark" values attached to the points, and a description of the spatial region or "window" in which the pattern was observed. Objects of class "ppp" can be created using the function ppp, converted from other data using the function as.ppp, or obtained in a variety of other ways.

In our current implementation, the mark attached to each point must be a *single* value (which may be numeric, character, complex, logical, or factor). Figure 3(a) shows an example where the mark is a positive real number. A multitype point pattern is represented as a marked point pattern for which the mark is a categorical variable (a "factor" in R). Figure 3(b) shows an example where the mark is a categorical variable with two levels (i.e. a bivariate point pattern).

If X is a point pattern object then typing X or print(X) will print a short description of the point pattern; summary(X) will print a longer summary; and plot(X) will generate a plot of the point pattern on a correct scale. Numerous facilities are available for manipulating point pattern datasets.

3 Data Analysis in spatstat

3.1 Data Input

Point pattern datasets (objects of class "ppp") can be entered into spatstat in various ways. We may create them from raw data using the function ppp,

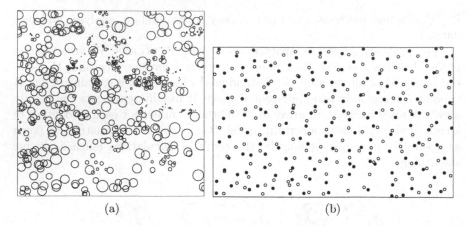

(a)   (b)

Fig. 3. Examples of marked point patterns. (a) continuous marks. Mark values (tree diameters) represented as radii of circles. The Longleaf Pines dataset, available as `longleaf`. (b) categorical marks. Mark values (cell types) represented as different graphical symbols. Hughes' amacrine cell dataset, available as `amacrine`

convert data from other formats (including other packages) using `as.ppp`, read data from a file using `scanpp`, manipulate existing point pattern objects using a variety of tools, or generate a random pattern using one of the simulation routines.

Suppose, for example, that we have data for a point pattern observed in the rectangle $[0, 10] \times [0, 4]$. Assume the Cartesian coordinates of the points are stored in R as vectors x and y. Then the command

```
X <- ppp(x, y, c(0,10), c(0,4))
```

creates a point pattern object containing this information.

3.2 Initial Inspection of Data

Chatfield [15] emphasises the importance of careful initial inspection of data. The same principles apply to point pattern data. A point pattern dataset should be inspected for the following: omission of data points; transcription errors; data file format violations; incorrect scaling of the coordinates; flipping of the axes; errors in delimiting the boundary; errors in inclusion/exclusion of points near the boundary; incorrect interpretation of the data type of the marks (e.g. categorical or continuous); inconsistency with plots of the same data in the original source publication; coarse rounding of the Cartesian coordinates; use of values such as 99 or −1 to indicate a missing value; incorrect software translation of the levels of a factor; and duplicated points.

Inspection can be accomplished in `spatstat` mainly with the commands `plot`, `print`, `summary`, `identify`, `hist` (to examine values of the Cartesian coordinates) and `nndist` (to detect duplicated points).

3.3 Exploratory Data Analysis

Before stochastic modelling of a point pattern dataset is attempted, and certainly before any formal hypothesis testing is contemplated, the data should be subjected to exploratory data analysis (EDA). General principles of EDA are outlined in [15, 18, 24, 71, 72]. Numerous tools for exploratory analysis of spatial point pattern data are surveyed in [22, 26, 28, 59, 61, 67, 68, 72].

In particular, the assumption of stationarity ("spatial homogeneity") is an essential requirement for many of the classical methods for spatial point pattern analysis [22, 26, 28, 58, 59, 61, 67, 68]. It seems clear that many real point patterns cannot be described as stationary [54, 69], and the use of the classical methods on such data would be invalid. Hence it is extremely important that the homogeneity of a point pattern dataset be critically evaluated. Techniques for analysing nonstationary ("spatially inhomogeneous") patterns are less developed [3, 52, 53, 54, 55, 69].

An exploratory analysis should typically begin with an assessment of spatial inhomogeneity using tools such as the kernel smoothed estimate of intensity [25] (available in spatstat as ksmooth.ppp), or LISA (Local Indicators of Spatial Association) methods [1, 21, 20]. The dataset could also be partitioned manually using the subset operator [] or the commands cut and split.

If a simple form of spatial inhomogeneity (such as a gradient from left to right) is suspected, this trend can be fitted using parametric methods as described in Sect. 6.

If the data are judged to be spatially homogeneous, the next step would be exploratory analysis using standard summary statistics such as Ripley's K-function. A wide choice of summary statistics is now available [22, 28, 61, 68, 67]. In spatstat the available choices include Kest, which estimates the K-function [58, 59],[68, Chap. 15]; Fest, estimating the empty space function F [59, 61] also known as the contact distribution function [68, Chap. 15] and point-event distance function [26, Sect. 2.4]; Gest, estimating the nearest neighbour distance distribution function G [26, Sect. 2.3],[68, Chap. 15]; pcf, the pair correlation function [68, Chap. 15]; Jest, the function $J(r) = (1 - G(r))/(1 - F(r))$ of [48]; Kmeasure, the reduced second moment measure [9, 10], [68, pp. 245, 247], [67]; and analogues of these functions for multitype and marked point patterns [49].

However if the data are judged to be spatially inhomogeneous, then at present there is limited scope for further exploratory analysis. One exception is the inhomogeneous K-function [3] implemented in spatstat as Kinhom.

4 Point Process Models

The spatstat package can fit parametric models of spatial point processes to point pattern data. This section describes the relevant class of models, and the next section explains how to fit them using spatstat.

A typical realisation of a point pattern X in the bounded region $W \subset \mathbb{R}^2$ will be denoted

$$\mathbf{x} = \{x_1, \ldots, x_n\}$$

where $x_i \in W$ are the individual points of the process, and the total number of points $n \geq 0$ is not fixed.

4.1 Formulation of Models

The point process models fitted in spatstat are Gibbs point processes, cf. [5, 47, 51]. The scope of possible models is very wide: they may include spatial trend, dependence on covariates, interpoint interactions of any order (i.e. we are not restricted to pairwise interactions), and dependence on marks.

Each model will be specified in terms of its *conditional intensity* rather than its likelihood. This turns out to be an intuitively appealing way to formulate point process models, as well as being necessary for technical reasons.

The (Papangelou) conditional intensity is a function $\lambda(u, \mathbf{x})$ of spatial location $u \in W$ and of the entire point pattern $\mathbf{x}$. Roughly speaking, if we consider an infinitesimal region around the point u of area du, then the conditional probability that the point process contains a point in this infinitesimal region, given the position of all points outside this region, is $\lambda(u, \mathbf{x}) \, du$. See [5, 17] and the excellent surveys by Ripley [61, 62].

For example, the homogeneous Poisson process (complete spatial randomness, CSR) has conditional intensity

$$\lambda(u, \mathbf{x}) = \beta \tag{1}$$

where β is the intensity (expected number of points per unit area). The inhomogeneous Poisson process with local intensity function $\beta(u)$, $u \in \mathbb{R}^2$, has conditional intensity

$$\lambda(u, \mathbf{x}) = \beta(u). \tag{2}$$

The Strauss process, a simple model of dependence between points, has conditional intensity

$$\lambda(u, \mathbf{x}) = \beta \gamma^{t(u, \mathbf{x})} \tag{3}$$

where $t(u, \mathbf{x})$ is the number of points of the pattern $\mathbf{x}$ that lie within a distance r of the location u. Here γ is the interaction parameter, satisfying $0 \leq \gamma \leq 1$, and $r > 0$ is the interaction radius.

The conditional intensity is a useful modelling tool because its functional form has a straightforward interpretation. The simplest form is a constant, $\lambda(u, \mathbf{x}) \equiv \beta$, which corresponds to "complete spatial randomness" (a uniform Poisson process). In most applications, this would be the null model. A conditional intensity $\lambda(u, \mathbf{x})$ which depends only on the location u, say $\lambda(u, \mathbf{x}) = \beta(u)$, corresponds to an inhomogeneous Poisson process with intensity function $\beta(u)$. In this case the functional form of $\beta(u)$ indicates the type of inhomogeneity (or "spatial trend").

A conditional intensity $\lambda(u, \mathbf{x})$ which depends on the point pattern $\mathbf{x}$, as well as on the location u, corresponds to a point process which exhibits stochastic dependence between points. For example, in the Strauss process (3) with $\gamma < 1$, dependence between points is reflected in the fact that the conditional probability of finding a point of the process at the location u is reduced if other points of the process are present within a distance r. In the special case $\gamma = 0$, the conditional probability of finding a point at u is zero if there are any other points of the process within a distance r of this location.

4.2 Scope of Models

Our technique [5] fits any model for which the conditional intensity is of the loglinear form

$$\lambda(u, \mathbf{x}) = \exp(\psi^{\mathsf{T}} B(u) + \varphi^{\mathsf{T}} C(u, \mathbf{x})) \tag{4}$$

where $\theta = (\psi, \varphi)$ are the parameters to be estimated. Both ψ and φ may be vectors of any dimension, corresponding to the dimensions of the vector-valued statistics $B(u)$ and $C(u, \mathbf{x})$ respectively.

The term $B(u)$ depends only on the spatial location u, so it represents "spatial trend" or spatial covariate effects. The term $C(u, \mathbf{x})$ represents "stochastic interactions" or dependence between the points of the random point process. For example $C(u, \mathbf{x})$ is absent if the model is a Poisson process.

Gibbs models may require reparametrisation in order to conform to (4). For example, the Strauss process conditional intensity (3) satisfies (4) if we set $B(u) \equiv 1$ and $C(u, \mathbf{x}) = t(u, \mathbf{x})$, and take the parameters to be $\psi = \log \beta$ and $\varphi = \log \gamma$.

In practice there is an additional constraint that the terms $B(u)$ and $C(u, \mathbf{x})$ must be implemented in software. Some point process models which belong to the class of Gibbs processes have a conditional intensity which is difficult to evaluate. Notable examples include Cox processes [8]. For these models, other approaches should be used [51].

4.3 Model-fitting Algorithm

Our software currently fits models by the method of maximum pseudolikelihood (in Besag's sense [13]), using a computational device developed for Poisson models by Berman & Turner [12] which we adapted to pseudolikelihoods of general Gibbs point processes in [5]. Although maximum pseudolikelihood may be statistically inefficient [42, 43], it is adequate in many practical applications [63] and it has the virtue that we can implement it in software with great generality. Future versions of spatstat will implement the Huang-Ogata improvement to maximum pseudolikelihood [39] which is believed to be highly efficient.

Let the point pattern dataset $\mathbf{x}$ consist of n points $x_1, \ldots, x_n$ in a spatial region $W \subseteq \mathbb{R}^d$. Consider a point process model governed by a parameter

θ and having conditional intensity $\lambda_\theta(u, \mathbf{x})$. The *pseudolikelihood* [13] of the model is

$$\mathrm{PL}(\theta; \mathbf{x}) = \prod_{i=1}^{n} \lambda_\theta(x_i; \mathbf{x}) \exp\left(-\int_W \lambda_\theta(u; \mathbf{x})\, du\right) \qquad (5)$$

The *maximum pseudolikelihood estimate* of θ is the value which maximises $\mathrm{PL}(\theta; \mathbf{x})$. Now discretise the integral in (5) to obtain

$$\int_W \lambda_\theta(u; \mathbf{x})\, du \approx \sum_{j=1}^{m} \lambda_\theta(u_j; \mathbf{x}) w_j. \qquad (6)$$

where $u_j \in W$ are "quadrature points" and $w_j \geq 0$ the associated "quadrature weights" for $j = 1, \ldots, m$. The quadrature scheme should be chosen so that (6) is a good approximation.

The Berman-Turner [12] device involves choosing a set of quadrature points $\{u_j\}$ which *includes all the data points* x_i as well as some other ("dummy") points. Let z_j be the indicator which equals 1 if u_j is a data point, and 0 if it is a dummy point. Then the logarithm of the pseudolikelihood can be approximated by

$$\log \mathrm{PL}(\theta; \mathbf{x}) \approx \sum_{j=1}^{m} [z_j \log \lambda_\theta(u_j; \mathbf{x}) - w_j \lambda_\theta(u_j; \mathbf{x})]$$

$$= \sum_{j=1}^{m} w_j(y_j \log \lambda_j - \lambda_j) \qquad (7)$$

where $y_j = z_j/w_j$ and $\lambda_j = \lambda_\theta(u_j, \mathbf{x})$. The key to the Berman-Turner device is to recognise that the right hand side of (7) has the same functional form as the log likelihood of m independent Poisson random variables Y_j with means λ_j and responses y_j. This enables us to maximise the pseudolikelihood using standard statistical software for fitting generalised linear models.

Given a point pattern dataset and a model of the form (4), our algorithm constructs a suitable quadrature scheme $\{(u_j, w_j)\}$, evaluates the vector valued sufficient statistic $s_j = (B(u_j), C(u_j, \mathbf{x}))$, forms the indicator variable z_j and the pseudo-response $y_j = z_j/w_j$, then calls standard R software to fit the Poisson loglinear regression model $Y_j \sim \mathsf{Poisson}(\lambda_j)$ where $\log \lambda_j = \theta s_j$. The fitted coefficient vector $\widehat{\theta}$ given by this software is returned as the maximum pseudolikelihood estimate of θ. For further explanation see [5]. Advantages of using existing software to compute the fitted coefficients include its numerical stability, reliability, and most of all, its flexibility.

5 Model-fitting in spatstat

5.1 Overview

The model-fitting function is called ppm and is strongly analogous to lm or glm. In simple usage, it is called in the form

```
ppm(X, trend, interaction, ...)
```

where X is the point pattern dataset, trend describes the spatial trend (the function $B(u)$ in equation (4)) and interaction describes the stochastic dependence between points in the pattern (the function $C(u, \mathbf{x})$ in equation (4)). Other arguments to ppm may provide covariates, select edge corrections, and control the fitting algorithm.

For example

```
ppm(X, ~1, Strauss(r=0.1), ....)
```

fits the stationary Strauss process (3) with interaction radius $r = 0.1$. The spatial trend formula ~1 is a constant, meaning the process is stationary. The argument Strauss(r=0.1) is an object representing the interpoint interaction structure of the Strauss process with interaction radius $r = 0.1$.

Similarly

```
ppm(X, ~x + y, Poisson())
```

fits the non-stationary Poisson process with a *loglinear* intensity of the form

$$\beta(x, y) = \exp(\theta_0 + \theta_1 x + \theta_2 y)$$

where $\theta_0, \theta_1, \theta_2$ are (scalar) parameters to be fitted, and x, y are the Cartesian coordinates.

5.2 Spatial Trend Terms

The trend argument of ppm describes any spatial trend and covariate effects. It must be a formula expression in the R language, and serves a role analogous to the formula for the linear predictor in a generalised linear model. See e.g. [72, Sect 6.2].

The right hand side of trend specifies the function $B(u)$ in equation (4) following the standard R syntax for a linear predictor. The terms in the formula may include the reserved names x, y for the Cartesian coordinates. Spatial covariates may also appear in the trend formula as we explain in Sect. 7.

Effectively, the function $B(u)$ in (4) is treated as the "systematic" component of the model. Note that the link function is always the logarithm, so the model formula in a ppm call is always a description of the **logarithm** of the conditional intensity.

The default trend formula is ~1, which indicates $B(u) \equiv 1$, corresponding to a process without spatial trend or covariate effects. The formula ~x indicates the vector statistic $B((x, y)) = (1, x)$ corresponding to a spatial trend of the form $\exp(\psi B((x, y))) = \exp(\alpha + \beta x)$, where α, β are coefficient parameters to be estimated, while ~x + y indicates $B((x, y)) = (1, x, y)$ corresponding to $\exp(\psi B((x, y))) = \exp(\alpha + \beta x + \gamma y)$.

A wide variety of model terms can easily be constructed from the Cartesian coordinates. For example

```
ppm(X, ~ ifelse(x > 2, 0, 1), Poisson())
```

fits an inhomogeneous Poisson process with different, constant intensities on each side of the line $x = 2$.

spatstat provides a function polynom which generates polynomials in 1 or 2 variables. For example

```
~ polynom(x, y, 2)
```

represents a polynomial of order 2 in the Cartesian coordinates x and y. This would give a "log-quadratic" spatial trend.[3]
Similarly

```
~ harmonic(x, y, 2)
```

represents the most general *harmonic* polynomial of order 2 in x and y.

Other possibilities include B-splines and smoothing splines, fitted with bs and s respectively. These terms introduce smoothing penalties, and thus provide an implementation of "penalised maximum pseudolikelihood" estimation (cf. [30]).

The special term offset can also be used in the trend formula. It has the same role in ppm as it does in other model-fitting functions, namely to add to the linear predictor a term which is not associated with a parameter. For example

```
~ offset(x)
```

will fit the model with log trend $\beta + x$ where β is the only parameter to be estimated.

Observed spatial covariates may also be included in the trend formula; see Sect. 7 below.

5.3 Interaction Terms

The dependence structure or "interpoint interaction" in a point process model is determined by the function $C(u, \mathbf{x})$ in (4). This term is specified by the interaction argument of ppm, which is strongly analogous to the family argument to glm. Thus, interpoint interaction is regarded as a "distributional" component of the point process model, analogous to the distribution family in a generalised linear model.

The interaction argument is an object of a special class "interact". The user creates such objects using specialised spatstat functions, similar to those which create the family argument to glm. For example, the command

---

[3] We caution against using the standard function poly for the same purpose here. For a model formula containing poly, prediction of the fitted model can be erroneous, for reasons which are well-known to R users. The function polynom provided in spatstat does not exhibit this problem.

```
Strauss(r=0.1)
```

will create an object of class "interact" representing the interaction function $C(u, \mathbf{x})$ for the Strauss process (3) with interaction radius r. This object is then passed to the model-fitting function ppm, usually in the direct form

```
ppm(cells, ~1, Strauss(r=0.1))
```

The following functions are supplied for creating interpoint interaction structures; details of these models can be consulted in [5].

Poisson . . . . Poisson process
Strauss . . . . Strauss process
StraussHard . . . . . . Strauss process with a hard core
Softcore . . . . Pairwise soft core interaction
PairPiece . . . . Pairwise interaction, step function potential
DiggleGratton . . . Diggle-Gratton potential
LennardJones . . . . Lennard-Jones potential
Geyer . . . . Geyer's saturation process
OrdThresh . . . . Ord's process, threshold on cell area

Note that ppm estimates only the "canonical" parameters of a point process model. These are parameters θ such that the conditional intensity is log-linear in θ, as in equation (4). Other so-called "irregular" parameters (such as the interaction radius r of the Strauss process) cannot be estimated by the Berman-Turner device, and their values must be specified a priori, as arguments to the interaction function. Estimation of irregular parameters is discussed in Sect. 9.

For more advanced use, the following functions will accept "user-defined potentials" in the form of an arbitrary R language function. They effectively allow arbitrary point process models of these three classes.

Pairwise . . . . Pairwise interaction, user-supplied potential
Ord . . . . Ord model, user-supplied potential
Saturated . . . Saturated pairwise model, user-supplied potential

6 Fitted Models

The value returned by ppm is a "fitted point process model" of class "ppm". It can be stored, inspected, plotted, predicted and updated. The following would be typical usage:

```
fit <- ppm(X, ~1, Strauss(r=0.1), ...)
fit
plot(fit)
pf <- predict(fit)
coef(fit)
```

Methods are provided for the following generic operations applied to "ppm" objects:

print   Print basic information
summary Print extensive summary information
coef   Extract fitted model coefficients
plot   Plot fitted intensity
fitted   Compute fitted conditional intensity or trend at data points
predict Compute predictions (spatial trend, conditional intensity)
update   Update the fit

Printing the fitted object fit will produce text output describing the fitted model. Plotting the object will display the spatial trend and the conditional intensity, as perspective plots, contour plots and image plots.

6.1 Interpretation of Fitted Coefficients

The easiest way to interpret a fitted point process model is to print it at the terminal. The print method attempts to produce a comprehensible description. For example,

```
> ppm(swedishpines, ~1, Strauss(7))
Stationary Strauss process
First order term:
        beta
0.01823799
Interaction: Strauss process
interaction distance:    7
Fitted interaction parameter gamma:
   [1] 0.2472
```

Thus the fitted model is the stationary Strauss process (3) with parameters $\beta = 0.01823799$ and $\gamma = 0.2472$.

Alternatively the coefficients of the fitted model may be extracted using coef. These should be interpreted as the canonical parameters $\theta = (\psi, \varphi)$ appearing in (4). For example

```
> u <- ppm(swedishpines, ~1, Strauss(7))
> coef(u)
(Intercept) Interaction
  -4.004248   -1.397759
```

Comparing (4) with (3) we see that the usual parameters β, γ of the Strauss process are $\beta = \exp \psi$ and $\gamma = \exp \varphi$, so typing

```
> exp(coef(u))  .
(Intercept) Interaction
 0.01823799  0.24715026
```

shows that the fitted parameters are $\beta = 0.0182$ and $\gamma = 0.2472$.

If the model includes a spatial trend, then the fitted *canonical* coefficients of the trend will be presented using the standard R conventions. For example

```
> ppm(swedishpines, ~x)
Nonstationary Poisson process
Trend formula: ~x
Fitted coefficients for trend formula:
(Intercept)          x
-5.13707319  0.00462614
```

indicates that the fitted model is a Poisson process with intensity function

$$\beta((x,y)) = \exp(-5.13707319 + 0.00462614x).$$

In more complex models, the interpretation of the fitted coefficients may depend on the choice of contrasts for the coefficients of linear models. For example, if the treatment contrasts [72, sect 6.2] are in force, then a model involving a factor will be printed as follows:

```
> ppm(swedishpines, ~factor(ifelse(x < 50, "left", "right")))
Nonstationary Poisson process
Trend formula: ~factor(ifelse(x < 50, "left", "right"))
Fitted coefficients for trend formula:
                                         (Intercept)
                                             -5.075
factor(ifelse(x < 50, "left", "right"))right
                                          0.331
```

The explanatory variable is a factor with two levels, left and right. By default the levels are sorted alphabetically. Since we are using the treatment contrasts, the value labelled "(Intercept)" is the fitted coefficient for the first level (left), while the value labelled "right" is the estimated treatment contrast (to be *added* to the intercept) for the level right. This indicates that the fitted model is a Poisson process with intensity $\exp(-5.075) = 0.00625$ on the left half of the dividing line $x = 50$, and intensity $\exp(-5.075 + 0.331) = 0.00871$ on the right side.

6.2 Invalid Models

For some values of the parameters, a point process model with conditional intensity (4) may be "invalid" or "undefined", in the sense that the corresponding probability density is not integrable. For example, the Strauss process (3) is defined only for $0 \leq \gamma \leq 1$ (equivalently for $\varphi \leq 0$); the density is not integrable if $\gamma > 1$, as famously announced in [44].

A point process model fitted by ppm may sometimes be invalid in this sense. For example, a fitted Strauss process model may sometimes have a value of γ greater than 1. This happens because, in the Berman-Turner device, the conditional intensity (4) is treated as if it were the mean in a Poisson loglinear regression model. The latter model is well-defined for all values of the linear predictor, so the software does not constrain the values of the canonical parameters ψ, φ in (4).

The spatstat package has internal procedures for deciding whether a fitted model is valid, and for mapping or "projecting" an invalid model to the nearest valid model. Currently these procedures are invoked only when we simulate a realisation of the fitted model. They are *not* invoked when a model is printed or when it is returned from ppm, so that the printed output from ppm may represent an invalid model.

6.3 Predicting and Plotting a Fitted Model

The predict method for a fitted point process model computes either the fitted spatial trend

$$\tau(u) = \exp(\widehat{\psi}B(u)) \tag{8}$$

or the fitted conditional intensity

$$\lambda_{\widehat{\theta}}(u, \mathbf{x}) = \exp(\widehat{\psi}B(u) + \widehat{\varphi}C(u, \mathbf{x})) \tag{9}$$

at arbitrary locations u. Note that $\mathbf{x}$ is always taken to be the observed data pattern to which the model was fitted.

The default behaviour is to produce a pixel image of both trend and conditional intensity, where these are nontrivial. A typical example is the following:

```
data(cells)
m <- ppm(cells,~polynom(x,y,2),Strauss(0.05), rbord=0.05)
trend <- predict(m,type="trend",ngrid=100)
cif   <- predict(m,type="cif",ngrid=100)
```

The resulting objects trend and cif are pixel images. One could then plot the resulting surfaces with calls like

```
persp(trend)
persp(cif, theta=-30,phi=40,d=4,ticktype="detailed",zlab="z")
```

We caution again that the result of predict may be incorrect if the trend formula of the point process model contains one of the functions poly, bs, lo, or ns.

The plot method (plot.ppm) will take a fitted point process model and plot the trend and/or the conditional intensity. By default this surface is calculated at a 40×40 grid of points on the (enclosing rectangle of) the observation window. The plots may be produced as perspective plots, images, or contour plots. For example

```
plot(fit,cif=FALSE,how="persp")
```

will generate a perspective plot of the fitted trend, where fit is the fitted model.

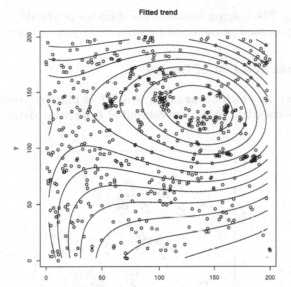

Fig. 4. Contour plot of fitted log-cubic trend for Longleaf Pines data (locations only) obtained using `plot.ppm`

7 Models with Covariates

In order to model the dependence of a point pattern on a spatial covariate, there are several requirements. First, the covariate must be a quantity $Z(u)$ observable (in principle) at each location u in the window (e.g. altitude, soil pH, or distance to another spatial pattern). There may be several such covariates, and they may be continuous-valued or factors. Second, the values $Z(x_i)$ of Z at each point of the data point pattern must be available. Thirdly, the values $Z(u)$ at *some* other points u in the window must be available. The accuracy of the algorithm depends on the number of these additional points and on their spatial arrangement. For a good approximation to the pseudo-likelihood, the density of the additional points should be high throughout the window.

The argument `covariates` to the function `ppm` specifies the values of the spatial covariates. It may be either a data frame or a list of pixel images.

(a) If `covariates` is a list of pixel images, then each image is assumed to contain the values of a spatial covariate at a fine grid of spatial locations. The names of the list entries should be the names of the covariates used in the trend formula when you call `ppm`.

(b) If `covariates` is a data frame, then the ith row of the data frame is expected to contain the covariate values for the ith "quadrature point" (see below). The column names of the data frame should be the names of the covariates used in the trend formula when you call ppm.

7.1 Covariates in a List of Images

The format (a), in which covariates is a list of images, would typically be used when the covariate values are computed from other data.

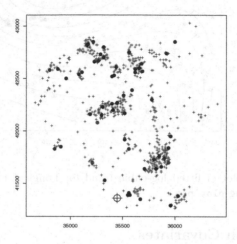

Fig. 5. The Chorley-Ribble data [27]. Cases of cancer of the larynx (•) and lung (+) in the Chorley-Ribble region of Lancashire, England, and the location of a disused industrial incinerator (⊕)

For example, Fig. 5 shows a spatial epidemiological dataset containing a point pattern X of disease cases, and another point pattern Y of control cases. We want to model X as a point process with intensity proportional to the local density ϱ of the susceptible population. We estimate ϱ by taking a kernel-smoothed estimate of the intensity of Y. Thus

```
rho.hat <- ksmooth.ppp(Y, sigma=1.2)
ppm(X, ~offset(log(rho)), covariates=list(rho=rho.hat))
```

The first line computes the values of the kernel-smoothed intensity estimate at a fine grid of pixels, and stores them in the pixel image object rho.hat (plotted in Fig. 6). The second line fits the Poisson process model with log intensity

$$\log \lambda(u) = \psi + \log \varrho(u) \tag{10}$$

where ψ is an unknown parameter; that is, it fits the Poisson model with intensity

$$\lambda(u) = \mu \, \varrho(u) \tag{11}$$

where $\mu = e^{\psi}$ is the only parameter to be estimated. Note that covariates must be a list, even though there is only one covariate. The variable name rho in the model formula must match the name rho in the list.

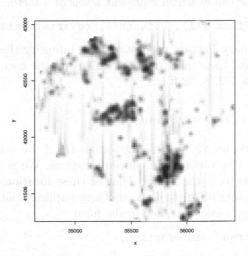

Fig. 6. Kernel smoothed intensity estimate $\widehat{\varrho}(u)$ of the lung cancer data from Fig. 5, which serves as a spatial covariate for modelling the laryngeal cancer data

Typical output is as follows:

```
> ppm(X, ~offset(log(rho)), Poisson(), data=list(rho=rho.hat))
Nonstationary Poisson process
Trend formula:  ~offset(log(rho))
Fitted coefficients for trend formula:
(Intercept)
   -2.889
```

This indicates that the estimate of the parameter ψ in (10) is $\widehat{\psi} = -2.889$. Equivalently the estimate of μ in (11) is $\widehat{\mu} = e^{-2.889} = 0.056$.

More complex models may be fitted to explore other effects by adding terms to the trend formula. For example

```
ppm(X, ~ x + offset(log(rho)), data=list(rho=rho.hat))
```

would fit a nonstationary Poisson model with intensity

$$\beta((x,y)) = e^{\psi + \varphi x} \varrho((x,y)).$$

Covariates represented by pixel images in spatstat may have values that are numerical, complex, logical, or character strings. Unfortunately a pixel image in spatstat cannot have categorical (factor) values, because R refuses to create a factor-valued matrix. In order to represent a categorical variate as a pixel image, the categorical values should be encoded as integers (for efficiency's sake) and assigned to an integer-valued pixel image. Then the model formula should invoke the `factor` command on this image. For example if `fim` is an image with integer values which represent levels of a factor, then

```
ppm(X, ~factor(f), Poisson(), covariates=list(f=fim))
```

fits the nonstationary Poisson process with an intensity that depends on the levels of this factor. Care must be taken to ensure the correct interpretation of the factor levels [72, p. 22 ff.].

7.2 Covariates in a Data Frame

Typically we would use the data frame format (b) if the values of the spatial covariates can only be observed at certain locations. We need to force `ppm` to use these locations to fit the model. That is, these locations must be used as the quadrature points u_j in the Berman-Turner approximation (6).

The function `ppm` may be called in the form

```
ppm(Q, trend, interaction, ...)
```

where `Q` is a "quadrature scheme" and the other arguments are unchanged. A quadrature scheme in `spatstat` is an object of a special class `"quad"` which comprises both *"data points"* (the points of the observed point pattern) and *"dummy points"* (some other locations in the window). It is usually created using the function `quadscheme`.

In the present context we will need to create a quadrature scheme based on the spatial locations where the covariate Z has been observed. Then the values of the covariate at these locations are passed to `ppm` through the data frame `covariates`.

For example, suppose that `X` is the observed point pattern and we are trying to model the effect of soil acidity (pH). Suppose we have measured the values of soil pH at the points x_i of the point pattern, and stored them in a vector `XpH`. Suppose we have measured soil pH at some other locations u in the window, and stored the results in a data frame `U` with columns `x`, `y`, `pH`. Then do as follows:

```
Q <- quadscheme(data=X, dummy=list(x=U$x, y=U$y))
df <- data.frame(pH=c(XpH, U$pH))
```

Then the rows of the data frame `df` correspond to the quadrature points in the quadrature scheme `Q`. To fit just the effect of pH, we type

```
ppm(Q, ~ pH, Poisson(), covariates=df)
```

where the term `pH` in the formula `~ pH` agrees with the column label `pH` in the argument `covariates = df`. This will fit an inhomogeneous Poisson process with intensity that is a loglinear function of soil pH. We could can also try (say)

```
ppm(Q, ~ pH, Strauss(r=1), covariates=df)
ppm(Q, ~ factor(pH > 7), Poisson(), covariates=df)
ppm(Q, ~ polynom(x, 2) * factor(pH > 7), covariates=df)
```

8 Fitting Models to Multitype Point Patterns

The function ppm will also fit models to multitype point patterns. A multitype point pattern is a point pattern in which the points are each classified into one of a finite number of possible types (e.g. species, colours, on/off states). In spatstat a multitype point pattern is represented by a "ppp" object X whose marks are a factor. Fig. 3(b) shows an example.

Currently, ppm will not fit models to a marked point pattern if the marks are not a factor.

8.1 Conditional Intensity

A multitype point process in a region $W \subset \mathbb{R}^2$, with a set $\mathcal{M}$ of possible types, may be regarded as a point process in $W \times \mathcal{M}$. Let $\mathbf{y} = \{(x_1, m_1), \ldots, (x_n, m_n)\}$ denote a typical realisation of the process, where $x_i \in W$ are the locations and $m_i \in \mathcal{M}$ the corresponding marks (types).

The conditional intensity is now of the form $\lambda((u, m), \mathbf{y})$, where $u \in W$ and $m \in \mathcal{M}$. It has the interpretation that $\lambda((u, m), \mathbf{y}) \, du$ is the conditional probability of finding a point *of type* m in an infinitesimal neighbourhood of the point u, given that the rest of the process coincides with $\mathbf{y}$.

This introduces some subtleties. A conditional intensity function which is constant,

$$\lambda((u, m), \mathbf{y}) = \beta \tag{12}$$

corresponds to a process in which **the points of each type** $m \in \mathcal{M}$ constitute a uniform Poisson process with intensity β. By standard properties of the Poisson process [45], this is equivalent to a marked Poisson process of total intensity $M\beta$ (where M is the number of possible types $M = |\mathcal{M}|$) in which the points have independent random marks, with equal probability $1/M$ for each possible type.

A conditional intensity function which depends only on the marks,

$$\lambda((u, m), \mathbf{y}) = \beta_m \tag{13}$$

where β_m, $m \in \mathcal{M}$ are constants, is a marked Poisson process of total intensity $\mu = \sum_m \beta_m$, in which the points have independent random marks, with probability $p_m = \beta_m/\mu$ for type m.

The most general multitype Poisson process has conditional intensity

$$\lambda((u, m), \mathbf{y}) = \beta_m(u) \tag{14}$$

where $\beta_m(u), m \in \mathcal{M}$ are arbitrary nonnegative integrable functions. This process has total intensity $\mu(u) = \sum_m \beta_m(u)$. The marks are independent but not identically distributed: a point at location u has conditional mark distribution $p_m(u) = \beta_m(u)/\mu(u)$ for $m \in \mathcal{M}$.

8.2 Multitype Models

Trend Component

In order to represent the dependence of the trend on the marks, the trend formula passed to ppm may involve the reserved name marks.

The trend formula ~1 states that the trend is constant and does not depend on the marks, as in (12). The formula ~marks indicates that there is a separate, constant intensity for each possible mark, as in (13). If a uniform multitype Poisson process is to be fitted to data, the usual intention is to allow for different intensities for each mark, so the appropriate call would be

```
ppm(X, ~ marks, Poisson())
```

The result of fitting this model to the data in Fig. 3(b) yields the following output.

```
Stationary multitype Poisson process
Possible marks: off on
Intensity: Trend formula: ~marks
Fitted intensities:
beta_off  beta_on
88.68302 94.92830
```

This indicates that the fitted model is a multitype Poisson process with intensities 88.7 and 94.9 for the points of type "off" and "on" respectively.

In more elaborate cases, the trend formula may involve both the marks and the spatial locations or spatial covariates. For example the trend formula ~marks + polynom(x,y,2) signifies that the first order trend is a log-quadratic function of the Cartesian coordinates, multiplied by a constant factor depending on the mark. The formulae

```
~ marks * polynom(x,2)
~ marks + marks:polynom(x,2)
```

both specify that, for each mark, the first order trend is a different log-quadratic function of the Cartesian coordinates. The second form looks "wrong" since it includes a "marks by polynom" interaction without having polynom in the model, but since polynom is a covariate rather than a factor this is is allowed, and makes perfectly good sense. As a result the two foregoing models are in fact mathematically equivalent. However, the fitted model objects will give slightly different output.

For example, the first model ~marks * polynom(x,2) fitted to the data in Fig. 3(b) gives the following output (assuming options("contrasts") is set to its default, namely the "treatment" contrasts):

```
Nonstationary multitype Poisson process
Trend formula: ~marks * polynom(x, 2)
Fitted coefficients for trend formula:
              (Intercept)                    markson
              4.3127945                    0.2681231
      polynom(x, 2)[x]             polynom(x, 2)[x^2]
              0.4651860                   -0.2363352
  markson:polynom(x, 2)[x]  markson:polynom(x, 2)[x^2]
             -0.6781045                    0.4023491
```

This form of the model gives two quadratic functions: a "baseline" quadratic

$$P_0(x, y) = 4.3127945 + 0.4651860x - 0.2363352x^2$$

and a quadratic associated with the mark level "on",

$$P_{on}(x, y) = 0.2681231 - 0.6781045x + 0.4023491x^2.$$

The baseline quadratic is the logarithm of the fitted trend for the points of type off, since off is the first level of the factor marks. For points of type on, since we are using the treatment contrasts, the log trend is

$$P_0(x, y) + P_{on}(x, y) = 4.580918 - 0.2129185x + 0.1660139x^2.$$

On the other hand, when the model ~marks + marks:polynom(x,2)) is fitted to the same dataset, the output is

```
Nonstationary multitype Poisson process
Trend formula: ~marks + marks:polynom(x, 2)
Fitted coefficients for trend formula:
              (Intercept)                    markson
              4.3127945                    0.2681231
  marksoff:polynom(x, 2)[x]   markson:polynom(x, 2)[x]
              0.4651860                   -0.2129185
  marksoff:polynom(x, 2)[x^2]  markson:polynom(x, 2)[x^2]
             -0.2363352                    0.1660138
```

This says explicitly that the log trend for points of type off is

$$Q_{off}(x, y) = 4.3127945 + 0.4651860x - 0.2363352x^2$$

while for points of type on it is

$$Q_{on}(x, y) = 4.580918 - 0.2129185x + 0.1660139x^2.$$

Hence the two fitted models are mathematically identical.

Interaction Component

For the interaction component of the model, any of the interactions listed above for unmarked point processes may be used. However these interactions do not depend on the marks. We have additionally defined two interactions which do depend on the marks:

MultiStrauss   multitype Strauss process

MultiStraussHard   multitype Strauss/hard core

For the multitype Strauss process, a matrix of "interaction radii" must be specified. If there are m distinct levels of the marks, we require a matrix r in which r[i,j] is the interaction radius r_{ij} between types i and j. For the multitype Strauss/hard core model, a matrix of "hardcore radii" must be supplied as well. These matrices will be of dimension $m \times m$ and must be symmetric.

9 Irregular Parameters

As explained in Sect. 4.2, our model-fitting technique [5] estimates the parameters θ which appear in loglinear form (4) in the conditional intensity. We call these "regular" parameters, while other model parameters are called "irregular". Most of the familiar point process models have irregular parameters controlling the scale or range of interaction: an example is the interaction radius r of the Strauss process (3). Irregular parameters cannot be estimated directly using our algorithm, and must be given a fixed value in any call to ppm.

Very little theory is available about the estimation of irregular parameters. An exception is the case of hard-core radii. For example, consider the classical hard-core process, which is the special case of the Strauss process (3) with $\gamma = 0$. It can easily be shown [64] that the maximum likelihood and maximum pseudolikelihood estimate of r is

$$\widehat{r} = \min_i \min_{j \neq i} ||x_i - x_j||,$$

the minimum interpoint distance in the point pattern $\mathbf{x}$.

Some irregular parameters can be determined from the pair correlation function or the K-function [28, 29, 34]. For the Strauss process with $\gamma < 1$, the pair correlation function has a jump at r. This leads to a useful procedure for estimating r called the "cusp method" [34, 66], [68, p. 333]. These methods are not yet implemented in spatstat.

One general strategy available in spatstat for estimating irregular parameters is *profile pseudolikelihood* [5, Sect. 8.2]. Let θ and ψ denote the regular and irregular parameters respectively, and write the pseudolikelihood as $\mathrm{PL}(\theta, \psi; \mathbf{x})$. Define the profile pseudolikelihood for ψ to be

$$\mathrm{PPL}(\psi; \mathbf{x}) = \max_{\theta} \mathrm{PL}(\theta, \psi; \mathbf{x}), \qquad (15)$$

the maximum value of pseudolikelihood obtained by maximising over the regular parameters with ψ held fixed. Then the maximum pseudolikelihood estimate of ψ is the value which maximises the profile pseudolikelihood,

$$\widehat{\psi} = \mathrm{argmax}_{\psi} \, \mathrm{PPL}(\psi; \mathbf{x}). \qquad (16)$$

In spatstat the profile pseudolikelihood (15) can be evaluated for any given value of ψ by fitting the model with this value of ψ using ppm. The resulting fitted model object has a component named maxlogpl which gives the maximised log pseudolikelihood.

For example, the following code computes the maximum pseudolikelihood estimate of the interaction radius r in the Strauss process model for the cells dataset.

```
data(cells)
rval <- seq(0.01, 0.2, by=0.01)
prof <- numeric(length(rval))
for(i in seq(rval)) {
    fit <- ppm(cells, ~1, Strauss(r=rval[i]),
                         correction="translate")
    prof[i] <- fit$maxlogpl
}
iopt <- min(which(prof == max(prof)))
rhat <- rval[iopt]
```

Note that the same edge correction must be used to fit each model in order that the pseudolikelihood values be comparable. The example above shows the translation edge correction. If the border edge correction is used, the correction distance rbord should be fixed at the maximum interaction radius of all models to be fitted.

For diagnostic purposes the profile pseudolikelihood prof should be plotted against the r argument rval to verify that the function has a unique global maximum. In the example shown above, there is an unambiguous peak in profile likelihood at $\widehat{r} = 0.1$. Section 12.2 gives an example where more care is required.

10 Model Validation

Having fitted a point process model to data, it is important to "validate" the model, i.e. to check formally or informally that the model is a good fit to the data, and that all terms in the model are appropriate [2, 16, 24, 72], [50, Chap. 12].

10.1 Residuals and Diagnostics

Residuals from the fitted model are an important diagnostic tool in other areas of applied statistics, but in spatial statistics they have only recently been developed [7, 46, 65]. The residuals and diagnostic plots introduced in [7] are available in spatstat. The function diagnose.ppm is the analogue of plotting the residuals against the covariates in a linear model, while qqplot.ppm is the analogue of a Q–Q plot of the residuals in a linear model.

These techniques are particularly well suited to detecting spatial inhomogeneity. For example, Fig. 7 (left) shows a point pattern simulated from the Poisson process with intensity $\lambda(x, y) = 300 \exp(-3|x - 0.5|)$ in the unit square. We then fitted the incorrect model, a *uniform* Poisson process, to these data. The right side of Fig. 7 shows the result of diagnose.ppm for this incorrect model. The striking deviations of the plots from their nominal (constant) values indicate clearly that the model is inappropriate, and suggest the form of departure from the model. For detailed information and examples, see [7] and the help files for these functions.

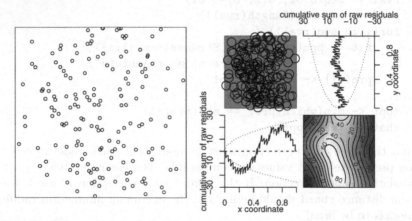

Fig. 7. Demonstration of diagnostic tools in spatstat. (*Left*) data point pattern, a realisation of an inhomogeneous Poisson process; (*Right*) diagnostic plots (generated by diagnose.ppm) for an incorrect model, a uniform Poisson process, fitted to the data

10.2 Formal Inference

Techniques

Formal hypothesis tests are often applied in spatial statistics for the following purposes:

1. to test whether the point pattern is a realisation of a uniform Poisson process (complete spatial randomness or CSR);

2. to assess the goodness-of-fit of a point process model that has been fitted to the point pattern data;
3. to select models (e.g. to decide whether a particular term in a point process model may be omitted).

The first type of test is the most popular in applications, following Ripley [57, 59]. However, this test may often be inappropriate or uninformative: we usually know that the data are not completely random, and a formal confirmation of this statement is not scientifically informative since it does not indicate the kind of departure from complete randomness. Normally the second and third types of tests are more useful in modelling.

Statistical theory of parameter estimation and hypothesis testing for spatial point processes is rather limited. See the recent surveys of Møller and Waagepetersen [51, Chaps. 8 10], Diggle [28, Chap. 2] and van Lieshout [47, Chap. 3].

Techniques available for formal inference depend on the class of models envisaged. For **Poisson processes** (homogeneous or inhomogeneous), much of the classical theory of maximum likelihood is applicable, including the likelihood ratio test. Goodness-of-fit tests based on the χ^2 distribution are also possible after discretisation (binning) of the data. For **Cox processes**, estimation methods include minimum contrast [51, p. 182] and maximum likelihood in special cases [51, Sect. 10.3]. In the latter case, a likelihood ratio test is applicable. For **Gibbs processes**, we may use Monte Carlo maximum likelihood [51, Sect. 9.1.4] which provides *approximate* maximum likelihood estimates, confidence intervals and likelihood ratio tests. Bayesian inference has also also been developed [51, Sects. 9.3, 10.4]. For **Gibbs processes**, we may also use maximum pseudolikelihood [51, Sect. 9.2]. Maximum pseudolikelihood estimates are known to be consistent and asymptotically normal in some contexts [42, 43], but at the time of writing there is no statistical theory for hypothesis tests based on the pseudolikelihood. For **general point processes** which can be simulated, some elementary *simulation-based inference* is feasible [51, Chap. 8]. The canonical example is the Monte Carlo test [14, 37, 59]. Monte Carlo tests based on envelopes of simulations of the K-function (and other summary functions) are very popular [28].

It is not known whether maximum likelihood estimation is optimally efficient [23, Sect. 8.5.8], [60]. Little is known about the power of the various tests mentioned above [28, p. 28]. The distributional information required for statistical inference can often be obtained only by using Monte Carlo methods. For example, simple formulae for the variance of estimators are available only for the uniform Poisson process [61, Sect. 3.3], but the Fisher information matrix for a general Gibbs process can be estimated by MCMC methods [32, 33], [47, p. 103].

Implementation in spatstat

Spatstat includes the following support for formal inference.

For Poisson processes, since maximum pseudolikelihood is equivalent to maximum likelihood, models fitted using ppm can be compared using the likelihood ratio test. The function anova.ppm performs analysis of deviance and reports p-values for the likelihood ratio test.

For example, suppose we wish to test the null hypothesis that the point process is a homogeneous Poisson process, against the alternative that it is an inhomogeneous Poisson process with intensity of the form

$$\lambda(x, y) = \exp(a + bx) \, .$$

The likelihood ratio test is performed as follows, assuming X is the point pattern dataset.

```
fit0 <- ppm(X, ~1, Poisson())
fit1 <- ppm(X, ~x, Poisson())
anova(fit0,fit1)
```

For Gibbs processes, spatstat provides a simulation algorithm rmh, an implementation of the Metropolis-Hastings algorithm. This implementation will simulate a wide range of models, including models fitted to data by ppm. The algorithm handles arbitrary first-order trends, using a renormalisation technique [6, Sect. 10.4]. Trends may be specified as symbolic functions, as pixel images, or using a fitted model object. However, due to the high computational load, interpoint interaction terms in the conditional intensity are calculated in Fortran. This restricts the range of models that can be simulated. Currently the available interaction terms include Poisson, Strauss, Strauss/hard core, soft core, Geyer saturation process, multitype Strauss, multitype Strauss/hard-core, and the general stationary pairwise interaction with step-function potential.

Here are two examples of the use of rmh(); see the help file in spatstat for a plethora of other examples.

```
m <- list(cif="strauss",par=c(beta=2,gamma=0.2,r=0.7),
          w=c(0,10,0,10))
X1 <- rmh(model=m,start=list(n.start=80),
          control=list(nrep=5e6,nverb=1e5))
fit <- ppm(cells, ~1, Strauss(0.1))
X2 <- rmh(fit,start=list(n.start=200),
          control=list(nrep=1e5,nverb=5000))
```

The user may exploit rmh to perform simulation-based inference. Currently, inferential techniques must be implemented by hand: an example is given in Sect. 11.4. Future extensions of the package will include basic support for simulation-based inference.

Pseudolikelihood Ratio and Monte Carlo Tests

A reasonable substitute for the likelihood ratio test statistic for a general Gibbs process is based on the log *pseudolikelihood* ratio. Consider a null hypothesis H_0 and an alternative H_1, and suppose H_0 is contained in H_1. Denote the point pattern dataset by **x**. Let $\widehat{\theta}_0 = \widehat{\theta}_0(\mathbf{x})$ be the estimate of the canonical parameters (by maximum pseudolikelihood) under H_0, and $\widehat{\theta}_1 = \widehat{\theta}_1(\mathbf{x})$ the estimate under H_1. The test statistic will be twice the log pseudolikelihood ratio

$$\Delta = \Delta(\mathbf{x}) = 2\left(\log \mathrm{PL}(\widehat{\theta}_1(\mathbf{x}); \mathbf{x}) - \log \mathrm{PL}(\widehat{\theta}_0(\mathbf{x}); \mathbf{x})\right) \qquad (17)$$

analogous to the deviance in likelihood theory. The following simple function calculates the quantity Δ from two fitted model objects in spatstat.

```
delta <- function(model0, model1)
    2 * (model1$maxlogpl - model0$maxlogpl)
```

However we emphasise again that there is no statistical theory available to support inferential interpretations of Δ. We explore a distributional approximation for Δ in Sect. 11. Alternatively one may simply use Δ as the test statistic in a Monte Carlo test. Suppose for example that H_0 is a simple hypothesis (i.e. in which θ_0 is fixed). Generate m independent realisations $\mathbf{x}^{(1)}, \dots, \mathbf{x}^{(m)}$ from the null hypothesis. Compute the corresponding values of the test statistic, say $\Delta_i = \Delta(\mathbf{x}^{(i)})$ for $i = 1, \dots, m$. Compute the rank of Δ in the set of values $\{\Delta_1, \dots, \Delta_m\} \cup \{\Delta\}$, that is, $R = 1 + \sum_{i=1}^{m} \mathbf{1}\{\Delta_i > \Delta\}$. Then under H_0, the rank R is uniformly distributed on $\{1, 2, \dots, m+1\}$, assuming there are no ties. Hence, the test which rejects H_0 when $R \le k$ has size $\alpha = k/(m+1)$ exactly, if H_0 is simple. The associated p-value is

$$p = \frac{R}{m+1}. \qquad (18)$$

Gamma Approximation to Distribution of Pseudolikelihood Ratio

Another possibility is to approximate the null distribution of the log pseudolikelihood ratio statistic Δ by a Gamma distribution. The Gamma family is chosen simply because it is a flexible class of distributions, and because it includes the χ^2 distribution, which is the asymptotic null distribution of the likelihood ratio test statistic.

Given some realisations from the null distribution of Δ, we fit a Gamma distribution using the method of moments, then calculate a critical value or p-value for the observed Δ statistic based on this fitted Gamma distribution. The p-value so obtained will be called the "gamma p-value", in contrast to the "Monte Carlo p-value" given by (18). The gamma approximation offers a substantial economy in the number of replicates used in the simulations. Of course this economy comes at the cost of placing trust in the approximation.

Some minimal experimentation indicates that the fit is generally good in the upper tail. See Fig. 8.

11 Harkness-Isham Ants' Nests Data

11.1 Description of Data

Figure 9 shows a point pattern data set recorded by Professor R.D. Harkness at a site in northern Greece, and described and analysed in [35]. The points record the locations of two species of ants: 68 nests of *Messor wasmanni* and 29 nests of *Cataglyphis bicolor*, in an irregular region 425 feet in diameter. Covariate information is also provided: the bold diagonal line in the Fig. 9 indicates a boundary between vegetation types, "field" and "scrub", while the two closely-spaced parallel lines delimit a foot track.

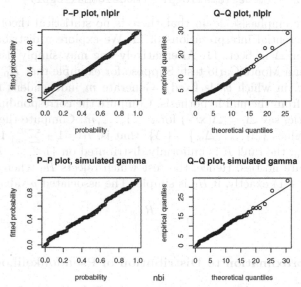

Fig. 8. Assessment of the gamma distribution approximation to the null distribution of the log pseudolikelihood ratio statistic Δ. Calculated for the test for between-species interaction described in Sect. 11. (*Top*) P–P and Q–Q plots comparing the empirical null distribution of Δ (from simulations of H_0) with the gamma distribution; (*Bottom*) analogous plots for i.i.d. random Gamma variates

Interest in these data focuses on whether there is evidence of spatial inhibition between *Messor* nests, and of a tendency for *Cataglyphis* nests to be situated close to *Messor* nests. Harkness and Isham suggested that the two species have a relationship similar to that of predator and prey. *Messor* is a harvester which collects seeds for food and builds nests composed mainly of seed husks. *Cataglyphis* is a forager which eats dead insects and other arthropods, and, while not preying upon the *Messor* ants, feeds upon dead *Messors* which have been killed by a predatory spider.

Rectangular subsets of the data were analysed in [5, 35, 36, 41, 70] and [63, Sect. 5.3]. Most of these analyses have used the dashed rectangles labelled

A and B in Fig. 9, which were defined by Harkness and Isham. The current analysis is, to our knowledge, the first to treat the full data set in its original (polygonal) window.

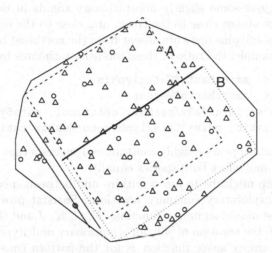

Fig. 9. Harkness-Isham ants' nests data [35, Fig. 1]. Locations of nests of two species of ants, *Messor wasmanni* (△) and *Cataglyphis bicolor* (○) in an irregular region 425 feet in diameter. Additional markings explained in the text. North at top of figure. Data reproduced by kind permission of Profs. R.D. Harkness and V. Isham

The nest locations (in units of half-feet) were kindly provided by Professor V. Isham. The polygonal window and the extra features (foot track, field-scrub boundary, rectangles A and B) were digitised by the first author from Fig. 1 in [35]. The full dataset is now available in spatstat as ants. Dr A. Särkkä also kindly provided a version of the subset in rectangle A which was analysed in her work [63].

Harkness and Isham [35] concluded from their analysis of rectangular subsets A and B that there is spatial dependence in the location of the nests, both within- and between-species. Results for subsets A and B were similar, suggesting that the field-scrub boundary has no effect. Särkkä [63] concluded from an analysis of subset A that there was strong inhibition among *Cataglyphis* nests, but obtained conflicting conclusions (depending on the choice of technique) about any dependence between species.

11.2 Exploratory Analysis

Our analysis of the full dataset in spatstat begins with exploratory methods (Sect. 3.3). Kernel-smoothed intensity estimates for the nests of each species (see Fig. 10) are plotted by the code at the top of the next page.

```
smoothants <- lapply(split(ants), ksmooth.ppp)
plot(smoothants$Cataglyphis, main="Cataglyphis nests")
plot(smoothants$Messor, main="Messor nests")
```

The results suggest some slightly nonstationary trends in nest abundance. Messor nests are absent close to the track, and close to the eastern corner of the polygon. Cataglyphis nests are absent from the northeast border. We may also plot, for example, the ratio of these intensity estimates by

```
cata <- smoothants$Cataglyphis
mess <- smoothants$Messor
ratio <- im(cata$v/mess$v, cata$xcol, cata$yrow)
plot(ratio, main="Cataglyphis-to-Messor ratio")
```

A plot of this ratio appears roughly constant and suggests that the same slight inhomogeneity may affect both species equally.

The next step might be to assess within- and between-species interaction by computing exploratory summary functions. Spatstat provides multitype versions of the standard summary functions F, G, K, J and the pair correlation function. In the notation of [49], for a stationary multitype point process, F_i denotes the empty space function F for the pattern consisting solely of points of type i, while $F_{\bullet}$ is the ordinary empty space function of the process of all points regardless of type. G_{ij} is the distribution function of the distance from a typical point of type i to the nearest point of type j, while $G_{i\bullet}$ is the distribution function of the distance from a typical point of type i to the nearest point regardless of type. Similarly K_{ij} is the K-function based on distances from points of type i to points of type j only, while $K_{i\bullet}$ is the K-function for distances from points of type i to points of any type. The pair correlation function ϱ_{ij} is defined by $\varrho_{ij}(t) = [(d/dt)K_{ij}(t)]/(2\pi t)$, analogously to the univariate case, and similarly for $\varrho_{i\bullet}$. Finally the J functions are defined [49] by $J_{ij}(t) = (1 - G_{ij}(t))/(1 - F_i(t))$ and $J_{i\bullet}(t) = (1 - G_{i\bullet}(t))/(1 - F_i(t))$. Diagnostic interpretation of these functions is described in [49, 68].

The spatstat function alltypes will compute these statistics and return an array of functions. For example

```
antsF <- alltypes(ants, "F")
plot(antsF)
```

computes the functions F_i (for i = Cataglyphis and i = Messor) and plots them. Similarly

```
antsG <- alltypes(ants, "G")
plot(antsG)
```

computes the functions G_{ij} for each i, j and plots them as a 2×2 array of panels, shown in Fig. 11. Similarly for the functions K_{ij} and J_{ij}. Algebraic transformations of these functions can be plotted easily using the R syntax for formulas.

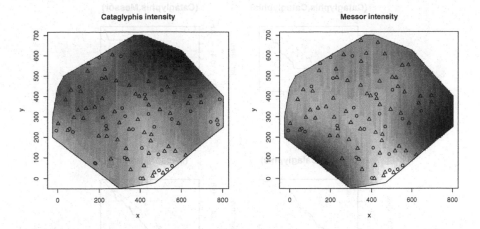

Fig. 10. Kernel smoothed intensity estimates for the two species of ants' nests

For example the corresponding L functions $L_{ij}(t) = \sqrt{K_{ij}(t)/\pi}$ may be plotted by

```
antsK <- alltypes(ants, "K")
plot(antsK, sqrt(trans/pi) ~ r)
```

The last line invokes `plot.fasp`. The second argument to `plot.fasp` is a model formula representing the variables which should be plotted. Here `trans` refers to the translation-correction estimate of K_{ij}. These plots all appear to evince some indication of between-species attraction and of within-species repulsion, at least over certain distance ranges. Plots based on the rectangular subset used by Särkkä are reasonably consistent in their appearance with those plots based on the full data set.

The pair correlation functions ϱ_{ij} are obtained from the K_{ij} estimates using `pcf`:

```
antsK <- alltypes(ants, "K")
antspcf <- pcf(antsK)
plot(antspcf)
```

This plot, shown in Fig. 12, tells a somewhat different story. It suggests that there is strong inhibition between *Messor* nests at all scales, while there is inhibition between *Cataglyphis* and *Messor* nests up to 10 half-feet and no interaction at longer distances. Between *Cataglyphis* nests there is a suggestion of short-scale inhibition and medium-scale attraction. For comparison we also show in Fig. 13 the pair correlation plot for rectangular subset A. This suggests inhibition for all combinations of nests.

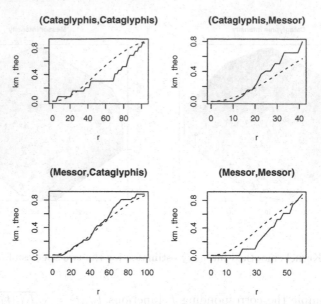

Fig. 11. Estimates of the cross-type nearest neighbour functions G_{ij} for the full ants' nests data

11.3 Modelling

Takacs & Fiksel [70] and Särkkä [63, Sect. 5.3] fitted a Strauss/hard core model (amongst other models) to assess the evidence of within- and between-species dependence. Let $\mathbf{y} = \{(x_1, m_1), \ldots, (x_n, m_n)\}$ denote a typical realisation of the process, where $x_i \in W$ are the locations and $m_i \in \mathcal{M}$ the corresponding marks (types). The Strauss/hard-core model has conditional intensity

$$\lambda((u, k), \mathbf{y}) = \beta_m \prod_i g(k, m_i, \|u - x_i\|) \tag{19}$$

where

$$g(k, m, d) = \begin{cases} 0 & \text{if } d < h_{km} \\ \gamma_{mk} & \text{if } h_{km} \leq d \leq r_{km} \\ 1 & \text{if } d > r_{km} \end{cases}$$

Here $\beta_m > 0$ are parameters influencing the intensity of the process, and $\gamma_{km} > 0$ are interaction parameters similar to the Strauss interaction parameter γ and satisfying $\gamma_{mk} = \gamma_{km}$. The parameters $h_{km} > 0$ are "hard-core distances" satisfying $h_{mk} = h_{km}$, while $r_{km} > 0$ are "interaction distances" analogous to the Strauss interaction radius r, and satisfying $r_{mk} = r_{km}$ and $r_{km} > h_{km}$. The process is well-defined and integrable provided either that $h_{mm} > 0$ for all m, or that $\gamma_{km} \leq 1$ for all k, m.

This model is chosen for its simplicity and flexibility in allowing for both negative and positive association within- and between-species. It is certainly a tentative model, and indeed pairwise interaction models such as the

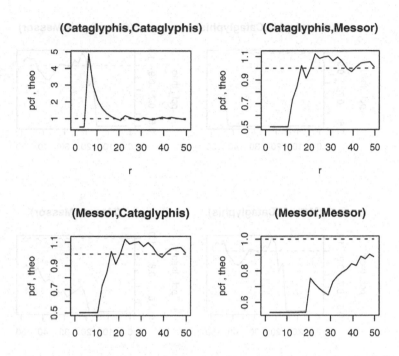

Fig. 12. Estimates of the cross-type pair correlation functions ϱ_{ij} for the full ants' nests data

Strauss/hard core model are sometimes regarded as inadequate for describing clustering.

Denoting the two types Cataglyphis and Messor by C and M respectively, the model has 5 regular parameters (the intensities β_C, β_M and interaction parameters $\gamma_{CC}, \gamma_{MM}, \gamma_{CM}$) and 6 irregular parameters (the hard core distances h_{ij} and interaction distances r_{ij}).

To reduce the computational load we estimate the hard core distances by their maximum likelihood (and maximum pseudolikelihood) estimates, which are the corresponding minimum interpoint distances, obtained by

```
d <- pairdist(ants)
mks <- ants$marks
tapply(d, list(mks[row(d)], mks[col(d)]), min)
```

Note that if rounding is performed, then these values must be rounded downward, to ensure that the model still has nonzero likelihood. The resulting values are $\hat{h}_{MM} = 18.7$, $\hat{h}_{CC} = 4.9$ and $\hat{h}_{CM} = 12.2$ (in half-feet). For the values of the Strauss interaction radii we adopted the same values as Takacs & Fiksel [70] and Särkkä [63], namely $r_{MM} = r_{CC} = r_{CM} = 90$ half-feet.

The model was fitted to the full dataset as follows.

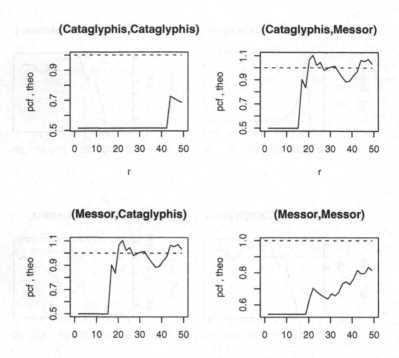

Fig. 13. Estimates of the pair correlation functions for Särkkä's version of the ants' nests data

```
rr   <- matrix(90,2,2)
hh   <- matrix(c(5.0,12.2,12.2,18.7),2,2)
types <- levels(ants$marks)
Int <- MultiStraussHard(types, rr, hh)
fit <- ppm(ants, ~marks, Int, correction="border", rbord=90)
```

Note that the trend formula must be ~marks in order to allow different intensity values β_M, β_C for the two species.

Printing the fitted model object fit shows the fitted values of all parameters. It is necessary to select a value for the correction argument specifying the edge correction for the pseudolikelihood [5]. Here we fitted the model using the "border" correction. Alternative choices of edge correction yield different fitted parameter values, as shown in the following table.

EDGE CORRECTION	$\beta_C \times 10^4$	$\beta_M \times 10^4$	γ_{CC}	γ_{MM}	γ_{CM}
Border	0.92	8.69	0.50	0.68	1.12
Translation	0.72	2.35	0.75	0.90	1.11
None	0.71	2.33	0.72	0.88	1.13

Evidence for between-species dependence is quantified by the Strauss interaction parameter γ_{CM}, which is slightly greater than unity, suggesting moderate positive association. The estimates of within-species interaction γ_{CC}, γ_{MM} are below unity, suggesting inhibition.

These conclusions should be compared to those of Särkkä's [63] analysis of the subset of data in the rectangle labelled A in Fig. 9. She obtained $\hat{\gamma}_{CM} = 0.88$, which would indicate *inhibition* between species. For the sake of direct comparison, we also fitted the Strauss/hard core model to the data used by Dr Särkkä and kindly supplied by her. Several small differences can be observed between Särkkä's dataset and the subset of our data indicated by rectangle A. These may be attributed to slight differences in digitising Fig. 1 of [35]. In Särkkä's dataset the minimum interpoint distance between *Cataglyphis* and *Messor* nests is 11.2 half-feet rather than 12.2.

Changing the inter-species hard core distance to $h_{CM} = 11.2$, we fitted the Strauss-hard core model to Särkkä's version of the data in rectangle A. This also allows us to compare four different edge corrections for the pseudolikelihood [5] which are implemented for rectangular windows, namely the border, periodic and translation edge corrections, and Ripley's isotropic correction.

The choice of edge correction appeared to have a substantial impact upon the results. (As a matter of convenience Särkkä used a periodic edge correction in her analysis, and a stochastic approximation to the pseudolikelihood.) Our estimates of γ_{CM}, based upon Särkkä's data, are 1.37 (border correction), 0.99 (periodic edge correction), 1.20 (translation correction) and 1.00 (Ripley isotropic correction). These estimates are larger than Särkkä's value of 0.88, and two of them are larger than unity, consistent with between-species *attraction*.

We also fitted the same model (i.e. with smaller $h_{CM} = 11.2$) to the complete data set, in its polygonal window, resulting in γ_{CM} estimates of 1.33 (border correction) and 1.12 (translation correction). In this case both estimates are greater than 1, perhaps substantially greater. The evidence at this point is thus somewhat contradictory. The exploratory summary functions F, G, K, J suggest interspecies attraction, while the pair correlation function exhibits no sign of between-species interaction. Four of the six estimates of γ_{CM} are larger than unity, again suggesting interspecies attraction. Formal methods may be useful at this point.

11.4 Formal Inference

We conducted formal hypothesis tests for the presence of inter-species interaction using the methods described in Sect. 10.2. The null hypothesis of no inter-species interaction can be formulated as

$$H_0: \quad \gamma_{CM} = 1, \quad h_{CM} = 0$$

which implies that the nests of the two species are independent point processes of Strauss-hard core type.

The Strauss interaction radii were held fixed at 90 half-feet in all instances. Under the null hypothesis, the cross-species interaction radius r_{CM} is not identifiable, since it plays no role in the model when $\gamma_{CM} = 1$. Hence r_{CM} should not be estimated from simulations of H_0. The within-species interaction parameters γ_{CC}, γ_{MM} were held fixed to reduce computational load, but they could have been estimated instead.

Under the null model, there are no interaction terms between nests of different species. In spatstat, assigning a value of NA to an irregular parameter will cause the interpoint interaction term associated with this parameter to be omitted from the analysis. Thus our null model is represented by assigning NA values to every off-diagonal entry in the matrices of hard core distances and of Strauss interaction distances. The following code fits the null and alternative hypotheses to the data and evaluates the log pseudolikelihood ratio statistic Δ:

```
Str1 <- matrix(c(90, 90, 90, 90), 2,2)
Str0 <- matrix(c(90, NA, NA, 90), 2,2)
Hard1 <- matrix(c(5.0, 12.2, 12.2, 18.7), 2,2)
Hard0 <- matrix(c(5.0, NA,   NA,   18.7), 2,2)
Int0 <- MultiStraussHard(types, Str0, Hard0)
Int1 <- MultiStraussHard(types, Str1, Hard1)
fit0 <- ppm(ants, ~marks, Int0, correction="translate")
fit1 <- ppm(ants, ~marks, Int1, correction="translate")
dobs <- 2 * (fit1$maxlogpl - fit0$maxlogpl)
```

To generate 99 realisations from the null distribution of Δ we proceed as follows:

```
dvalues <- numeric(99)
for(i in 1:99) {
    Xsim <- rmh(fit0)
    hc1 <- nnd(Xsim)
    hc0 <- matrix(NA, 2, 2)
    diag(hc0) <- diag(hc1)
    Int0sim <- MultiStraussHard(types, Str0, hc0)
    Int1sim <- MultiStraussHard(types, Str1, hc1)
    fit0sim <- ppm(Xsim, ~marks, Int0sim,
                        correction="translate")
    fit1sim <- ppm(Xsim, ~marks, Int1sim,
                        correction="translate")
    dvalues[i] <-  2 * (fit1sim$maxlogpl - fit0sim$maxlogpl)
}
```

where nnd is a small function to compute the minimum nearest-neighbour distances between each pair of types:

```
nnd <- function(X) {
    mks <- X$marks
```

```
          d <- pairdist(X)
          tapply(d, list(mks[row(d)], mks[col(d)]), min)
     }
```

The resulting Monte Carlo test for between-species interaction gave a p-value (18) of 0.18, based on 99 simulations from the null model as above. Using the gamma approximation (Sect. 10.2) based on a separate set of only 30 simulated realisations from the null model, the approximate p-value obtained was 0.1885. (Validity of this approximation is confirmed by Fig. 8 in Sect. 10.2.) Thus there appears to be no evidence of between-species interaction.

We then checked whether there was evidence of any interaction at all. In this case the null model simply consists of two independent Poisson processes, of different intensities. This is fitted by calling ppm with the trend given as ~marks and the interaction as Poisson. The alternative model was taken to be the full model, including both between- and within-species interactions. We obtained a Monte Carlo p-value of 0.03, and a gamma approximation p-value of 0.0402, thus providing evidence that some sort of interaction is present.

If we eliminate between-species interaction from the model, we can test for within-species interaction either for both species simultaneously, or in the context of univariate models fitted to each species separately. The p-values for the simultaneous test were 0.03 (Monte Carlo) and 0.0021 (gamma) indicating some evidence of within-species interaction. The test based on univariate models gave p-values for the Messor ants of 0 (Monte Carlo) and 0.004 (gamma) and for the Cataglyphis ants of 0.64 (Monte Carlo) and 0.6279 (gamma), suggesting that there is within-species interaction among the Messor ants, but not among the Cataglyphis ants.

Finally, as a check on the absence of between-species interaction, we performed a test in terms of a univariate model fitted to the Messor ants *conditional upon* the Cataglyphis ants. This model used the Strauss/hard core interaction as before, but added a trend term, the trend being a log-linear function of distance to the nearest Cataglyphis nest. The null model was formed simply by omitting the trend term. The empirical p-values for this test were 0.18 (Monte Carlo) and 0.1108 (gamma), which are again consistent with the hypothesis of no between-species interaction.

11.5 Incorporation of Covariates

In addition to recording the locations of the ants' nests, Harkness [35] noted a boundary between "field" and "scrub" crossing the middle of the study region, and a foot track running close to the perimeter. The relevance of these geographical features to the ants' nests pattern can easily be assessed using spatstat. Here we demonstrate the use of the modelling software to formulate and fit point process models which depend on covariates (Sect. 7).

A very simple model for "field/scrub" effect is one in which the intensity of the process is a different, constant value on each side of the field/scrub boundary. In more complicated models, the intensity might also depend on distance from the field/scrub boundary.

It is convenient to use the function `fsdistance(x,y)` which is displayed on p. 60. This function computes the *signed* distance from any location (x, y) to the field/scrub boundary. A point `(x,y)` belongs to the field region if `fsdistance(x,y) > 0`.

The simplest sensible model, in which intensity of each species is a different, constant value on each side of the field/scrub boundary, can be fitted by including a covariate which is a two-level factor, indicating whether the point in question is in field or scrub. One way to do this is by means of the function `fsfac(x,y)` shown below.

```
fsdistance <- function(x,y) {
  ends <- ants.extra$fieldscrub
  para <- c(diff(ends$x),diff(ends$y))
  perp <- c(para[2], -para[1])
  unit <- perp/sqrt(sum(perp^2))
  cbind(x,y) %*% unit - (ends$x[1] * unit[1] +
                                  ends$y[1] * unit[2])
}

fsfac <- function(x,y) {
  factor(ifelse(fsdistance(x,y) > 0,  "field", "scrub"))
}
```

The desired model can then be fitted via:

```
ppm(ants, ~ marks * fsfac(x,y), Poisson())
```

Note carefully that the variable names x and y in the call to `ppm` above, are reserved names which refer to the Cartesian coordinates in the quadrature scheme. The code above exploits the fact that the chosen covariate can be expressed as a function of the Cartesian coordinates. If this is not true, then the covariates must be supplied either as pixel images or as columns in a data frame, as explained in Sect. 7.

The fitted model output (after rounding) is

```
Nonstationary multitype Poisson process
Trend formula: ~marks * fs(x, y)
Fitted coefficients
(Intercept)    marksMessor  fsfac(x,y)scrub
-9.35          0.52           -0.77
               marksMessor:fsfac(x,y)scrub
                         0.97
```

Since `field` is the first level of the factor `fsfac(x,y)`, this output indicates that the fitted intensities are as follows:

	Cataglyphis	*Messor*
FIELD	$\exp(-9.35)$	$\exp(-9.35 + 0.52)$
	$= 0.9 \times 10^{-4}$	$= 1.5 \times 10^{-4}$
SCRUB	$\exp(-9.35 - 0.77)$	$\exp(-9.35 + 0.52 - 0.77 + 0.97)$
	$= 0.4 \times 10^{-4}$	$= 1.7 \times 10^{-4}$

These values could also have been obtained by geometrically dividing the study region into two subregions and counting the numbers of nests of each species in each subregion. They show that *Cataglyphis* has a marked preference for the field region, while *Messor* nests have approximately equal intensity in field and scrub regions. This finding was reported by Harkness & Isham [35].

However, these differences are not significant according to the (asymptotic) likelihood ratio test for a field/scrub effect. Typing

```
fit1 <- ppm(ants, ~ marks * fsfac(x,y), Poisson())
fit0 <- ppm(ants, ~ marks,              Poisson())
anova(fit0, fit1, test="Chi")
```

yields a p-value of 0.12 (with reference to the χ^2_2 distribution).

Interpoint interaction may be incorporated, and probably should be incorporated, even into the simplest model. For example, we may fit

```
ppm(ants, ~ marks * fsfac(x,y), Int1)
```

where Int1 is the interaction object representing the multitype Strauss/hard core model, constructed in the previous section using MultiStraussHard. The fitted intensity parameters β are as follows.

	Cataglyphis	*Messor*
FIELD	1.0×10^{-4}	2.0×10^{-4}
SCRUB	0.3×10^{-4}	2.0×10^{-4}

This strengthens the earlier suggestion that *Cataglyphis* nests have an affinity for field over scrub while *Messor* nests are indifferent.

Extending the model further, we might fit a trend (in either or both of the types) depending on the *distance* from the field/scrub boundary, as well as on the distinction between field and scrub. Assuming that the dependence on distance is loglinear, the model can be fitted by

```
fsdist <- function(x,y) { abs(fsdistance(x,y)) }
ppm(ants, ~ marks * fsdist(x,y) * fsfac(x,y), Int1)
```

This trend is essentially the simplest which can be fitted and which makes full use of all the variables of interest. It is admittedly arbitrary, but should have a reasonable chance of revealing a trend dependent upon the field/scrub dichotomy if such a trend exists.

We tested for a trend of the specified form, first in terms of a model allowing for both between- and within-species interactions, and then in terms of the model which appears most appropriate in the light of the tests previously conducted, namely a model in which there is interaction within the

Messor species only. Testing for the trend yielded empirical p-values of 0.32 (Monte Carlo) and 0.3692 (gamma) when the full multivariate Strauss/hard core interaction was used, and of 0.50 (Monte Carlo) and 0.5709 (gamma) when the within-Messor-only interaction was used. Thus there is no evidence of a field/scrub effect, at least as described by a model of this form. Models involving smooth intensity functions can also be fitted in the same style.

12 The Queensland Copper Data

12.1 Data and Previous Analyses

The Queensland copper data, shown in Fig. 14, were introduced and analysed by Berman [11]. They consist of a point pattern of 67 copper ore deposits, and a line segment pattern of 146 geological features, called 'lineaments', obtained from an intensive geological survey of a 70×158 km region in central Queensland, Australia. It is of interest to find any association between the copper deposits and the lineaments. Since the lineaments are visible on satellite images, they might be used to guide the search for copper deposits, by predicting regions of high intensity for the copper points.

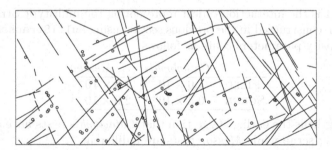

Fig. 14. Copper ore deposits (○) and lineaments (—) in a region of central Queensland. North at top of frame. Reproduced by kind permission of Dr A Green, Dr J Huntington, Dr M Berman and the Royal Statistical Society

Berman [11] developed formal tests for dependence of the points upon the lineaments, based on measuring the distance from each point to the nearest lineament. The points are assumed to constitute an inhomogeneous Poisson process, with an intensity that depends on distance to the nearest lineament. The null hypothesis is that the intensity is constant.

Let X denote the copper point process and L the lineament process. All analysis will be performed conditionally upon L. In [11] it is assumed that X is conditionally Poisson given L, with intensity function of the form

$$\lambda_{X|L}(u) = \varrho(d(u, L)) \tag{20}$$

where $d(u, L)$ denotes the shortest distance from the point u to the nearest lineament, and ϱ is an unknown function. Under this assumption, the *observed* distances $d_i = d(x_i, L)$ for all points $x_i \in X$ are i.i.d. The null hypothesis, that ϱ is constant, corresponds to assuming a known distribution for the d_i (determined by the geometry of L) and hence can be tested using the Kolmogorov-Smirnov or other tests of goodness-of-fit. For details see [11].

For geological reasons, lineaments lying in different spatial orientations have typically been created at different epochs. Hence Berman [11] also considered the possibility that the intensity of the points depends only upon distance to lineaments lying in a particular subset of orientations. This subset consists of those lineaments having an angle (measured in the anticlockwise direction from the horizontal, with 0° pointing east) between 120° and 160°. He also considered the subset whose angles lie between 10° and 40°, but found the results from this latter set not to differ from the results for all lineaments.

Berman concluded that there is some evidence of dependence of the intensity of points upon the lineaments, when the entire window is considered, but speculated that this dependence might be a spurious artifact due to the scarcity of points in the northern half of the window. When he restricted attention to the southern half of the window (shown in Fig. 15) he found no evidence of association between points and lineaments.

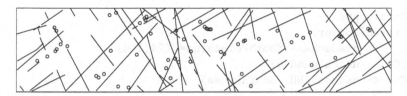

Fig. 15. Southern half window of the Queensland copper data

The data in the southern half window were re-analysed in [7, 31]. Both analyses concluded that there is no evidence of dependence.

12.2 Analysis

In this work we re-visit these data, making use of the spatstat package. The convenient model-fitting and simulation facilities of spatstat make it easy to conduct tests of association between the points and lineaments, and to explore other aspects of the nature of these data. In particular we investigate the assumption that the points are conditionally Poisson. Attention is mainly restricted to the southern half window, but a further analysis of the entire window is discussed briefly.

We test for dependence of the points on the lineaments, using three simple parametric loglinear models for the intensity of the points given the lines:

$$\lambda(u) = \beta \exp\{\alpha d\} \tag{21}$$

$$\lambda(u) = \beta \exp\{\alpha_1 d + \alpha_2 d \cos\theta + \alpha_3 d \sin\theta\} \tag{22}$$

$$\lambda(u) = \exp\{\alpha_{0,i} + \alpha_{1,i}d\}, \ i = 1, \ldots, 3 . \tag{23}$$

These models are expressed in terms of $d = d(u, L)$, the distance from a point u to the nearest lineament, and $\theta = \theta(u, L)$, the spatial orientation of the lineament closest to u (measured as an anticlockwise angle from the east-pointing direction). Model (23) is obtained by dividing the angle θ into classes, with breakpoints determined by the lineament subsets investigated by Berman in [11]. These breakpoints are $10°$, $40°$, $120°$, and $160°$. There were no lineaments in the intervals $[0°, 10°]$ nor $[160°, 180°]$ so there are effectively three classes. The index $i = 1, \ldots, 3$ is determined by the class in which the angle θ falls.

Under the assumption that the points, given the lineaments, are a realisation of an (inhomogeneous) Poisson processes we may apply the likelihood ratio test. The three models (and the null model comprising a constant intensity Poisson process) are fitted as follows. First we construct a data frame Cov containing the desired covariates: it has columns d (the distance to the nearest lineament), angle (the angle made by this nearest lineament with the horizontal) and cat.ang (the categorical variable or factor resulting from classifying angle into three groups).

```
data(copper)
attach(copper)
Q   <- quadscheme(SouthPoints,nd=c(24,106))
UQ  <- union.quad(Q)
Cov <- makecov(UQ, SouthLines)
```

The function makecov is a one-off utility which performs the analytic geometry of computing distances between points and line segments. The implementation of such calculations will change shortly. Interested readers should contact the authors for further information.

The three models (21)–(23) can then be fitted, along with the null model, as follows

```
FO  <- ppm(Q)
F1  <- ppm(Q, ~d, covariates=Cov)
F2  <- ppm(Q, ~d + I(d * sin(angle)) + I(d * cos(angle)),
            covariates=Cov)
F3  <- ppm(Q, ~d * cat.ang, covariates=Cov)
```

These models are Poisson by default. Note that expressions like d*sin(angle) must be protected by I() within a call to ppm() to ensure that * is interpreted as multiplication.

The likelihood ratio test for each successive pair of models can now be performed using anova.ppm, or by hand as indicated in Sect. 10. The resulting p-values are as follows.

Model	Statistic	p-value
1	0.5613	0.4538
2	6.0080	0.1112
3	3.7561	0.5850

Another subdivision of the lineament orientations with breakpoints of $0°$, $60°$, $120°$ and $180°$, resulted in a likelihood ratio statistic of 7.7873 with p-value 0.1684.

There is thus no evidence of dependence of the points upon the lineaments, at least in the forms of these models. Thus our conclusions here are in agreement with the previous analyses [7, 11, 31].

However, the foregoing analyses assume that the copper points form an (inhomogeneous) Poisson process given the lineaments. This assumption should be validated. One possibility is to use the residual plots described in Sect. 10.1; an analysis of these data is reported in [7].

Alternatively we may use the inhomogeneous version of the K-function, $K_{\mathrm{inhom}}(r)$ introduced in [3]. This requires the intensity function of the copper point process, evaluated at the data points. We estimated the intensity function in four ways: from the three foregoing parametric models, and also non-parametrically.

The parametric estimates are straightforward. The fitted intensity at the data points is provided by `fitted.ppm`. Thus for example

```
lambda <- fitted(F1)
K1 <- Kinhom(SouthPoints, lambda)
plot(K1)
```

computes the inhomogeneous K-function based on the intensity function λ estimated under the model (21), and plots the result.

For the non-parametric estimate of the intensity function, we assume (20) holds, where the form of the function ϱ is not specified. We make use of the following relationship [31]. Suppose X and L are jointly stationary. Let F_L be the empty space function for the L process, that is, $F_L(t)$ is the cumulative distribution function of the distance from an arbitrary point in the plane to the nearest lineament in L. Let G_{XL} be the cumulative distribution function of the distance from a typical point of the process X to the nearest line segment in L. Then if (20) holds, we have

$$G_{XL}(t) = \frac{\int_0^t \varrho(s)dF_L(s)}{\int_0^\infty \varrho(s)dF_L(s)} \qquad (24)$$

From this it follows that

$$\varrho(t) = \mu \frac{dG_{XL}(t)/dt}{dF_L(t)/dt} \tag{25}$$

where

$$\mu = \int_0^\infty \varrho(s) dF_L(s) = \mathbb{E}(\varrho(d(u,L)))$$

is the intensity of X. The moment estimator of μ is $\hat{\mu} = n(X)/|W|$.

To estimate $dG_{XL}(t)/dt$ and $dF(t)/dt$ we can compute empirical estimates of F and G, fit smoothing splines, and take the derivatives of the splines. Estimates of F and G can be computed using standard methods. At the time of writing, these methods must be implemented by hand for line segment patterns. Future extensions of spatstat will include support for these calculations. The graph of $\hat{\varrho}(t)$ is shown in Fig. 16.

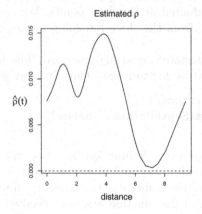

Fig. 16. Estimate of the function $\varrho(t)$ in (20) obtained using by substituting spline estimates in (25)

There were very substantial differences in the appearance of the intensity surfaces computed by different parametric methods, and by the nonparametric methods. Despite this, the four estimates of the inhomogeneous K-function turned out to be virtually identical to each other, and to the estimated conventional K-function, for each window. Plots of one of the inhomogeneous K-function estimates and of the conventional K-function estimate are shown in Fig. 17. The explanation is that the inhomogeneous K-function depends only on the estimated intensity values at the points of the point pattern. These intensity values were approximately constant for these data.

The K functions suggest that there is positive association between the copper deposits, conditioned on the lineaments. Again, we would like to be able

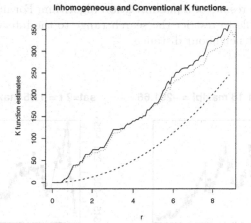

Fig. 17. Inhomogeneous K-function (*solid line*), conventional K-function (dotted line) and theoretical K-function under CSR (*dashed line*) for the Queensland copper data. Inhomogeneous version computed using parametric model (22)

to test this formally, and this requires a class of models which allow for positive association or clustering. One convenient choice is the Geyer saturation model [32].

In order to fit the Geyer model we need to estimate the "irregular" parameters of the model, namely the interaction radius $r > 0$ and the saturation number s. Rough estimates may be found by searching over a small set of integer values for s (1 to 5 inclusive) using profile pseudolikelihood. Note that the maximum over r of the log pseudolikelihood (for a fixed value of s) must occur at one of the interpoint distances of the observed pattern.

The values obtained for the estimates of the irregular parameters were $\hat{r} = 1.18$ and $\hat{s} = 2$ respectively. Similar estimates ($\hat{r} = 1.05$ and $\hat{s} = 2$) were obtained when we also included in the model a trend of the form (22) along with the Geyer interaction. Sample plots of the profiles, for the trend-included setting, are shown in Fig. 18. The profile over r for $s = 5$ is very similar to that for $s = 4$ and is omitted to save space.

The profile log pseudolikelihood in Fig. 18 is shown only for $r \leq 10$ km. Localised sharp peaks occur for some larger values of r, and in fact the overall unconstrained maximum occurs at $r = 10.62$ and $s = 1$. However, this value of r is not credible. The associated estimate of γ is 94.272, which would cause immensely strong clustering. Plots of the estimated G and K functions and the pair correlation function suggest an interaction range between 1 and 3 km. It seems plausible that the value of $r = 10.62$ is a numerical artifact, since it is just slightly larger than the *maximum* nearest neighbour distance in the data, and the observation window has a width of only 35 km. We therefore decided to dismiss the profile peaks for $r > 10$ as anomalies. This example illustrates the delicacy of estimating irregular parameters and the need to check the results of a maximisation algorithm. For interaction radii it is probably sensible to restrict the search range to the interval from 0 to the maximum nearest neighbour distance.

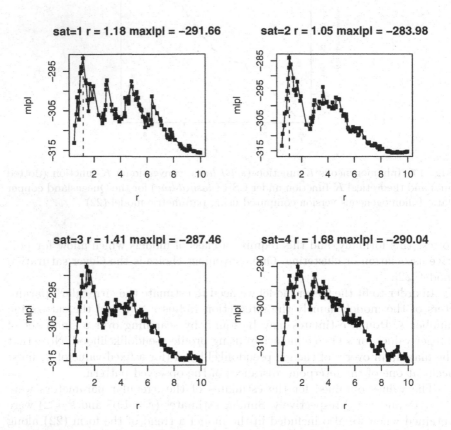

Fig. 18. Profile log pseudolikelihood for a Geyer model, as a function of interaction radius r, for several values of the saturation parameter s. Southern half window. Trend from model (22) included in the fit

We tested the model with trend given by model (22) and interaction given by `Geyer(1.05,2)` against the null model with trend only. We also tested a stationary model with `Geyer(1.18,2)` interaction against a completely null (i.e. constant intensity Poisson) model.

In the first case we obtained a log pseudolikelihood ratio statistic $\Delta = 57.35$, a Monte Carlo p-value of 0.01, and a gamma approximation p-value of 4.6×10^{-6}. In the second case $\Delta = 59.36$, with Monte Carlo p-value 0.01, and gamma p-value of 4.8×10^{-11}.

These tests appear to confirm the impression given by the K-function plots that there is positive association between the points. Given that there is attraction, the tests for dependence of the points upon the lineaments, based upon inhomogeneous Poisson models, cannot be considered valid. However we can now conduct tests allowing for the apparent interaction. The test of trend (given by model (22)) plus Geyer(1.05,2) against Geyer interaction only gave $\Delta = 4.75$, with a Monte Carlo p-value of 0.12, and a gamma p-value of 0.144. Thus when interaction is allowed for, the evidence of dependence of the points upon the lines is still "insignificant" and is in fact slightly weaker than if we assume the points to arise from an (inhomogeneous) Poisson process.

12.3 North-South Effect

We now briefly consider the complete data set rather than the southern half window. In particular we focus on Berman's conjecture that the apparent dependence of points on lineaments, when the entire window is considered, might be a spurious artifact due to the scarcity of points in the northern half of the window. If this is indeed the case, then it may be possible to adjust for the low intensity in the northern half window by introducing a trend depending upon the spatial covariates x and y.

One convenient class of models uses the smoothing term s in the trend formula. For example

```
Q  <- quadscheme(copper$points, nd=c(34,75))
F0 <- ppm(Q,~s(y),use.gam=TRUE)
```

fits a Poisson model with a smooth trend in the y coordinate (Northing) only. We may test this null model against more elaborate models such as

```
UQ  <- union.quad(Q)
Cov <- makecov(UQ, copper$lines)
F1  <- ppm(Q, ~s(y) + d + I(d * sin(angle)) +
           I(d * cos(angle)), covariates=Cov, use.gam=TRUE)
F2  <- ppm(Q, ~s(y) + d * cat.ang, covariates=Cov,
           use.gam=TRUE)
```

The likelihood ratio test of models F1 and F2 against model F0 turned out to have values of 18.22 and 9.96 on 3 and 9 degrees of freedom respectively. (Note that for the full window, all five angle categories are non-empty.) The corresponding p-values are 0.0003 and 0.3537. Thus there appears to remain an indication of dependence of the points on the lineaments via model (22) (although not via model (23)) for the full data set, even after a spatial trend (depending on the x and y coordinates) is allowed for.

The foregoing likelihood ratio test may be criticised since we had already demonstrated an interpoint interaction in the southern half window. Instead we should be conducting Monte Carlo tests involving an interpoint interaction term. The Geyer model irregular parameters may be estimated for the full window by profiling, yielding $\hat{r} = 1.18$ and $\hat{s} = 2$ (as for the southern half window when no trend is included. The estimates of the irregular parameters were the same for the full window whether a lineaments-dependent trend was allowed for or not.) We might thus set out to test the null model

```
FO <- ppm(Q, ~s(y), Geyer(1.18,2), covariates=Cov,
          use.gam=TRUE)
```

against (for instance)

```
F1 <- ppm(Q,~ s(y) + d * cat.ang,  Geyer(1.18,2),
          covariates=Cov, use.gam=TRUE)
```

Notice that here, as elsewhere, we propose to conduct inference *conditionally* on the fitted values of the irregular parameters. This is done mainly to save computational time. A definitive formal analysis should also look at the effect of estimating the irregular parameters.

References

[1] L. Anselin. Local indicators of spatial association – LISA. *Geographical Analysis*, 27:93–115, 1995.

[2] A.C. Atkinson. *Plots, Transformations and Regression*. Number 1 in Oxford Statistical Science Series. Oxford University Press/ Clarendon, 1985.

[3] A.J. Baddeley, J. Møller, and R.P. Waagepetersen. Non- and semiparametric estimation of interaction in inhomogeneous point patterns. *Statistica Neerlandica*, 54(3):329–350, 2000.

[4] A.J. Baddeley and R. Turner. spatstat: an R package for analyzing spatial point patterns. *Journal of Statistical Software*, 12(6):1–42, 2005. URL: http://www.jstatsoft.org.

[5] A.J. Baddeley and R. Turner. Practical maximum pseudolikelihood for spatial point patterns (with discussion). *Australian and New Zealand Journal of Statistics*, 42(3):283–322, 2000.

[6] A.J. Baddeley and R. Turner. Spatstat: an R package for analyzing spatial point patterns. Research Report 2004/13, School of Mathematics and Statistics, University of Western Australia, September 2004.

[7] A.J. Baddeley, R. Turner, J. Møller and M. Hazelton. Residual analysis for spatial point processes. *Journal of the Royal Statistical Society (series B)*, 67:1–35, 2005.

[8] A.J. Baddeley, M.N.M. van Lieshout and J. Møller. Markov properties of cluster processes. *Advances in Applied Probability*, 28:346–355, 1996.

[9] M.S. Bartlett. The spectral analysis of two-dimensional point processes. *Biometrika*, 51:299–311, 1964.

[10] M.S. Bartlett. *The statistical analysis of spatial pattern*. Chapman and Hall, London, 1975.

[11] M. Berman. Testing for spatial association between a point process and another stochastic process. *Applied Statistics*, 35:54–62, 1986.

[12] M. Berman and T.R. Turner. Approximating point process likelihoods with GLIM. *Applied Statistics*, 41:31–38, 1992.

[13] J.E. Besag. Statistical analysis of non-lattice data. *The Statistician*, 24:179–195, 1975.

[14] J.E. Besag and P.J. Diggle. Simple Monte Carlo tests for spatial pattern. *Applied Statistics*, 26:327–333, 1977.

[15] C. Chatfield. *Problem solving: a statistician's guide*. Chapman and Hall, 1988.

[16] D. Collett. *Modelling Binary Data*. Chapman and Hall, London, 1991.

[17] D.R. Cox and V. Isham. *Point processes*. Chapman and Hall, London, 1980.

[18] D.R. Cox and E.J. Snell. *Applied Statistics: principles and examples*. Chapman and Hall, 1981.

[19] The Comprehensive R Archive Network. URL `http://www.cran.r-project.org`.

[20] N.A.C. Cressie and L.B. Collins. Analysis of spatial point patterns using bundles of product density LISA functions. *Journal of Agricultural, Biological and Environmental Statistics*, 6:118–135, 2001.

[21] N.A.C. Cressie and L.B. Collins. Patterns in spatial point locations: local indicators of spatial association in a minefield with clutter. *Naval Research Logistics*, 48:333–347, 2001.

[22] N.A.C. Cressie. *Statistics for Spatial Data*. John Wiley and Sons, New York, 1991.

[23] N.A.C. Cressie. *Statistics for Spatial Data*. John Wiley and Sons, New York, 1993. Revised edition.

[24] A.C. Davison and E.J. Snell. Residuals and diagnostics. In D.V. Hinkley, N. Reid, and E.J. Snell, editors, *Statistical theory and modelling (in honour of Sir David Cox FRS)*, Chap. 4, pp. 83–106. Chapman and Hall, London, 1991.

[25] P.J. Diggle. A kernel method for smoothing point process data. *Journal of the Royal Statistical Society, series C (Applied Statistics)*, 34:138–147, 1985.

[26] P.J. Diggle. *Statistical analysis of spatial point patterns*. Academic Press, London, 1983.

[27] P.J. Diggle. A point process modelling approach to raised incidence of a rare phenomenon in the vicinity of a prespecified point. *Journal of the Royal Statistical Society, series A*, 153:349–362, 1990.

[28] P.J. Diggle. *Statistical Analysis of Spatial Point Patterns*. Arnold, second edition, 2003.

[29] P.J. Diggle, D.J. Gates, and A. Stibbard. A nonparametric estimator for pairwise-interaction point processes. *Biometrika*, 74:763–770, 1987.

[30] F. Divino, A. Frigessi and P.J. Green. Penalised pseudolikelihood estimation in Markov random field models. *Scandinavian Journal of Statistics*, 27(3):445–458, 2000.

[31] R. Foxall and A.J. Baddeley. Nonparametric measures of association between a spatial point process and a random set, with geological applications. *Applied Statistics*, 51(2):165–182, 2002.

[32] C.J. Geyer. Likelihood inference for spatial point processes. In O.E. Barndorff-Nielsen, W.S. Kendall, and M.N.M. van Lieshout, editors, *Stochastic Geometry: Likelihood and Computation*, number 80 in Monographs on Statistics and Applied Probability, Chap. 3, pp. 79–140. Chapman and Hall / CRC, Boca Raton, Florida, 1999.

[33] C.J. Geyer and J. Møller. Simulation procedures and likelihood inference for spatial point processes. *Scandinavian Journal of Statistics*, 21(4):359–373, 1994.

[34] K.-H. Hanisch and D. Stoyan. Remarks on statistical inference and prediction for a hard-core clustering model. *Statistics*, 14:559–567, 1983.

[35] R.D. Harkness and V. Isham. A bivariate spatial point pattern of ants' nests. *Applied Statistics*, 32:293–303, 1983.

[36] H. Högmander and A. Särkkä. Multitype spatial point patterns with hierarchical interactions. *Biometrics*, 55:1051–1058, 1999.

[37] A.C.A. Hope. A simplified Monte Carlo significance test procedure. *Journal of the Royal Statistical Society, series B*, 30:582–598, 1968.

[38] K. Hornik. The R FAQ: Frequently asked questions on R. URL http://www.ci.tuwien.ac.at/~hornik/R/. ISBN 3-901167-51-X.

[39] F. Huang and Y. Ogata. Improvements of the maximum pseudolikelihood estimators in various spatial statistical models. *Journal of Computational and Graphical Statistics*, 8(3):510–530, 1999.

[40] R. Ihaka and R. Gentleman. R: A language for data analysis and graphics. *Journal of Computational and Graphical Statistics*, 5(3):299–314, 1996.

[41] V.S. Isham. Multitype Markov point processes: some approximations. *Proceedings of the Royal Society of London, Series A*, 391:39–53, 1984.

[42] J.L. Jensen and H.R. Künsch. On asymptotic normality of pseudo likelihood estimates for pairwise interaction processes. *Annals of the Institute of Statistical Mathematics*, 46:475–486, 1994.

[43] J.L. Jensen and J. Møller. Pseudolikelihood for exponential family models of spatial point processes. *Annals of Applied Probability*, 1:445–461, 1991.

[44] F.P. Kelly and B.D. Ripley. On Strauss's model for clustering. *Biometrika*, 63:357–360, 1976.

[45] J.F.C. Kingman. *Poisson Processes*. Oxford University Press, 1993.

[46] A.B. Lawson. A deviance residual for heterogeneous spatial Poisson processes. *Biometrics*, 49:889–897, 1993.

[47] M.N.M. van Lieshout. *Markov Point Processes and their Applications*. Imperial College Press, 2000.

[48] M.N.M. van Lieshout and A.J. Baddeley. A nonparametric measure of spatial interaction in point patterns. *Statistica Neerlandica*, 50:344–361, 1996.

[49] M.N.M. van Lieshout and A.J. Baddeley. Indices of dependence between types in multivariate point patterns. *Scandinavian Journal of Statistics*, 26:511–532, 1999.

[50] P. McCullagh and J.A. Nelder. *Generalized Linear Models*. Chapman and Hall, second edition, 1989.

[51] J. Møller and R.P. Waagepetersen. *Statistical Inference and Simulation for Spatial Point Processes*. Chapman and Hall/CRC, Boca Raton, 2003.

[52] Y. Ogata and M. Tanemura. Estimation of interaction potentials of spatial point patterns through the maximum likelihood procedure. *Annals of the Institute of Statistical Mathematics*, B 33:315–338, 1981.

[53] Y. Ogata and M. Tanemura. Likelihood analysis of spatial point patterns. *Journal of the Royal Statistical Society, series B*, 46:496–518, 1984.

[54] Y. Ogata and M. Tanemura. Likelihood estimation of interaction potentials and external fields of inhomogeneous spatial point patterns. In I.S. Francis, B.J.F. Manly, and F.C. Lam, editors, *Pacific Statistical Congress*, pp. 150–154. Elsevier, 1986.

[55] A. Penttinen. *Modelling Interaction in Spatial Point Patterns: Parameter Estimation by the Maximum Likelihood Method*. Number 7 in Jyväskylä Studies in Computer Science, Economics and Statistics. University of Jyväskylä, 1984.

[56] R Development Core Team. *R: A language and environment for statistical computing*. R Foundation for Statistical Computing, Vienna, Austria, 2004. ISBN 3-900051-00-3.

[57] B.D. Ripley. The second-order analysis of stationary point processes. *Journal of Applied Probability*, 13:255–266, 1976.

[58] B.D. Ripley. Modelling spatial patterns (with discussion). *Journal of the Royal Statistical Society, Series B*, 39:172–212, 1977.

[59] B.D. Ripley. *Spatial Statistics*. John Wiley and Sons, New York, 1981.

[60] B.D. Ripley. Spatial statistics: developments 1980–3. *International Statistical Review*, 52:141–150, 1984.

[61] B.D. Ripley. *Statistical Inference for Spatial Processes*. Cambridge University Press, 1988.

[62] B.D. Ripley. Gibbsian interaction models. In D.A. Griffiths, editor, *Spatial Statistics: Past, Present and Future*, pp. 1–19. Image, New York, 1989.

[63] A. Särkkä. *Pseudo-likelihood approach for pair potential estimation of Gibbs processes*. Number 22 in Jyväskylä Studies in Computer Science, Economics and Statistics. University of Jyväskylä, 1993.

[64] B.W. Silverman and T.C. Brown. Short distances, flat triangles and poisson limits. *Journal of Applied Probability*, 15:815–825, 1978.

74   Adrian Baddeley and Rolf Turner

[65] D. Stoyan and P. Grabarnik. Second-order characteristics for stochastic structures connected with Gibbs point processes. *Mathematische Nachrichten*, 151:95–100, 1991.

[66] D. Stoyan and P. Grabarnik. Statistics for the stationary Strauss model by the cusp point method. *Statistics*, 22:283–289, 1991.

[67] D. Stoyan, W.S. Kendall and J. Mecke. *Stochastic Geometry and its Applications*. John Wiley and Sons, Chichester, second edition, 1995.

[68] D. Stoyan and H. Stoyan. *Fractals, random shapes and point fields*. Wiley, 1995.

[69] D. Stoyan and H. Stoyan. Non-homogeneous Gibbs process models for forestry — a case study. *Biometrical Journal*, 40:521–531, 1998.

[70] R. Takacs and T. Fiksel. Interaction pair-potentials for a system of ants' nests. *Biometrical Journal*, 28:1007–1013, 1986.

[71] J. Tukey. *Exploratory Data Analysis*. Addison-Wesley, Reading, Mass., 1977.

[72] W.N. Venables and B.D. Ripley. *Modern Applied Statistics with S-Plus*. Springer, second edition, 1997.

Theoretical and Methodological Advances in
Spatial Point Processes

Part II

Theoretical and Methodological Advances in
Spatial Point Processes

Strong Markov Property of Poisson Processes and Slivnyak Formula

Sergei Zuyev

Department of Statistics and Modelling Science, University of Strathclyde, Glasgow, G1 1XH, UK, sergei@stams.strath.ac.uk

Summary. We discuss strong Markov property of Poisson point processes and the related stopping sets. Viewing Poisson process as a set indexed random field, we demonstrate how the martingale technique applies to establish the analogues of the classical results: Doob's theorem, Wald identity in this multi-dimensional setting. In particular, we show that the famous Slivnyak-Mecke theorem characterising the Poisson process is a consequence of the strong Markov property.

Key words: Gamma-type result, Poisson point process, Slivnyak formula, Strong Markov property

1 Filtrations and Stopping Sets

To outline the idea of this paper, let us start with an example of a temporal stochastic process, i. e. a random function $\xi_\bullet(\omega) = \{\xi_t(\omega)\}_{t\geq 0}$, $\omega \in \Omega$ indexed by one-dimensional parameter $t \geq 0$ which we refer as *time*. Surely, this map from sample space Ω into the appropriate function space over $[0, \infty)$ should be measurable with respect to a suitably chosen σ-algebra. However, such a definition is usually too general as it does not describe the temporal evolution of $\xi_\bullet$. Therefore it is useful to define a growing sequence of σ-algebras $\mathcal{F}_{[0,s]}$ of subsets of Ω representing the process' history up to time s, and impose the condition that the restriction of $\xi_\bullet$ onto time interval $[0, s]$, i. e. the function $\{\xi_t(\omega)\}_{t\in[0,s]}$, should be $\mathcal{F}_{[0,s]}$-measurable for all $s \geq 0$. Of course, this is a stronger notion of measurability for the random function which is called *progressive measurability*. The system of growing σ-algebras $\mathcal{F}_{[0,s]}$ is called *filtration*.

One of the central notions for temporal processes is the *stopping time*. It is a random variable τ such that event $\{\omega \in \Omega : \tau(\omega) \leq s\}$ is $\mathcal{F}_{[0,s]}$-measurable for all $s \geq 0$. In words, the fact that τ is observed before time s is defined only by the history $\mathcal{F}_{[0,s]}$ up to time s only. With every stopping time one may associate the corresponding stopping σ-algebra which is the collection of events

$$\mathcal{F}_\tau = \{\Sigma \in \mathcal{F}_{[0,\infty]} : \ \Sigma \cap \{\omega : \ \tau(\omega) \le s\} \in \mathcal{F}_{[0,s]} \text{ for all } s \ge 0\}. \qquad (1)$$

The main object of our study here are point processes in a general space. We shall see how far we can mimic the above objects in this intrinsically multidimensional setting. We treat point processes as random countable measures and as we will see, their usual definition actually assumes the progressive measurability. Specifically, let X be a locally compact separable topological space (LCS-space) which we call a *phase space* of the process and $\mathcal{B}$ be its Borel σ-algebra. X plays the role of the index set $[0,\infty)$ above – we typically consider $X = \mathbb{R}^d$ for simplicity. Let $\mathcal{N}$ be a set of counting measures on $\mathcal{B}$, so that a measure $\phi \in \mathcal{N}$, if $\phi(B) \in \{0,1,2,\dots\} = \mathbb{Z}_+$ for any Borel B. Any such measure can be represented as the sum of unit masses: $\phi = \sum_i \delta_{x_i}$, where x_i are not necessarily different and $\delta_x(B) = \mathbf{1}_B(x)$. We call the support points *particles*.

A point process $N = N(\omega)$ is a $[\mathcal{F}, \Xi]$-measurable mapping from some abstract probability space $(\Omega, \mathcal{F}, \mathbf{P})$ into the measurable space $[\mathcal{N}, \Xi]$ of counting measures. σ-algebra Ξ is generated by the sets of the type $\{\phi \in \mathcal{N} : \phi(B) = k\}$, $B \in \mathcal{B}$, $k \in \mathbb{Z}_+$. This is a natural definition of measurability for point processes as this makes the events of type $\{\omega \in \Omega : N(\omega, B) = k\}$ measurable. Often $[\Omega, \mathcal{F}]$ is taken to be $[\mathcal{N}, \Xi]$ itself and N is identity mapping. Such processes are called *canonically defined*. From now on we consider canonically defined processes and write ϕ (a point configuration) instead of ω to stress that and give up notation Ξ in favour of $\mathcal{F}$.

The *intensity measure* of a point process $N = N(\phi)$ defined on Borel $B \in \mathcal{B}$ as $\lambda_N(B) = \mathbf{E}N(B)$. The *Campbell measure* is a measure $\mathcal{C}(\mathrm{d}\phi \, \mathrm{d}x)$ on $\mathcal{F} \otimes \mathcal{B}$ defined on $\Sigma \times B$ as $\mathcal{C}(\Sigma \times B) = \mathbf{E}N(\phi, B)\mathbf{1}_\Sigma(\phi)$. We observe that $\mathcal{C}(\Sigma \times \bullet)$ as a measure on $\mathcal{B}$ is absolutely continuous with respect to $\lambda_N(\mathrm{d}x)$, thus there exists a Radon-Nikodym derivative $\mathbf{P}_N^x(\Sigma)$ which is a measurable function of $x \in X$, but which can also be chosen to be a probability measure on $[\mathcal{N}, \mathcal{F}]$ called the *Palm distribution* corresponding to N at x. By definition the following identity called *refined Campbell theorem* holds:

$$\mathbf{E} \int F(\phi, x) \, N(\mathrm{d}x) = \int \mathbf{E}_N^x F(\phi, x) \, \lambda_N(\mathrm{d}x) \qquad (2)$$

for any measurable function F. The Palm measure $\mathbf{P}_N^x$ is concentrated on configurations ϕ such that $\phi(\{x\}) > 0$ and can be regarded as a distribution of a random configuration conditioned on having a particle at x.

Let $\mathbb{F}, \mathbb{K}$ be the system of closed and compact subsets of X respectively. Then for every $K \in \mathbb{K}$ one may define the σ-algebra $\mathcal{F}_K$ which is generated by the sets $\{\phi \in \mathcal{N} : \phi(B \cap K) = k\}$, $B \in \mathcal{B}$, $k \in \mathbb{Z}_+$. Similarly to one-dimensional case, the following properties allow us to call the system $\{\mathcal{F}_K\}$, $K \in \mathbb{K}$ a *filtration*:

- monotonicity: $\mathcal{F}_{K_1} \subseteq \mathcal{F}_{K_2}$ for any two compact $K_1 \subseteq K_2$;
- continuity from above: $\mathcal{F}_K = \cap_{n=1}^\infty \mathcal{F}_{K_n}$ if $K_n \downarrow K$.

By construction, the restriction of the point process N onto K is $\mathcal{F}_K$-measurable, so N is automatically progressively measurable and $\{\mathcal{F}_K\}$, $K \in \mathbb{K}$ is thus the *natural filtration* associated with the process. We see a complete analogy when one-dimensional parameter t – time is replaced now by a compact set K. To pursue this analogy we need a notion of a random compact set which supersedes a random time.

A *random closed set* $\mathcal{N}$ is a measurable mapping $\mathcal{N} : [\mathcal{N}, \mathcal{F}] \mapsto [\mathbb{F}, \sigma_f]$, where σ_f is the σ-algebra generated by the system $\{F \in \mathbb{F} : F \cap K \neq \emptyset\}$, $K \in \mathbb{K}$.

A random compact set $S = S(\phi)$ is called a *stopping set* (more precisely, $\{\mathcal{F}_K\}$-stopping set) if the event $\{\phi : S(\phi) \subseteq K\}$ is $\mathcal{F}_K$ measurable for all $K \in \mathbb{K}$. It is a natural generalisation of the notion of a stopping time: knowing the configuration of $N(\phi)$ inside a compact K is sufficient to conclude whether $S(\phi) \subseteq K$ or not.

Similarly to (1), with each stopping set S there is an associated *stopping σ-algebra*:

$$\mathcal{F}_S = \{\Sigma \in \mathcal{F} : \Sigma \cap \{\phi : S(\phi) \subseteq K\} \in \mathcal{F}_K \text{ for all } K \in \mathbb{K}\}.$$

It can be shown that

$$S(\phi) = S(\psi|_{S(\phi)}) \quad \text{and} \quad F(\phi) = F(\phi|_{S(\phi)}) \tag{3}$$

if F is $\mathcal{F}_S$-measurable. Here and afterwards, $\phi|_B(\bullet) = \phi(B \cap \bullet)$ denotes restriction of a counting measure ϕ onto B This stems from [7, Prop. 3] on the structure of the stopping σ-algebra and reflects the fact that to decide whether S is a stopping set or not, one only needs to know configuration in S itself. Since non-random compacts are also stopping sets, then (6) also covers (4).

Perhaps, the simplest of stopping set is based on the stopping time: if τ is a finite stopping time in 1D case, then the set $[0, \tau]$ is a compact $\{\mathcal{F}_{[0,s]}\}$-stopping set. More complex examples. Assume that X is a metric space and $N(X) \geq k$ almost surely for some $k \geq 1$. Then the smallest closed ball $B(x_0)$ centred in a given point x_0 containing k points of the process inside is a stopping set. Indeed, given realisation $N(\phi)$, start "growing" a ball from x_0 increasing its radius from 0 to infinity and stop when it first accumulates k points (or maybe more at once, when the process points are not always in a general position or may overlap). Then whatever compact K is considered, either we stop before this growing ball touches the complement K^c, so that $B(x_0) \subseteq K$, or we reach K^c and thus $B(x_0) \not\subseteq K$. Either way, we only used point configuration inside K to decide whether or not $B(x_0) \subseteq K$, i.e. this event is $\mathcal{F}_K$-measurable.

This observation actually shows a very useful way to establish the stopping property: if there is a one-parameter sequence of growing compact sets which eventually leads to construction of the random compact, then this compact is a stopping set. Consider $X = \mathbb{R}^2$ and $N(\phi)$ containing almost surely at least one particle in each of the four quadrants. Assume also that N does not contain multiple points and that all the particles are in a general position (no three points are aligned and no four points lie on a circle). Construct the Voronoi cell centred in the origin O with respect to $N(\phi) \cup \{O\}$. It consists of the points which are closer to O than to any particle from $N(\phi)$. Its vertices are the centres of the balls which have the origin and exactly two particles of $N(\phi)$ on their boundaries and no point of $N(\phi)$ inside. The union $F(\phi)$ of these balls is known as the *Voronoi flower* or fundamental region as the geometry of the cell is completely determined by F. Let us show that F is a stopping set.

Let S_0 be the largest disk centred on the positive x-axis passing through the origin and one of the particles (call it x_1), *and* not having any particles in its interior (see Fig. 1). The right bisector of O and x_1 can be seen on the figure; it is the side of the Voronoi polygon cut by the positive x-axis. Now consider the continuum of disks passing through O and x_1, with centre moving upward along this right bisector. Stop when this "growing" disk first hits another particle (which is labelled x_2). This disk is B_1. In a similar fashion, we move a circle-centre along the next right-bisector, stopping the growing disk (which passes through O and x_2) when it hits another particle, x_3. The last of these constructions stops when x_1 is encountered by a growing disk. This algorithm successfully constructs the Voronoi flower $F = S_0 \cup B_1 \cup \ldots B_n$, if the cell has n sides.

A *Poisson process* with intensity measure $\lambda(dx)$ is a point process Π with the following two properties: the variables $\Pi(B_1), \ldots, \Pi(B_k)$ are mutually independent for disjoint $B_1, \ldots, B_k$ for any k; and $\Pi(B)$ follows Poisson distribution with parameter $\lambda(B)$. As a result, for any Borel set B and any functional $F(\phi)$, $\phi \in \mathcal{N}$ one has:

$$\int F(\phi) \mathbf{P}(d\phi) = \int F(\phi|_B + \phi|_{B^c}) \mathbf{P}(d\phi)$$

$$= \iint F(\phi|_B + \phi'|_{B^c}) \mathbf{P}(d\phi) \mathbf{P}(d\phi')$$

$$= \iint F(\phi + \phi') \mathbf{P}_B(d\phi) \mathbf{P}_{B^c}(d\phi'), \qquad (4)$$

where $\mathbf{P}_B$ is the restriction of $\mathbf{P}$ onto the σ-algebra $\mathcal{F}_B$. The property (4) reflects complete independence of the Poisson process distribution due to which $\mathbf{P} = \mathbf{P}_B \otimes \mathbf{P}_{B^c}$. In particular, a Poisson process is a Markov process. Therefore it also possesses the *strong Markov property*:

$$\int F(\phi) \mathbf{P}(d\phi) = \iint F(\phi|_{S(\phi)} + \phi'|_{S^c(\phi)}) \mathbf{P}(d\phi) \mathbf{P}(d\phi') \qquad (5)$$

for every compact stopping set S, see [5, Thm. 4].

Relation (5) can also be expressed as

$$\mathbf{E}[F(\Pi) \mid \mathcal{F}_S](\phi|_{S(\phi)}) = \mathbf{E}_{S^c(\phi)} F(\phi|_{S(\phi)} + \Pi) \qquad (6)$$

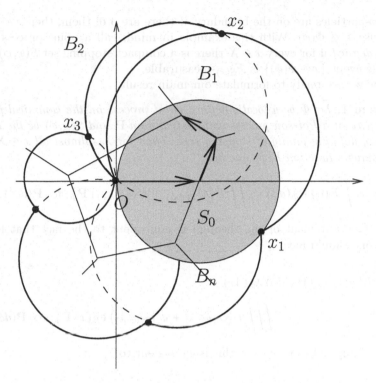

Fig. 1. Incremental construction of the Voronoi flower. Stopping set S_0 is shaded. Direction of the circle-centre move is shown by arrows

(more exactly, this is one of versions of the conditional expectation).

2 Slivnyak Theorem for Locally Defined Processes

It is common in stochastic geometry and other applications to have another point process Φ which is defined as a function of the reference process N. For instance, Φ may be the process of vertices of the Voronoi tessellation constructed with respect to planar process N. The way this process is constructed uses only local information to decide where the positions of Φ-particles are. Assume for simplicity that N is simple, i.e. with probability 1 it does not contain multiple particles. Then, given a configuration ϕ of the reference process N, the points of $\Phi(\phi)$ have the following identifying property: $x \in \Phi(\phi)$ if and only if there is a ball centred at x which contains exactly 3 ϕ-particle on its boundary and no ϕ-particle inside. A way to establish if $x \in \Phi(\phi)$ is simple: start "blowing" a ball centred at x until it hits a ϕ-particles. Call that inflated ball with at least one particle on the boundary $S(x, \phi)$. As we already discussed above, $S(x, \phi)$ is a stopping set. Then we just count how many ϕ-particles are on the boundary, if there are 3 of them, then $x \in \Phi(\phi)$, otherwise $x \notin \Phi(\phi)$. With this example in mind, call a point process $\Phi(\phi)$ *locally defined* if for every $x \in X$ there is a compact stopping set $S(x, \phi)$ such that the event $\{x \in \Phi(\phi)\}$ is $\mathcal{F}_{S(x)}$-measurable.

Now we are ready to formulate our main result.

Theorem 1. *Let Φ be a locally defined point process on the canonical probability space of a Poisson process with distribution $\mathbf{P}$ and $S(x, \phi)$ be the corresponding defining family of stopping sets. Then for λ_Φ-almost all $x \in X$ and a measurable function $F(\phi)$ one has*

$$\mathbf{E}_\Phi^x F = \int F(\phi) \, \mathbf{P}_\Phi^x(d\phi) = \iint F(\phi|_{S(x,\phi)} + \phi'|_{S^c(x,\phi)}) \, \mathbf{P}_\Phi^x(d\phi) \, \mathbf{P}(d\phi'). \quad (7)$$

Proof. The statement of the theorem is equivalent to the fact that for all $B \in \mathcal{B}$ one should have

$$\iint F(\phi) \mathbf{1}_B(x) \, \mathbf{P}_\Phi^x(d\phi) \, \lambda_\Phi(dx)$$

$$= \iiint F(\phi|_{S(x,\phi)} + \phi'|_{S^c(x,\phi)}) \mathbf{1}_B(x) \, \mathbf{P}_\Phi^x(d\phi) \, \mathbf{P}(d\phi').$$

By the Campbell theorem (2), this is equivalent to

$$\iint F(\phi) \mathbf{1}_B(x) \, \Phi(\phi, dx) \, \mathbf{P}(d\phi)$$

$$= \iiint F(\phi|_{S(x,\phi)} + \phi'|_{S^c(x,\phi)}) \mathbf{1}_B(x) \, \Phi(\phi, dx) \, \mathbf{P}(d\phi) \, \mathbf{P}(d\phi'). \quad (8)$$

Apply identity (5) to the left hand side of (8). By the local definition of Φ and by (3) one has $\Phi(\phi|_{S(x,\phi)} + \phi'|_{S^c(x,\phi)}) = \Phi(\phi|_{S(x,\phi)}) = \Phi(\phi)$. The result is indeed the right hand side, and the proof is complete.

A few remarks are now in order.

A result similar to (7) was first established in [4] for the above example of the nodes of the Voronoi tessellation constructed with respect to a stationary Poisson process. The proof there uses particular geometric properties of the empty Delaunay disks ($S(x)$ in our notation) and cannot be ported to our general setting. In this above form, the result was shown in [1] for the case of stationary processes. In the stationary case the Palm distribution is just no longer a function of x, so it is covered by the same identity (7).

Consider the case when the Poisson process Π is simple and Φ coincides with Π itself. It is trivially locally defined: the stopping sets $S(x)$ are just the singletons $\{x\}$. Now the formula (7) transforms into

$$\int F(\phi) \, \mathbf{P}^x(d\phi) = \int F(\delta_x + \phi') \, \mathbf{P}(d\phi') = \int F(\phi + \delta_x) \, \mathbf{P}(d\phi) \quad (9)$$

which is exactly the Slivnyak's theorem, see [3, 6]. So this Slivnyak-Mecke characterising formula is no more than another face of the Strong Markov property of the Poisson process.

The proof of the theorem used only the strong Markov property (5) of the Poisson process distribution **P** which, in turn, was a consequence of the complete independence property (4). Thus Theorem 1 also holds for completely independent point processes. Such processes are, in fact, a superposition of two independent point processes: a counting measure concentrated on a non-random at most countable set of atoms and a Poisson process with a diffuse intensity measure, see [2, Thm. 2.4.VIII]. This Poisson process is thus simple. We saw, however, that when the first component is absent, the theorem implies identity (9) which *characterises* Poisson point process distribution, as was proved in [3]. Thus, as a by-product we have shown that there is no simple complete independent point process other than Poisson.

Let us also mention another generalisation of the idea of locally defined point processes to higher dimensional random sets. For simplicity of formulations, we only deal with the phase space $X = \mathbb{R}^d$.

Consider an n-dimensional $(n < d)$ random fiber process, i.e. a random closed set Φ on the Poisson process' probability space $[\mathcal{N}, \Xi]$ such that its n-dimensional *intensity measure* $\lambda_\Phi(\bullet) = \mathbf{E}H^n(\bullet \cap \Phi)$ is non-trivial and σ-finite (H^n is the n-dimensional Hausdorff measure in $\mathbb{R}^d$).

As above, call Φ locally defined if for every $x \in X$ there is a compact stopping set $S(x, \phi)$ such that the event $\{x \in \Phi(\phi)\}$ is $\mathcal{F}_{S(x)}$-measurable. A visual example may provide the collection of n-dimensional edges of the Voronoi cells constructed with respect to the particles of the process. A point x belongs to n-dimensional edge if and only if the glowing ball centred at x will hit at least $d - n + 1$ particles at once, see Fig. 2.

Similarly to point process case, one may introduce the Campbell measure $\mathcal{C}(\Sigma \times B) = \mathbf{E}\mathbf{1}_\Sigma H^n(B \cap \Phi)$ and its Radon-Nikodym derivative

$$\mathbf{P}_\Phi^x(\Sigma) = \frac{d\mathcal{C}(\Sigma \times \bullet)}{d\lambda_\Phi}(x), \quad \Sigma \in \Xi,$$

which is called the Palm probability (more exactly, its version which is a probability measure on Ξ).

Now, the proof of Theorem 1 can be carried through to give us a similar result:

Theorem 2. *Assume that the fiber process Φ is locally defined. Then Formula (7) holds for Φ and its Palm distribution.*

Acknowledgements

The author is grateful to Günter Last for his enlightening questions on the nature of local definiteness.

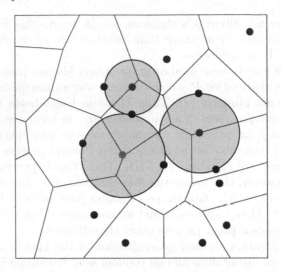

Fig. 2. Edges of the Voronoi cells and the corresponding defining stopping sets

References

[1] R. Cowan, M. Quine and S. Zuyev. Decomposition of Gamma-distributed domains constructed from poisson point processes. *Advances in Applied Probability*, 35(1):56–69, 2003.

[2] D.J. Daley and D. Vere-Jones. *An Introduction to the Theory of Point Processes*. Springer, New York, 1988.

[3] J. Mecke. Stationäre zufällige Masse auf localcompakten Abelischen Gruppen. *Z. Wahrscheinlichkeitsth*, 9:36–58, 1967.

[4] J. Mecke and L. Muche. The Poisson Voronoi tessellation I. basic identity. *Mathematische Nachrichten*, 176:199–208, 1995.

[5] Yu. A. Rozanov. *Markov random fields*. Springer, New York, 1982.

[6] I.M. Slivnyak. Some properties of stationary flows of homogeneous random events. *Teoriya Veroyatnostei i ee Primeneniya*, pages 347–352, 1962. (In Russian). English translation: *Theory of Probability and its Applications* 7:336–341.

[7] S. Zuyev. Stopping sets: Gamma-type results and hitting properties. *Advances in Applied Probability*, 31(2):355–366, 1999.

Bayesian Analysis of Markov Point Processes

Kasper K. Berthelsen and Jesper Møller

Aalborg University, Department of Mathematical Sciences,
F. Bajers Vej 7G, DK-9220 Aalborg, Denmark
kkb@math.aau.dk, jm@math.aau.dk

Summary. Recently Møller, Pettitt, Berthelsen and Reeves [17] introduced a new MCMC methodology for drawing samples from a posterior distribution when the likelihood function is only specified up to a normalising constant. We illustrate the method in the setting of Bayesian inference for Markov point processes; more specifically we consider a likelihood function given by a Strauss point process with priors imposed on the unknown parameters. The method relies on introducing an auxiliary variable specified by a normalised density which approximates the likelihood well. For the Strauss point process we use a partially ordered Markov point process as the auxiliary variable. As the method requires simulation from the "unknown" likelihood, perfect simulation algorithms for spatial point processes become useful.

Key words: Bayesian inference, Markov chain Monte Carlo, Markov point process, Partially ordered Markov point process, Perfect simulation, Spatial point process, Strauss process

1 Introduction

Markov point processes [14, 16, 19] are models for point processes with interacting points, and they constitute one of the most important classes of spatial point process models. The basic problem with parametric inference for such point processes is the presence of a normalising constant which cannot be evaluated explicitly, cf. Chap. 9 in [19]. So far most work on parametric inference for Markov point processes have concentrated on parameter estimation based on maximum pseudo likelihood estimation [1, 5, 13] or approximate maximum likelihood estimation using Markov chain Monte Carlo (MCMC) algorithms [9, 10, 18, 19]. Apart from a few papers [3, 12], very little has been done on Bayesian inference for Markov point processes.

In this paper we consider the problem of simulating from a posterior density

$$\pi(\theta|y) \propto \pi(\theta)\pi(y|\theta) \tag{1}$$

when the likelihood

$$\pi(y|\theta) = q_\theta(y)/Z_\theta \tag{2}$$

is given by an unnormalised density $q_\theta(y)$ with an unknown normalising constant (or partition function) Z_θ. By "unknown", we mean that Z_θ is not available analytically and/or that exact computation is not feasible. Indeed this is the case when (2) is a likelihood function for a parametric family of Markov point process models, cf. [16, 19].

For example, consider a Strauss process defined on a region $S \subset \mathbb{R}^2$ of area $|S| \in (0, \infty)$. This has a density

$$\pi(y|\theta) = \frac{1}{Z_\theta} \beta^{n(y)} \gamma^{s_R(y)} \tag{3}$$

with respect to μ, which denotes a homogeneous Poisson point process on S with intensity one. Further, y is a point configuration, i.e. a finite subset of S; $\theta = (\beta, \gamma, R)$, with $\beta > 0$ (known as the chemical activity in statistical physics), $0 < \gamma \leq 1$ (the interaction parameter), and $R > 0$ (the interaction range); $n(y)$ is the cardinality of y; and

$$s_R(y) = \sum_{\{\xi,\eta\} \subseteq y : \xi \neq \eta} 1[\|\eta - \xi\| \leq R]$$

is the number of pairs of points in y within a distance R from each other. Figure 1 shows a realisation y of a Strauss point process, where $s_R(y)$ is given by the number of pairs of overlapping discs with diameter $R/2$ and centred at the points in y. For $\gamma = 1$, we obtain a homogeneous Poisson process on S with intensity β. For $\gamma < 1$, typical realisations look more regular than in the case $\gamma = 1$. This is due to inhibition between the points, and the inhibition gets stronger as γ decreases or R increases. The normalising constant is unknown when $\gamma < 1$, since

$$Z_\theta = e^{-|S|} + e^{-|S|} \sum_{n=1}^{\infty} \beta^n \int_S \cdots \int_S \gamma^{s_R(\{y_1,\ldots,y_n\})} \, dy_1 \cdots dy_n$$

where the n-fold integrals are unknown, cf. [14].

It is not straightforward to generate samples from (1) by MCMC algorithms: Consider a Metropolis-Hastings algorithm, see e.g. [20]. If θ is the current state of the chain generated by the algorithm, and if a proposal θ' with density $p(\theta'|\theta)$ is generated, then θ' is accepted as the new state with probability $\alpha(\theta'|\theta) = \min\{1, H(\theta'|\theta)\}$, and otherwise we retain θ. Here

$$H(\theta'|\theta) = \frac{\pi(\theta'|y)p(\theta|\theta')}{\pi(\theta|y)p(\theta'|\theta)}$$

is the Hastings ratio. By (2),

$$H(\theta'|\theta) = \frac{\pi(\theta')q_{\theta'}(y)p(\theta|\theta')}{\pi(\theta)q_\theta(y)p(\theta'|\theta)} \bigg/ \frac{Z_{\theta'}}{Z_\theta} \tag{4}$$

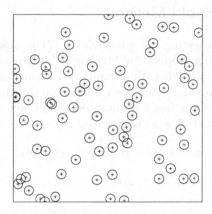

Fig. 1. Realisation of a Strauss point process on the unit square, with $(\beta, \gamma, R) = (100, 0.5, 0.05)$, and generated by perfect simulation algorithm (dominated CFTP, see[15]). Circles centred at points have radii 0.025.

is unknown, since it depends on the ratio of unknown normalising constants $Z_{\theta'}/Z_{\theta}$.

Because of their intractability, earlier Bayesian work on Markov point processes attempted to avoid algorithms involving unknown normalising constants. An example in connection to spatial point processes is Heikkinen and Penttinen [12], who instead of estimating the entire posterior distribution, focused on finding the maximum a posteriori estimate for the interaction function in a Bayesian model where the likelihood function is given by a pairwise interaction point processes (like the Strauss process) and its normalising constant is unknown. Recently, Berthelsen and Møller [3] performed a more detailed Bayesian MCMC analysis, using path sampling [8] or, as it is known in statistical physics, thermodynamic integration, for estimating the ratio of normalising constants.

Section 2 considers the approach introduced by Møller et al. [17] which avoids approximations of (ratios of) normalising constants such as those discussed above. Their approach consists in introducing an auxiliary variable x into a Metropolis-Hastings algorithm for (θ, x) so that ratios of normalising constants no longer appear but the posterior distribution for θ is retained. Access to algorithms for making perfect (or exact) simulations [2, 11, 15, 19] from (2) is an important ingredient as explained later. Section 3 applies this approach to a Bayesian analysis of a Strauss process. This section has earlier been published as a part of the research report [17]. Finally, Sect. 4 contains some concluding remarks.

2 Auxiliary Variable Method

Consider the general setting (1) when Z_θ in (2) is unknown. The method described in this section applies for Markov point process models as well as many other statistical models with an unknown normalising constant, cf. [17].

We introduce an auxiliary variable x defined on the same space as the state space of y. Assume that x has a normalised conditional density $f(x|\theta, y)$, so that the joint density of (θ, x, y) is given by

$$\pi(\theta, x, y) = f(x|\theta, y)\pi(y|\theta)\pi(\theta).$$

The posterior density with $\pi(y|\theta)$ given by (2),

$$\pi(\theta, x|y) \propto f(x|\theta, y)\pi(\theta)q_\theta(y)/Z_\theta$$

still involves the unknown Z_θ.

A Metropolis-Hastings algorithm for drawing from $\pi(\theta, x|y)$ has a Hasting ratio given by

$$
\begin{aligned}
H(\theta', x'|\theta, x) &= \frac{\pi(\theta', x'|y)p(\theta, x|\theta', x')}{\pi(\theta, x|y)p(\theta', x'|\theta, x)} \\
&= \frac{f(x'|\theta', y)\pi(\theta')q_{\theta'}(y)p(\theta, x|\theta', x')}{f(x|\theta, y)\pi(\theta)q_\theta(y)p(\theta', x'|\theta, x)} \bigg/ \frac{Z_{\theta'}}{Z_\theta}
\end{aligned}
$$

where $p(\theta', x'|\theta, x)$ is the proposal density for (θ', x'). The proposal density can be factorised as

$$p(\theta', x'|\theta, x) = p(x'|\theta', \theta, x)p(\theta'|\theta, x) \tag{5}$$

and the choice of proposal distribution is arbitrary from the point of view of the equilibrium distribution of the chain of θ-values. Hence we may take the proposal density for the auxiliary variable x' to be the same as the likelihood, but depending on θ', rather than θ,

$$p(x'|\theta', \theta, x) = p(x'|\theta') = q_{\theta'}(x')/Z_{\theta'}. \tag{6}$$

Then

$$H(\theta', x'|\theta, x) = \frac{f(x'|\theta', y)\pi(\theta')q_{\theta'}(y)q_\theta(x)p(\theta|\theta', x')}{f(x|\theta, y)\pi(\theta)q_\theta(y)q_{\theta'}(x')p(\theta'|\theta, x)} \tag{7}$$

does not depend on $Z_{\theta'}/Z_\theta$, and the marginalisation over x of the equilibrium distribution $\pi(\theta, x|y)$, gives the desired distribution $\pi(\theta|y)$. In contrast to (4) we now have a much simpler problem of finding the ratio of the distributions of the proposed and current auxiliary variable, $f(x'|\theta', y)/f(x|\theta, y)$, the other factors in (7) presenting no difficulty in evaluation.

Henceforth, for simplicity, we assume that

$$p(\theta'|\theta, x) = p(\theta'|\theta) \tag{8}$$

does not depend on x. For simulation from the proposal density (5) we suppose that it is straightforward to make simulations from $p(\theta'|\theta)$ but not necessarily from $p(x'|\theta', \theta, x)$; for $p(x'|\theta', \theta, x)$ given by (6) appropriate perfect simulation algorithms [2, 11, 15, 19] are used to avoid convergence questions of straightforward MCMC algorithms.

A critical design issue for the algorithm is to choose an appropriate auxiliary density $f(x|\theta, y)$ and proposal density $p(\theta'|\theta)$ so that the algorithm has good mixing and convergence properties. Assume for the moment that Z_θ is known and the algorithm based on (4) has good mixing properties. If we let $f(x|\theta, y) = q_\theta(x)/Z_\theta$, then by (8), (7) reduces to (4), and so the mixing and convergence properties of the two Metropolis-Hastings algorithms using (4) and (7) are the same. Furthermore, recommendations on how to tune Metropolis-Hastings algorithms to obtain optimal acceptance probabilities may exist in the case of (4). This suggests that the auxiliary distribution should approximate the distribution given by q_θ,

$$f(x|\theta, y) \approx q_\theta(x)/Z_\theta. \tag{9}$$

It is interesting to notice that if equality holds in (9), then the states from the chain for the auxiliary variable x can be interpreted as posterior predictions. Choices where (9) are satisfied will be discussed in the following. One particular choice is

$$f(x|\theta, y) = q_{\tilde{\theta}}(y)/Z_{\tilde{\theta}}, \tag{10}$$

where $\tilde{\theta}$ is fixed. This choice is expected to work well if the posterior distribution is concentrated around $\tilde{\theta}$.

3 The Strauss Process

The Strauss process (3) is an example of a so-called locally stable point process, and in fact most Markov point processes used in applications are locally stable [9, 19]. Locally stable point processes can be simulated perfectly by an extension of the Propp-Wilson CFTP algorithm, called dominated CFTP, see [15]. Maximum likelihood and maximum pseudo likelihood estimation for the Strauss process is well established [1, 3, 5, 9, 10, 13, 18, 19].

3.1 Specification of Auxiliary Point Processes

In Sect. 3.2 we consider results for three different kinds of auxiliary variables (referred to as auxiliary point processes) with densities $f = f_1, f_2, f_3$ with respect to μ. In the sequel, for simplicity, we fix R, though our method extends to the case of varying interaction radius, but at the expense of further calculations.

The simplest choice of an auxiliary point process is a homogeneous Poisson point process on S. We let its intensity be given by the MLE $n(y)/|S|$ based on the data y. This auxiliary point process has density

$$f_1(x|\theta, y) = e^{|S|-n(y)}(n(y)/|S|)^{n(x)}, \tag{11}$$

see e.g. [19]. We refer to (11) as the fixed Poisson process.

The second choice takes the interaction into account. Its density is given by

$$f_2(x|\theta, y) \propto \hat{\beta}^{n(x)}\hat{\gamma}^{s_R(x)} \tag{12}$$

where $(\hat{\beta}, \hat{\gamma})$ is the MLE based on y and approximated by MCMC methods (for details, see Sect. 3 in [3]). We refer to (12) as the fixed Strauss process and to $(\hat{\beta}, \hat{\gamma})$ as the MCMC MLE.

The densities f_1 and f_2 do not depend on the parameters β and γ, and they are both of the type (10). The third choice we consider takes both interaction and parameters into account, but not the data y. Its density is more complicated to present, but it is straightforward to make a simulation in a sequential way: Assume for instance that S is rectangular (the following easily extends to a general region S). Choose a subdivision C_i, $i = 1, \ldots, m$ of S into, say, rectangular cells C_i of equal size. The simulation is then done in a single sweep, where the cells are visited once in some order. Each visit to a cell involves updating the point configuration within the cell in a way that only depends on the point configuration within the cells already visited.

Specifically, let $I = \{1, \ldots, m\}$ be the index set for the subdivision and for each $i \in I$ let X_i be a point process on C_i. Furthermore, we introduce a permutation $\rho : I \mapsto I$ of I; we shall later let ρ be random but for the moment we condition on ρ. Then, let $X_{\rho(1)}$ be a homogeneous Poisson point process on $C_{\rho(1)}$ with intensity κ_1 and for $i = 2, \ldots, m$, conditional on $X_{\rho(1)} = x_1, \ldots, X_{\rho(i-1)} = x_{i-1}$, let $X_{\rho(i)}$ be a homogeneous Poisson point process on $C_{\rho(i)}$ with intensity κ_i, where κ_i may depend on $x_1, \ldots, x_{i-1}$ (which is the case below). Then $X = \cup_{i=1}^m X_i$ is a point process which is an example of a so-called partially ordered Markov model (POMM).

POMMs were introduced by Cressie and Davidson [6] who applied POMMs in the analysis of grey scaled digital images. POMMs have the attractive properties that their normalising constants are known (and equal one), and that they can model some degree of interaction. Cressie, Zhu, Baddeley and Nair [7] consider what they call directed Markov point processes (DMPP) as limits of POMM point processes. Such processes are similar to our POMM point process X.

When specifying κ_i, $i \in I$ we want to approximate a Strauss point process. To do so we introduce the following concepts and notation. To each cell C_i, $i \in I$ we associate a reference point $\xi_i \in C_i$. Two cells C_i and C_j, $i \neq j$, are said to be neighbour cells if $\|\xi_i - \xi_j\| \leq R_P$, where $R_P > 0$ is the POMM interaction range (to be specified below). Further, for a given point configuration $x \subset S$, let $n_i(x) = n(x \cap C_{\rho(i)})$ denote the number of points in cell $C_{\rho(i)}$, and let

$s_{i,R_P,\rho}(x) = \sum_{j \in I: j < i} n_j(x) \mathbf{1}[\|\xi_{\rho(j)} - \xi_{\rho(i)}\| \leq R_P]$ be the number of points in the cells $C_{\rho(j)}$, $j < i$, which are neighbours to $C_{\rho(i)}$ (setting $s_{1,R_P,\rho}(x) = 0$). Note that we have suppressed the dependence on $\{C_i : i \in I\}$ and $\{\xi_i : i \in I\}$ in the notation. Setting $\kappa_i = \beta_P \gamma_P^{s_{i,R_P,\rho}(x)}$ we have that X is a POMM point process with density

$$f_P(x|\beta_P, \gamma_P, R_P, \rho) = \exp\left(-\beta_P \sum_{i \in I} |C_{\rho(i)}| \gamma_P^{s_{i,R_P,\rho}(x)}\right) \beta_P^{n(x)} \prod_{i \in I} \gamma_P^{n_i(x) s_{i,R_P,\rho}(x)}$$

(13)

with respect to μ.

Cressie et al. [7] use a Strauss like DMPP which obviously suffers from directional effects (incidentally this does not show up in the examples they consider). In order to eliminate directional effects in our POMM point process we consider ρ as a random variable uniformly distributed over all permutations of I independent of (θ, y). Moreover, we assume that x given (θ, y, ρ) has density f_3 as specified below. Letting ρ be a random variable requires a slight modification of the auxiliary variable method: each Metropolis-Hastings update consists in first proposing new values of θ and ρ and then conditional on these proposals proposing a new value of x. Using a uniform proposal ρ' the Hastings ratio (7) is modified by replacing $f(x'|\theta', y)/f(x|\theta, y)$ with $f_3(x'|\theta', \rho', y)/f_3(x|\theta, \rho, y)$ when (θ, x, ρ) is the current state of the chain and (θ', x', ρ') is the proposal; for further details, see Appendix A.

It remains to specify f_3 and (β_P, γ_P, R_P) in terms of $\theta = (\beta, \gamma, R)$. Let $(\beta_P, \gamma_P, R_P) = g(\theta) \equiv (g_1(\theta), g_2(\theta), g_3(\theta))$ where $g : (0, \infty) \times (0, 1] \times (0, \infty) \mapsto (0, \infty) \times (0, 1] \times (0, \infty)$ is a function specified as follows. Conditional on (θ, ρ, y), the POMM auxiliary point process has density

$$f_3(x|\theta, \rho, y) = f_P(x|g(\theta), \rho).$$

(14)

When specifying g we note that for point configurations x (except for a null set with respect to a homogeneous Poisson process), $\sum_{i \in I} s_{i,R_P,\rho}(x)$ tends to $s_{R_P}(x)$ as $m \to \infty$. This motivates setting $g_3(\theta) = R$ when the cell size is small compared to R. We would like that

$$(g_1(\theta), g_2(\theta)) = \mathbb{E}[\text{argmax}_{(\tilde{\beta}, \tilde{\gamma})} f_P(Y|\tilde{\beta}, \tilde{\gamma}, R, \rho)]$$

(15)

where Y is a Strauss process with parameter $\theta = (\beta, \gamma, R)$ and ρ is uniformly distributed and independent of Y. As this expectation is unknown to us, it is approximated as explained in Appendix B. In Table 1, Sect. 3.2, we refer to (15) as the "MLE". For comparison, we also consider the identity mapping $g(\theta) = \theta$ in Sect. 3.2. In Table 1 we refer to this case as the "identity".

3.2 Results for the Auxiliary Variable Method

In our simulation study, the data y is given by the perfect simulation in Fig. 1, where $S = [0, 1]^2$, $\beta = 100$, $\gamma = 0.5$, $R = 0.05$, $n(y) = 75$, and $s_R(y) = 10$. For the MCMC MLE, we obtained $\hat{\beta} = 108$ and $\hat{\gamma} = 0.4$. A priori we assume that $R = 0.05$ is known and β and γ are independent and uniformly distributed on $(0, 150]$ and $(0, 1]$, respectively; perfect simulations for $\beta > 150$ can be slow [2, 3]. For the POMM point process we divide S into $m = N^2$ square cells of side length $1/N$. Below we consider the values $N = 50, 100, 200$, or in comparison with $R = 0.05$, $1/N = 0.02, 0.01, 0.005$. Further details on the auxiliary variable method can be found in Appendix A.

The results are summarised in Table 1 for the different auxiliary processes, and in the POMM case, for different choices of N, the function g in (14), and proposal distributions. Experiments with the algorithm for the fixed Poisson and Strauss and the POMM processes with smaller values of N showed that trace plots of $n(x)$ and $s_R(x)$ (not shown here) may exhibit seemingly satisfactory mixing properties for several million updates and then get stuck – sometimes for more than 100,000 updates. Therefore we consider the fraction of acceptance probabilities below $\exp(-10)$ as an indicator for the mixing properties of the chain. Table 1 also shows the mean acceptance probability and the lag 100 autocorrelation of β and γ.

Aux. proc.	g	Prop σ_β	Prop σ_γ	MAcP	Extr	c_β	c_γ
Fixed Poisson		2	0.05	0.128	0.151	0.88	0.53
POMM (N=100)	identity	2	0.05	0.171	0.127	0.86	0.54
POMM (N=200)	identity	2	0.05	0.213	0.064	0.85	0.47
POMM (N=50)	MLE	2	0.05	0.246	0.055	0.85	0.46
Fixed Strauss		2	0.05	0.393	0.031	0.79	0.46
POMM (N=100)	MLE	4	0.1	0.298	0.030	0.52	0.21
POMM (N=200)	MLE	4	0.1	0.366	0.014	0.41	0.14
POMM (N=100)	MLE	2	0.05	0.321	0.013	0.79	0.38
POMM (N=200)	MLE	2	0.05	0.406	0.002	0.75	0.33

Table 1. Empirical results: For each auxiliary process considered, one million updates were generated. "Aux. Proc." is the type of auxiliary process used; g is the type of mapping used for each POMM point process (see the end of Sect. 3.1); "Prop σ_β" and "Prop σ_γ" are the proposal standard deviations for β and γ; "MAcP" is the mean acceptance probability; "Extr" is the fraction of acceptance ratios below $\exp(-10)$; c_β and c_γ are the lag 100 autocorrelation for β and γ

The different cases of auxiliary processes in Table 1 are ordered by the values of "Extr" (the fraction of extremely low acceptance probabilities). Seemingly the results for the autocorrelations depend predominantly on the choice of proposal standard deviations for β and γ. Using the POMM point process with $N = 200$ and $g = $ MLE appears to give the best mixing. Fig. 2 shows the marginal and joint posterior distributions for β and γ when using the POMM process with $N = 200$, $g = $ MLE, and proposal standard deviations for β and γ equal to 2 and 0.05. From Fig. 2 it can be seen that the MCMC MLE

$(\hat{\gamma}, \hat{\beta}) = (0.4, 108)$ is not far from the approximative posterior mode obtained by simulation. This is of course to be expected since we have a uniform prior for (γ, β). The marginal posterior modes are close to the posterior mode, since the posterior has nearly elliptical contours.

Despite a seemingly fair number of points in the data, Fig. 2 shows a rather large degree of posterior uncertainty about β and γ. The posterior distribution of β suggests that the upper bound of 150 on β should be slightly increased, however we do not expect that increasing this bound would affect the overall picture.

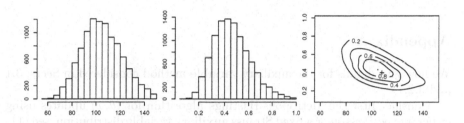

Fig. 2. Empirical marginal posterior distributions of β (left plot) and γ (centre plot) generated using a POMM auxiliary process with $N - 200$ and $g =$ MLE. Empirical joint posterior distribution of (β, γ) (right plot) where "." denotes the approximate posterior mode and "+" denotes the approximate MLE.

In conclusion, to obtain a significant improvement by using a POMM auxiliary process with $g =$ MLE compared to using a fixed Strauss process, a cell side length less than about $R/10$ is needed. Computer times show that using the POMM with $N = 100$ are not much slower than using the fixed Strauss process. For $N = 200$ the POMM takes twice as long as for $N = 100$.

4 Concluding Remarks

The technique used in this paper adds significantly to the ability of simulation-based Bayesian inference for Markov point processes, which previously have been subject to one or another approximate analysis. By using the auxiliary variable method presented here in conjunction with perfect sampling, we remove the need for estimating ratios of normalising constants.

We have demonstrated that a workable auxiliary variable distribution has the attribute of closely matching the unnormalised likelihood, while not requiring the computation of a normalising constant. Perhaps the most important consequence of this is that the proposal for the auxiliary variable is then very similar to its full conditional density, which we expect to promote good mixing. For the simulation study in Sect. 3 a POMM is a more appropriate choice of auxiliary variable than an auxiliary variable density based on the unnormalised likelihood evaluated at the MLE.

To the best of our knowledge, prior specification for Markov point processes has so far not been discussed much in the literature (however, see [12] and [3]). Here we have chosen uniform priors to keep things simple as our main purpose is to illustrate the auxiliary variable method for Markov point processes. Choosing another prior, our choice of proposal density $p(\theta'|\theta)$ may be different, but otherwise the method is the same.

In [4] we use the auxiliary variable method for a semi-parametric inhomogeneous Markov point process, using again a POMM auxiliary point process.

Appendix A

We now give details for the auxiliary variable method considered in Sects. 3.1 and 3.2.

Consider first the Metropolis-Hastings algorithm for (θ, x) updates using either a fixed Poisson or a fixed Strauss auxiliary variable distribution, see (11) and (12). Recall that $\theta = (\beta, \gamma, R)$ where $R = 0.05$ is fixed. As initial values we choose $\theta = (n(y), 1, 0.05)$ and x is a realisation of a Poisson point process on $S = [0, 1]^2$ with intensity $n(y)$. Then, if (θ, x) comprises the current state of the Metropolis-Hastings algorithm with $\theta = (\beta, \gamma, R)$, the next state is generated as follows with f in step 3 replaced by either f_1 (fixed Poisson case) or f_2 (fixed Strauss case).

1. Draw proposals β' and γ' from independent normal distributions with means β and γ.
2. Generate a realisation x' from a Strauss process specified by $\theta' = (\beta', \gamma', R)$ and using dominated CFTP.
3. With probability

$$\min\left\{1, \mathbf{1}[0 < \beta' \leq 150, 0 < \gamma' \leq 1] \times \left(\frac{\beta'}{\beta}\right)^{n(y)} \left(\frac{\gamma'}{\gamma}\right)^{s_R(y)} \frac{f(x'|y, \theta')}{f(x|y, \theta)} \frac{\beta^{n(x)} \gamma^{s_R(x)}}{\beta'^{n(x')} \gamma'^{s_R(x')}}\right\}$$

set $\theta = \theta'$ and $x = x'$, otherwise do nothing.

The standard deviations of the normal distributions in step 1 can be adjusted to give the best mixing of the chain.

Consider next using a POMM auxiliary process. Then an extra auxiliary variable, the random permutation ρ, and an additional step is required in the update above. If the current state consists of (β, γ), ρ, and x, then steps 1 and 2 above are followed by

3. Generate a uniform random permutation ρ'.

4. With probability

$$\min\left\{1, \mathbf{1}[0 < \beta' \le 150, 0 < \gamma' \le 1] \times\right.$$
$$\left.\left(\frac{\beta'}{\beta}\right)^{n(y)} \left(\frac{\gamma'}{\gamma}\right)^{s_R(y)} \frac{f_3(x'|y,\theta',\rho')}{f_3(x|y,\theta,\rho)} \frac{\beta^{n(x)}\gamma^{s_R(x)}}{\beta'^{n(x')}\gamma'^{s_R(x')}}\right\}$$

set $(\theta, \rho, x) = (\theta', \rho', x')$, otherwise do nothing.

Here f_3 is given by (14).

Appendix B

When the mapping g in Sects. 3.1 and 3.2 is not the identity, it is specified as follows.

Based on the range of the empirical posterior distributions in the fixed Strauss case (not shown here) we define a grid $G = \{50, 52, \ldots, 150\} \times \{0.1, 0.2, \ldots, 1.0\} \times \{0.05\}$. For each grid point $\theta = (\beta, \gamma, R) \in G$, using dominated CFTP, we generate 10 independent realisations $x^{(1)}, \ldots, x^{(10)}$ of a Strauss point process with parameter θ together with the generation of 10 independent random permutations $\rho^{(1)}, \ldots, \rho^{(10)}$. For $\theta \in G$, $g(\theta)$ is given by

$$(g_1(\theta), g_2(\theta)) = \frac{1}{10} \sum_{i=1}^{10} \operatorname{argmax}_{(\tilde{\beta}, \tilde{\gamma})} f_P(x^{(i)} | \tilde{\beta}, \tilde{\gamma}, R, \rho^{(i)}),$$

and $g_3(\theta) = R$. For $(\beta, \gamma, 0.05) \notin G$, we set $g(\beta, \gamma, 0.05) = g(\tilde{\beta}, \tilde{\gamma}, 0.05)$ where $(\tilde{\beta}, \tilde{\gamma}, 0.05) \in G$ is the grid point closest to $(\beta, \gamma, 0.05)$.

Fig. 3 shows $g_1(\beta, \gamma, R) - \beta$ and $g_2(\beta, \gamma, R) - \gamma$ for a range of β and γ values when $N = 200$. Results for $N = 50$ and $N = 100$ are almost identical to those for $N = 200$. In cases of strong interaction, i.e. for combinations of low values of γ and high values of β, the parameters $\beta_P = g_1(\beta, \gamma, R)$ and $\gamma_P = g_2(\beta, \gamma, R)$ in the POMM process are much smaller than β and γ in the Strauss process. This is explained by the fact that the interaction in the POMM auxiliary process is weaker than in the Strauss process when $(\beta_P, \gamma_P, R_P) = (\beta, \gamma, R)$.

Acknowledgements

The research of J. Møller was supported by the Danish Natural Science Research Council and the Network in Mathematical Physics and Stochastics (MaPhySto), founded by grants from the Danish National Research Foundation.

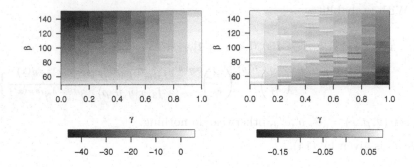

Fig. 3. Plot of difference between $g(\theta)$ and θ for $\theta \in G$: $g_1(\beta, \gamma, R) - \beta$ (left) and $g_2(\beta, \gamma, R) - \gamma$ (right).

References

[1] A.J. Baddeley and R. Turner. Practical maximum pseudolikelihood for spatial point patterns. *Australian and New Zealand Journal of Statistics*, 42:283–322, 2000.

[2] K.K. Berthelsen and J. Møller. A primer on perfect simulation for spatial point processes. *Bulletin of the Brazilian Mathematical Society*, 33:351–367, 2002.

[3] K.K. Berthelsen and J. Møller. Likelihood and non-parametric Bayesian MCMC inference for spatial point processes based on perfect simulation and path sampling. *Scandinavian Journal of Statistics*, 30:549–564, 2003.

[4] K.K. Berthelsen and J. Møller. Semi-parametric Bayesian inference for inhomogeneous Markov point processes. In preparation.

[5] J.E. Besag. Some methods of statistical analysis for spatial data. *Bulletin of the International Statistical Institute*, 47:77–92, 1977.

[6] N.A.C. Cressie and J.L. Davidson. Image analysis with partially ordered Markov models. *Computational Statistics and Data Analysis*, 29(1):1–26, 1998.

[7] N.A.C. Cressie, J. Zhu, A.J. Baddeley and M.G. Nair. Directed Markov point processes as limits of partially ordered Markov models. *Methodology and Computing in Applied Probability*, 2:5–21, 2000.

[8] A. Gelman and X.-L. Meng. Simulating normalizing contants: From importance sampling to bridge sampling to path sampling. *Statistical Science*, 13:163–185, 1998.

[9] C.J. Geyer. Likelihood inference for spatial point processes. In O. E. Barndorff-Nielsen, W. S. Kendall, and M. N. M. van Lieshout, editors, *Stochastic Geometry: Likelihood and Computation*, pp. 79–140, Boca Raton, Florida, 1999. Chapman & Hall/CRC.

[10] C.J. Geyer and J. Møller. Simulation procedures and likelihood inference for spatial point processes. *Scandinavian Journal of Statistics*, 21:359–373, 1994.

[11] O. Häggström, M.N.M. Van Lieshout and J. Møller. Characterisation results and Markov chain Monte Carlo algorithms including exact simulation for some spatial point processes. *Bernoulli*, 5:641–658, 1999.

[12] J. Heikkinen and A. Penttinen. Bayesian smoothing in the estimation of the pair potential function of Gibbs point processes. *Bernoulli*, 5:1119–1136, 1999.

[13] J.L. Jensen and J. Møller. Pseudolikelihood for exponential family models of spatial point processes. *Annals of Applied Probability*, 3:445–461, 1991.

[14] F.P. Kelly and B.D. Ripley. A note on Strauss' model for clustering. *Biometrika*, 63:357–360, 1976.

[15] W.S. Kendall and J. Møller. Perfect simulation using dominating processes on ordered spaces, with application to locally stable point processes. *Advances in Applied Probability*, 32:844–865, 2000.

[16] M.N.M. van Lieshout. *Markov Point Processes and their Applications*. Imperial College Press, London, 2000.

[17] J. Møller, A.N. Pettitt, K.K. Berthelsen and R.W. Reeves. An efficient MCMC method for distributions with intractable normalising constants. Research report r-2004-02, Department of Mathematical Sciences, Aalborg University, 2004.

[18] J. Møller and R.P. Waagepetersen. An introduction to simulation-based inference for spatial point processes. In J. Møller, editor, *Spatial Statistics and Computational Methods*, Lecture Notes in Statistics 173, pp. 143–198. Springer-Verlag, New York, 2003.

[19] J. Møller and R.P. Waagepetersen. *Statistical Inference and Simulation for Spatial Point Processes*. Chapman and Hall/CRC, 2003.

[20] L. Tierney. Markov chains for exploring posterior distributions. *Annals of Statistics*, 22:1701–1728, 1994.

Statistics for Locally Scaled Point Processes

Michaela Prokešová[1], Ute Hahn[2] and Eva B. Vedel Jensen[3]

[1] Charles University, Department of Probability, Sokolovská 83, 18675 Praha 8, Czech Republic, prokesov@karlin.mff.cuni.cz
[2] University of Augsburg, Department of Applied Stochastics, 86135 Augsburg, Germany, Ute.Hahn@Math.Uni-Augsburg.de
[3] University of Aarhus, The T.N. Thiele Centre of Applied Mathematics in Natural Science, Department of Mathematical Sciences, Ny Munkegade, 8600 Aarhus C, Denmark, eva@imf.au.dk

Summary. Recently, locally scaled point processes have been proposed as a new class of models for inhomogeneous spatial point processes. They are obtained as modifications of homogeneous template point processes and have the property that regions with different intensity differ only by a location dependent scale factor. The main emphasis of the present paper is on analysis of such models. Statistical methods are developed for estimation of scaling function and template parameters as well as for model validation. The proposed methods are assessed by simulation and used in the analysis of a vegetation pattern.

Key words: Inhomogeneous spatial point processes, Local scaling of point processes, Model validation and simulation

1 Introduction

The present paper deals with statistical analysis for inhomogeneous point processes that are obtained by local scaling. In these point processes, local geometry is constant, that is, subregions of the inhomogeneous process with different intensity appear to be scaled versions of the same homogeneous process. This property is characteristic of locally scaled point processes and not present in the other models for inhomogeneous point processes discussed in [9]. Such patterns occur for example in vegetation of dry areas, as shown in Fig. 1. Heterogeneity on a small scale is largely due to a patchy soil mosaic combining drier, sandy soils with clay textured soils of higher water capacity [1]. Where water or other resources are short, plants grow sparsely and keep larger distances between individuals than in regions with better supply.

Naturally there is no preference for a direction, and therefore the vegetation pattern is locally isotropic. Local scaling of an isotropic template process yields locally isotropic patterns in contrast to transformation of an isotropic template process [13].

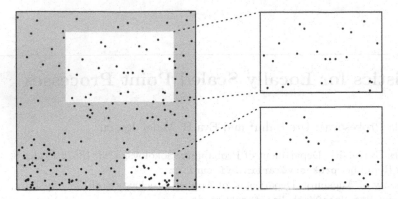

Fig. 1. (*Left part*) Map of 171 individuals of a *Scholtzia* aff. *involucrata* in Australian bush on a 22 × 22 m square; (*Right part*) Two rectangular subregions with different intensity were rescaled such that they have the same number of individuals by unit area. Data from [1]

Similar locally scaled structures are found in arrangements of solid bodies with constant shape but location dependent size, such as the sinter filter discussed in [9] or in sponges with constant porosity but small pore size close to the surface and large pore size in the interior.

Locally scaled point processes are finite point processes derived from a homogeneous template process which describes the interaction between points and is responsible for the local geometry of the resulting pattern. We will put the major focus on Markov template processes. Inhomogeneity is introduced through a location dependent function that gives the local scale, as explained in detail in Sect. 2 below.

Fitting a model to a given pattern thus consists of finding the parameters inherited from the template and choosing an appropriate scaling function. This can be achieved by simultaneous maximum (pseudo)likelihood estimation as discussed in Sect. 3. A less time consuming procedure is two step maximum likelihood estimation where the parameters of the template are estimated after having determined the scaling function. Sect. 4 is centred on two step estimation, which is assessed and demonstrated by a simulation study presented in Sect. 5. Sect. 6 addresses possibilities to estimate the scaling function by other methods.

A widely used popular method for model validation in the homogeneous case is to compare the empirical K-function with the theoretically known or simulated K-function of the fit. An inhomogeneous analogue of the K-function is proposed in Sect. 7. Furthermore we suggest an inhomogeneous version of the Q^2-statistic recently proposed by [8] for model validation of homogeneous point processes.

Finally, a statistical analysis of the point pattern in Fig. 1 is presented in Sect. 8.

2 Locally Scaled Point Processes

In this section, we introduce the locally scaled point processes and discuss some of their basic properties.

Let X be a finite point process, defined on a full-dimensional bounded subset $\mathcal{X}$ of $\mathbb{R}^k$. We suppose that X has a density f_X with respect to the restriction of the unit rate Poisson point process Π to $\mathcal{X}$. Let $\nu^* = (\nu^0, \ldots, \nu^k)$ be the set of d-dimensional volume measures (Hausdorff measures) ν^d in $\mathbb{R}^k$, $d = 0, 1, \ldots, k$. Let us suppose that f_X is of the following form

$$f_X(\mathbf{x}) \propto g(\mathbf{x}; \nu^*), \quad \mathbf{x} \subset \mathcal{X} \text{ finite} , \tag{1}$$

where the function g is scale-invariant, i.e.

$$g(c\,\mathbf{x}; \nu_c^*) = g(\mathbf{x}; \nu^*) , \tag{2}$$

for all $\mathbf{x}$ and $c > 0$. Here, $\nu_c^* = (\nu_c^0, \ldots, \nu_c^k)$ and $\nu_c^d(A) = \nu_d(c^{-1}A)$, $A \in \mathbb{B}_k$. The classical homogeneous point processes have densities with this property.

The process X will serve as a template process. In order to construct a locally scaled version of X with scaling function $c : \mathbb{R}^k \to \mathbb{R}_+$, we replace the d-dimensional volume measure ν^d in $\mathbb{R}^k$ with a locally scaled version

$$\nu_c^d(A) = \int_A c(u)^{-d} \nu^d(\mathrm{d}u) , \quad A \in \mathbb{B}_k ,$$

$d = 0, 1, \ldots, k$. In what follows, we assume that the scaling function c is bounded from below and from above, i.e. $0 < \underline{c} < c(u) < \overline{c}$, $u \in \mathbb{R}^k$. Furthermore, we will assume that $g(\cdot; \nu_c^*)$ is integrable with respect to the Poisson point process Π_c with ν_c^k as intensity measure. A locally scaled point process X_c on $\mathcal{X}$ with template process X is then a finite point process defined by the following density with respect to Π_c

$$f_{X_c}^{(c)}(\mathbf{x}) \propto g(\mathbf{x}; \nu_c^*) . \tag{3}$$

Note that the density of X_c with respect to Π is

$$f_{X_c}(\mathbf{x}) = \exp\left(-\int_{\mathcal{X}} [c(u)^{-k} - 1]\nu^k(\mathrm{d}u)\right) \prod_{x \in \mathbf{x}} c(x)^{-k} \times f_{X_c}^{(c)}(\mathbf{x}) . \tag{4}$$

Example 1. The Strauss process X with intensity parameter $\beta > 0$, interaction parameter $\gamma \in [0, 1]$ and interaction distance $\delta > 0$ is given by the density

$$f_X(\mathbf{x}) \propto \beta^{n(\mathbf{x})}\gamma^{s(\mathbf{x})}, \quad \mathbf{x} \subset \mathcal{X} \text{ finite} ,$$

where $n(\mathbf{x})$ is the number of points in $\mathbf{x}$ and $s(\mathbf{x})$ is the number of δ-close pairs, cf. [15]. The density is of the form (1) with

$$g(\mathbf{x}; \nu^*) = \beta^{\nu^0(\mathbf{x})}\gamma^{\sum_{\{u,v\} \subseteq \mathbf{x}}^{\neq} \mathbf{1}\{\nu^1([u,v]) \leq \delta\}} ,$$

where the superscript $\neq$ in the summation indicates that u and v are different. It is easy to check that this function is indeed scale-invariant. The locally scaled Strauss process X_c has density with respect to Π_c of the form

$$f_{X_c}^{(c)}(\mathbf{x}) \propto \beta^{n(\mathbf{x})} \gamma^{s_c(\mathbf{x})},$$

where

$$s_c(\mathbf{x}) = \sum_{\{u,v\} \subseteq \mathbf{x}}^{\neq} 1\{\nu_c^1([u,v]) \le \delta\}.$$

Fig. 2 shows locally scaled Strauss processes on $\mathcal{X} = [0,1]^2$ with scaling function of the exponential form

$$c_\theta(u) = \sqrt{\frac{1 - e^{-2\theta}}{2\theta}} \, e^{\theta u_1}, \quad u = (u_1, u_2) \in \mathcal{X}, \quad \theta \in \mathbb{R}^1, \tag{5}$$

for 4 different values of the inhomogeneity parameter $\theta \in \{0.25, 0.5, 1, 1.5\}$. The normalisation $\sqrt{\frac{1-e^{-2\theta}}{2\theta}}$ ensures that the 4 point patterns have approximately the same number of points (see Sect. 4.2 for details).

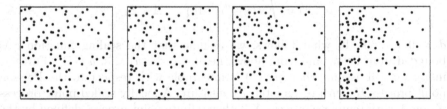

Fig. 2. Simulation of locally scaled Strauss processes on $[0,1]^2$ with exponential scaling function (5) for $\theta \in \{0.25, 0.5, 1, 1.5\}$ (*from left to right*) and template parameters $\beta = 250, \gamma = 0.3$ and $\delta = 0.05$

□

Example 2. The area-interaction point process with intensity parameter $\beta > 0$, interaction parameter $\gamma > 0$ and interaction distance $\delta > 0$ is given by the density

$$f(\mathbf{x}) \propto \beta^{n(\mathbf{x})} \gamma^{-\nu^2(U_\delta(\mathbf{x}))}, \quad \mathbf{x} \subset \mathcal{X} \text{ finite},$$

where $U_\delta(\mathbf{x}) = \bigcup_{x \in \mathbf{x}} b(x, \delta)$ is the union of balls with centres in $\mathbf{x}$ and radius δ. For $\gamma > 1$ the point pattern appears clustered, for $\gamma < 1$ regular, cf. [5]. The density is again of the form (1) with scale invariant

$$g(\mathbf{x}; \nu^*) = \beta^{\nu^0(\mathbf{x})} \gamma^{-\nu^2(\bigcup_{x \in \mathbf{x}}\{v \in \mathcal{X} : \nu^1([v,x]) \le \delta\})}.$$

The locally scaled area-interaction process has density with respect to Π_c of the form

$$f_{X_c}^{(c)}(\mathbf{x}) \propto \beta^{n(\mathbf{x})} \gamma^{-\nu_c^2(U_{c,\delta}(\mathbf{x}))} \, ,$$

where $U_{c,\delta} = \bigcup_{x \in \mathbf{x}} b_c(x, \delta)$ and $b_c(x, \delta) = \{v \in \mathcal{X} \; : \; \nu_c^1([v, x]) \leq \delta\}$ is the scaled ball. Fig. 3 shows locally scaled area-interaction processes with the same scaling function (5) as in Example 2. The value of the interaction parameter γ was chosen so that $\gamma^{-\pi\delta^2} \approx 0.1$ and the point patterns are visibly clustered.

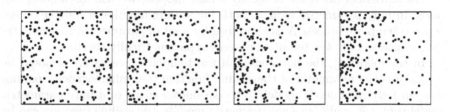

Fig. 3. Simulation of locally scaled area-interaction processes on $[0, 1]^2$ with exponential scaling function (5) for $\theta \in \{0.25, 0.5, 1, 1.5\}$ (*from left to right*) and template parameters $\beta - 180, \gamma = 6.7 \cdot 10^{31}$ and $\delta = 0.1$

$\square$

The Strauss process and the area-interaction process are examples of point processes from two large classes of homogeneous template processes, viz. the distance-interaction processes and the shot noise weighted processes. For these two classes, it has been shown in [9] that the Papangelou conditional intensities λ_{X_c} of the locally scaled process X_c and λ_X of the template process X, respectively, satisfy

$$\lambda_{X_c}(x \mid \mathbf{x}) = c_0^{-k} \lambda_X\left(\frac{x}{c_0} \mid \frac{\mathbf{x}}{c_0}\right), \tag{6}$$

if the scaling function c is constant and equal to c_0 in a scaled neighbourhood of x. If we let dx be an infinitesimal region around x and $\nu^k(dx)$ the k-dimensional volume (Lebesgue measure) of dx, then $\lambda(x \mid \mathbf{x})\nu^k(dx)$ can be interpreted as the conditional probability of finding a point from the process in dx given the configuration elsewhere is $\mathbf{x}$, cf. e.g. [10]. Since the right-hand side of (6) is the conditional intensity of a globally scaled template process with scaling factor c_0, it is expected that the locally scaled process appears as a scaled version of the template process if the scaling function is slowly varying compared to the interaction radius. The development of further formal reasoning, supporting this statement, seems very hard.

It is also of interest to study the unconditional intensity function $\lambda_c(x), x \in \mathcal{X}$, of the locally scaled process. Let us suppose that the template process X

is homogeneous with intensity λ_0 (X may, for instance, be defined on a torus with periodic boundary conditions). Then

$$\lambda_c(x) = c(x)^{-k}\lambda_0, \quad x \in \mathcal{X} , \tag{7}$$

holds if the template process is Poisson or the scaling function is constant. Also, (7) holds for any locally scaled distance-interaction process in $\mathbb{R}^1$, see the Appendix. The equality (7) is expected to hold approximately if the scaling function is slowly varying, compared to the interaction radius.

For statistical inference of locally scaled models, we will distinguish two cases. In fully parametric models, both the scaling function c and the homogeneous template process X are specified by a set of parameters. In semiparametric models only the template process is parametrically specified.

In the following, the parameters of the template process are denoted by ψ, and θ is the parameter of the scaling function (i.e. inhomogeneity parameter) in fully parametric models. The parameter space of a fully parametric model is $\Theta \times \Psi$, while, in semiparametric models, the scaling function can be any function in the space $\mathcal{C}^+$ of measurable positive functions, satisfying the regularity conditions mentioned above.

A particularly attractive parametric form of the scaling function is the exponential form

$$c_\theta(u) = \alpha(\theta)\, e^{\theta \cdot \tau(u)}, \quad u \in \mathbb{R}^k , \tag{8}$$

where $\theta \in \Theta \subseteq \mathbb{R}^l$, $\alpha(\theta) \in \mathbb{R}_+$, $\cdot$ indicates the inner product and $\tau(u) \in \mathbb{R}^l$. A locally scaled model with an exponential scaling function is called an exponentially scaled model. Note that if $\tau(u) = u$, then scaled distances can be calculated explicitly. Using the coarea formula we get

$$\begin{aligned}
\nu_c^1([u,v]) &= \int_{[u,v]} \alpha(\theta)^{-1} e^{-\theta \cdot x}\, \nu^1(\mathrm{d}x) \\
&= \int_0^1 \nu^1([u,v])\, \alpha(\theta)^{-1} e^{-\theta \cdot (u+t(v-u))}\, \mathrm{d}t \\
&= \nu^1([u,v]) \frac{[c_\theta(u)^{-1} - c_\theta(v)^{-1}]}{\theta \cdot (v-u)} , \qquad u,v \in \mathbb{R}^k .
\end{aligned}$$

3 Simultaneous Maximum (Pseudo)likelihood Estimation of Scaling Function and Template Parameters

In a fully parametric model, the likelihood factorizes as, cf. (4),

$$L(\theta, \psi; \mathbf{x}) = L_0(\theta; \mathbf{x}) \times L_1(\theta, \psi; \mathbf{x}) , \tag{9}$$

where L_0 is the likelihood of an inhomogeneous Poisson point process Π_c with intensity measure ν_c^k, and $L_1(\theta, \psi; \mathbf{x}) = f_{X_c}^{(c)}(\mathbf{x}; \psi)$ is the density of the scaled process X_c with respect to Π_c. Recall that the scaling function is parametrised by θ, i.e. $c = c_\theta$.

Maximum likelihood estimation is most feasible in exponential families, since it amounts to moment estimation there. Most popular homogeneous Markov point process models are partially exponential, and the set ψ splits into two components – the nuisance parameters and the remaining parameters, that form exponential family parameters given the nuisance parameters. Since the likelihood in Markov point processes is known only up to the normalizing constant, one has to resort to MCMC methods for MLE, cf. e.g. [11]. Whilst moment estimation in these models can be done relatively precisely with affordable effort, estimation of the normalizing constant entails numerical pitfalls and should be avoided as much as possible. This suggests that MLE should be done on a grid of nuisance parameters, since given this component, the remaining parameters are exponential family parameters. In locally scaled processes, the inhomogeneity parameter also acts as a nuisance parameter.

Usually, the point process X_c is observed in a sampling window $W \subseteq \mathcal{X}$. In such cases, a conditional likelihood may be used, based on the conditional density of $X_c \cap W$ given $X_c \cap W^c = \mathbf{x}_{W^c}$ where $\mathbf{x}_{W^c}$ is a finite subset of W^c. Since

$$f_{X_c}(\cdot \mid \mathbf{x}_{W^c}) \propto f_{X_c}(\cdot \cup \mathbf{x}_{W^c}),$$

It follows from (4) that (9) still holds for the conditional likelihoods. This result is mainly of interest for locally scaled Markov point processes.

A less computational demanding procedure is based on the pseudolikelihood function, see [4] and references therein. The pseudolikelihood function of a point process density f with respect to a Poisson point process with intensity measure μ, based on observation in W, is defined by

$$\exp\left(-\int_W [\lambda(u \mid \mathbf{x}) - 1]\,\mu(du)\right) \prod_{x \in \mathbf{x} \cap W} \lambda(x \mid \mathbf{x}\backslash\{x\}), \quad W \subseteq \mathcal{X},$$

where $\mathbf{x}$ is the realised point pattern in $\mathcal{X}$ and

$$\lambda(u \mid \mathbf{x}) = \frac{f(\mathbf{x} \cup \{u\})}{f(\mathbf{x})}, \quad u \notin \mathbf{x},$$

is the Papangelou conditional intensity associated with f.

Based on observation in W, let $PL_W(\theta, \psi; \mathbf{x})$ be the pseudolikelihood function for the density $f_{X_{c_\theta}}(\cdot\,; \psi)$ with respect to the unit rate Poisson point process and let $PL_{W,1}(\theta, \psi; \mathbf{x})$ be the pseudolikelihood for the density $f_{X_{c_\theta}}^{(c_\theta)}(\cdot\,; \psi)$ with respect to the Poisson point process with intensity measure $\nu_{c_\theta}^k$. Then, using (4), we find

$$PL_W(\theta, \psi; \mathbf{x}) = L_0(\theta; \mathbf{x} \cap W) \times PL_{W,1}(\theta, \psi; \mathbf{x}). \tag{10}$$

Note that

$$\lambda_{\theta,\psi}(u \mid \mathbf{x}) = c_\theta(u)^{-k} \lambda_{\theta,\psi}^{(c_\theta)}(u \mid \mathbf{x}), \quad u \notin \mathbf{x},$$

where $\lambda_{\theta,\psi}$ and $\lambda_{\theta,\psi}^{(c_\theta)}$ are the conditional intensities associated with $f_{X_{c_\theta}}$ and $f_{X_{c_\theta}}^{(c_\theta)}$, respectively.

A proof of (10) can be constructed as follows. ¿From (4), we get

$$PL_W(\theta, \psi; \mathbf{x})$$

$$= \exp(-\int_W [\lambda_{\theta,\psi}(u \mid \mathbf{x}) - 1]\nu^k(du)) \prod_{x \in \mathbf{x} \cap W} \lambda_{\theta,\psi}(x \mid \mathbf{x}\backslash\{x\})$$

$$= \exp(-\int_W [c_\theta(u)^{-k} - 1]\nu^k(du)) \prod_{x \in \mathbf{x} \cap W} c_\theta(x)^{-k}$$

$$\times \exp(-\int_W [\lambda_{\theta,\psi}^{(c_\theta)}(u \mid \mathbf{x}) - 1]c_\theta(u)^{-k}\nu^k(du)) \prod_{x \in \mathbf{x} \cap W} \lambda_{\theta,\psi}^{(c_\theta)}(x \mid \mathbf{x}\backslash\{x\})$$

$$= L_0(\theta; \mathbf{x}) \times PL_{W,1}(\theta, \psi; \mathbf{x}).$$

As the values of the scaled interaction statistics (e.g. $s_{c_\theta}(\mathbf{x})$ in the Strauss model) and subsequently the values of $\lambda_{\theta,\psi}^{(c_\theta)}(u \mid \mathbf{x})$ depend on the inhomogeneity parameter θ, the latter is a nuisance parameter also in the pseudolikelihood estimation. This means we have to evaluate the profile pseudolikelihood on a grid of nuisance parameters similarly to the maximum likelihood approach. However, this is much less computational intensive in maximum pseudolikelihood estimation than in maximum likelihood estimation, since PL_1 can be calculated directly without having to estimate an unknown normalizing constant by simulation as it is the case with L_1.

4 Two Step Maximum Likelihood Estimation of Scaling Parameters Prior to Template Parameters

The structural similarity of the full likelihood in locally scaled models and the full likelihood in transformation models for point processes suggests that partial likelihood inference as in the paper [14] will be successful also for locally scaled models. [14] estimated the inhomogeneity parameters by maximizing the Poisson part L_0 of the likelihood only, assuming no interaction. They chose an exponential model for the inhomogeneity function, since this largely simplifies calculations.

Below, this approach is followed for the locally scaled models. In Sect. 4.1, we find the maximum likelihood estimate $\widehat{\theta}_0$ of θ on the basis of L_0 and, in Sect. 4.2, it is shown that $\widehat{\theta}_0$ can be regarded as an approximate moment estimator. In Sect. 4.3, estimation of the template parameters is considered.

4.1 Estimation of Scaling Parameters, Using the Poisson Likelihood

We suppose that the scaling function is of the form

$$c(u) = \alpha\, e^{\theta \cdot \tau(u)}, \qquad u \in \mathbb{R}^k, \tag{11}$$

where $\theta \in \Theta \subseteq \mathbb{R}^l$ and $\alpha \in \mathbb{R}_+$. In addition to the inhomogeneity parameter θ, the scaling function contains a global scaling parameter α. For the moment, these two parameters vary in a product set $\Theta \times \mathbb{R}_+$.

Then, the Poisson part of the likelihood of the process X_c, observed in a set W, is

$$L_0(\theta, \alpha; \mathbf{x} \cap W) = \exp\left(-\int_W (\alpha^{-k} e^{-k\,\theta \cdot \tau(u)} - 1)\, \nu^k(\mathrm{d}u)\right) \times$$
$$\prod_{x \in \mathbf{x} \cap W} (\alpha^{-k} e^{-k\,\theta \cdot \tau(x)}) .$$

The log-likelihood becomes

$$l_0(\theta, \alpha; \mathbf{x} \cap W) = \int_W 1\, \nu^k(\mathrm{d}u) - \int_W \alpha^{-k} e^{-k\,\theta \cdot \tau(u)}\, \nu^k(\mathrm{d}u)$$
$$- k\, n(\mathbf{x} \cap W) \ln \alpha + \sum_{x \in \mathbf{x} \cap W} (-k\,\theta \cdot \tau(x)) .$$

Assume that $n(\mathbf{x} \cap W) > 0$ and $\|\tau(u)\|\, e^{\theta \cdot \tau(u)}$ is uniformly bounded in $u \in W$ and $\theta \in \Theta$. Then by differentiating we get $l+1$ equations

$$k\alpha^{-k-1} \int_W e^{-k\,\theta \cdot \tau(u)}\, \nu^k(\mathrm{d}u) = k\, n(\mathbf{x} \cap W)\alpha^{-1}$$
$$\alpha^{-k} \int_W k\tau_i(u) e^{-k\,\theta \cdot \tau(u)}\, \nu^k(\mathrm{d}u) = k \sum_{x \in \mathbf{x} \cap W} \tau_i(x), \quad i = 1, \dots, l .$$

Dividing the last l equations by the first equation we get the vector equation

$$\frac{t(\mathbf{x} \cap W)}{n(\mathbf{x} \cap W)} = m(\theta) , \tag{12}$$

where $t(\mathbf{x} \cap W) = \sum_{x \in \mathbf{x} \cap W} \tau(x)$ and

$$m(\theta) = \frac{\int_W \tau(u) e^{-k\,\theta \cdot \tau(u)}\, \nu^k(\mathrm{d}u)}{\int_W e^{-k\,\theta \cdot \tau(u)}\, \nu^k(\mathrm{d}u)} . \tag{13}$$

Thus the estimate of θ does not depend on the estimate of the constant α and furthermore the estimate depends only on the statistic $t(\mathbf{x} \cap W)/n(\mathbf{x} \cap W)$.

It turns out that we get exactly the same estimate of θ if we impose the following normalizing condition on c_θ

$$\int_W c_\theta(u)^{-k} \nu^k(du) = \nu^k(W) \,, \tag{14}$$

implying that

$$\alpha = \alpha(\theta) = \left[\int_W e^{-k\theta \cdot \tau(u)} \nu^k(du) / \nu^k(W) \right]^{1/k}. \tag{15}$$

To see this, note that under (15), the Poisson likelihood takes the form

$$L_0(\theta; \mathbf{x} \cap W) = \exp\left(-\int_W (\alpha(\theta)^{-k} e^{-k\theta \cdot \tau(u)} - 1) \nu^k(du) \right)$$
$$\times \prod_{x \in \mathbf{x} \cap W} (\alpha(\theta)^{-k} e^{-k\theta \cdot \tau(x)})$$
$$= \left(\alpha(\theta)^{-k} \exp\left(-k\theta \cdot \frac{t(\mathbf{x} \cap W)}{n(\mathbf{x} \cap W)} \right) \right)^{n(\mathbf{x} \cap W)}.$$

Taking the logarithm and differentiating with respect to θ, we again get the vector equation (12). As we shall see in Sect. 4.2, (14) appears to be a very natural condition.

The existence and uniqueness of a solution $\widehat{\theta}_0$ to (12) have been studied in [14] in a closely related set–up where the parameter of interest was $\tilde{\theta} = -k\theta$. The same type of arguments applies here. Using (15), it is seen that

$$\left\{ \frac{\alpha(\theta)^{-k}}{\nu^k(W)} e^{-k\theta \cdot \tau(u)} : \theta \in \Theta \right\}$$

is an exponential family of densities on W, with respect to ν^k. If the family is regular, then the function m in (13) is a bijection of Θ on int$S$ where S is the convex support of the family, cf. e.g. [6]. Thus, under these conditions, there is a unique solution to (12) if $n(\mathbf{x} \cap W) > 0$ and $t(\mathbf{x} \cap W)/n(\mathbf{x} \cap W) \in \text{int}S$.

Example 3. Let $\tau(u) = u$ and $W = [0,1]^k$. Then, $\Theta = \mathbb{R}^k$

$$c_\theta(u) = \alpha(\theta) e^{\theta \cdot u} \,,$$

$$\alpha(\theta) = \left(\prod_{i=1}^{k} \frac{1 - e^{-k\theta_i}}{k\theta_i} \right)^{1/k},$$

and $m(\theta) = (m_1(\theta), \ldots, m_k(\theta))$ where

$$m_i(\theta) = \frac{1 - e^{-k\theta_i} - k\theta_i e^{-k\theta_i}}{k\theta_i(1 - e^{-k\theta_i})}, \quad i = 1, \ldots, k \,.$$

$\square$

4.2 Statistical Properties of $\widehat{\theta}_0$

The estimator $\widehat{\theta}_0$ is the maximum likelihood estimator of θ if the template process is Poisson. It is also possible to give theoretical support to the use of $\widehat{\theta}_0$ for general template processes, as shown below.

Proposition 1. *Suppose that the intensity of the locally scaled process X_{c_θ} satisfies*

$$\lambda_{\theta,\psi}(u) = c_\theta(u)^{-k}\lambda_{0\psi} . \tag{16}$$

Then,

$$\frac{\mathbf{E}_{\theta,\psi}[t(X_{c_\theta} \cap W)]}{\mathbf{E}_{\theta,\psi}[n(X_{c_\theta} \cap W)]} = m(\theta) .$$

Proof. We use the following version of the Georgii-Nguyen-Zessin formula for X_{c_θ}, cf. [12],

$$\mathbf{E}_{\theta,\psi} \sum_{x \in X_{c_\theta}} h(x) = \int_{\mathbb{R}^k} h(x)\lambda_{\theta,\psi}(x)\nu^k(\mathrm{d}x) .$$

We get

$$\mathbf{E}_{\theta,\psi}[t(X_{c_\theta} \cap W)] = \mathbf{E}_{\theta,\psi} \sum_{x \in X_{c_\theta} \cap W} \tau(x)$$

$$= \int_W \tau(x)\lambda_{\theta,\psi}(x)\nu^k(\mathrm{d}x)$$

$$= \lambda_{0\psi} \int_W \tau(x)c_\theta(x)^{-k}\nu^k(\mathrm{d}x) .$$

In particular,

$$\mathbf{E}_{\theta,\psi}[n(X_{c_\theta} \cap W)] = \lambda_{0\psi} \int_W c_\theta(x)^{-k}\nu^k(\mathrm{d}x) . \tag{17}$$

The result now follows directly.   □

If (16) holds, $\widehat{\theta}_0$ can thus be regarded as a moment estimator.

As mentioned in Sect. 2, the equation (16) holds if the template process is homogeneous and the scaling function is constant. More interestingly, (16) holds for a not necessarily constant scaling function for distance-interaction processes in $\mathbb{R}^1$, see the Appendix. Generally, equation (16) is expected to hold approximately if the scaling function varies slowly compared to the interaction radius.

4.3 Estimation of the Template Parameters

Having estimated the scaling parameter θ we can proceed by the estimation of the template process parameters. We will here concentrate on the case where the pseudolikelihood $PL_{W,1}(\widehat{\theta}_0, \psi; \mathbf{x})$ from the decomposition (10) is used. In the following we discuss the practical implementation of this method for the locally scaled models. We consider general parametric scaling functions.

Recall that the pseudolikelihood $PL_{W,1}(\theta, \psi; \mathbf{x})$ for the density $f_{X_{c_\theta}}^{(c_\theta)}(\,\cdot\,; \psi)$ with respect to the Poisson point process with intensity measure $\nu_{c_\theta}^k$, based on observation in a window $W \subset \mathbb{R}^k$, is defined as follows

$$PL_{W,1}(\theta, \psi; \mathbf{x}) = \exp\left(-\int_W [\lambda_{\theta,\psi}^{(c_\theta)}(u \mid \mathbf{x}) - 1]\, \nu_{c_\theta}^k(\mathrm{d}u) \right) \times \qquad (18)$$

$$\prod_{x \in \mathbf{x} \cap W} \lambda_{\theta,\psi}^{(c_\theta)}(x \mid \mathbf{x}\backslash\{x\}) \,. \qquad (19)$$

In the second step of the two-step estimation procedure we fix the scaling parameter θ to $\widehat{\theta}_0$ and maximize $PL_{W,1}(\widehat{\theta}_0, \psi; \mathbf{x})\, \exp(-\nu_{c_{\widehat{\theta}_0}}^k(W))$ as a function of ψ. This can be done in a way similar to the procedure used in the homogeneous case, cf. [4]. We partition W into a finite number of cells C_i, each containing one dummy point $u_i, i = 1, \ldots, l$. The union of the dummy points and the points of the observed pattern is denoted $\{u_j : j = 1, \ldots, m\}$. Furthermore let $C_{i(j)}$ be the unique cell containing $u_j, j = 1, \ldots, m$, with dummy point $u_{i(j)}$. Then we approximate the integral in the pseudolikelihood by

$$\int_W \lambda_{\widehat{\theta}_0, \psi}^{c_{\widehat{\theta}_0}}(u \mid \mathbf{x})\nu_{c_{\widehat{\theta}_0}}^k(\mathrm{d}u) \approx \sum_{j=1}^m \lambda_{\widehat{\theta}_0, \psi}^{c_{\widehat{\theta}_0}}(u_j \mid \mathbf{x}\backslash\{u_j\})\, w_j \,,$$

where

$$w_j = \frac{\nu^k(C_{i(j)})}{c_{\widehat{\theta}_0}(u_{i(j)})^k} \frac{1}{(1 + n(\mathbf{x} \cap C_{i(j)}))} \approx \nu_{c_{\widehat{\theta}_0}}^k(C_{i(j)}) \frac{1}{(1 + n(\mathbf{x} \cap C_{i(j)}))} \,. \qquad (20)$$

Here, $n(\mathbf{x} \cap C_{i(j)})$ is the total number of observed points in the cell $C_{i(j)}$. $\nu^k(C_{i(j)}) \,/\, c_{\widehat{\theta}_0}(u_{i(j)})^k$ approximates $\nu_{c_{\widehat{\theta}_0}}^k(C_{i(j)})$ if the cells $C_{i(j)}$ are sufficiently small, such that the scaling function c is approximately constant in $C_{i(j)}$. Let us denote $\lambda_{\widehat{\theta}_0, \psi}^{c_{\widehat{\theta}_0}}(u_j \mid \mathbf{x}\backslash\{u_j\})$ by $\lambda_j, j = 1, \ldots, m$. The pseudolikelihood can then be approximated as a weighted likelihood of independent Poisson variables y_j with means λ_j and weights w_j

$$\log(PL_{W,1}(\widehat{\theta}_0, \psi; \mathbf{x})\exp(-\nu_{c_{\widehat{\theta}_0}}^k(W))) \approx \sum_{j=1}^m (y_j \log \lambda_j - \lambda_j)\, w_j \,, \qquad (21)$$

$$y_j = \frac{1}{w_j}\mathbf{1}\{u_j \in \mathbf{x}\} \,, \quad j = 1, \ldots, m \,. \qquad (22)$$

When the conditional intensity $\lambda_{\widehat{\theta}_0,\psi}^{c_{\widehat{\theta}_0}}$ is of exponential family form, (21) can easily be maximised, using standard software for generalised linear models.

5 Simulation Study

In order to further study the properties of the estimation procedure proposed in Sect. 4, a simulation study was carried out. The simulation experiment concerns the exponentially scaled Strauss point process with scaling function

$$c_\theta(u) = \sqrt{\frac{1 - e^{-2\theta}}{2\theta}}\, e^{\theta u_1}\ ,\quad u = (u_1, u_2) \in \mathbb{R}^2\ ,\qquad (23)$$

observed on the unit square $W = [0,1]^2$. We used four different values of the inhomogeneity parameter $\theta \in \{0.25, 0.5, 1, 1.5\}$. For the template Strauss process we fixed the interaction radius δ to 0.05 and used a dense set of γ-values in $\{0.01, 0.02, \ldots, 1.00\}$. For β, we used the two values of 250 and 100 to investigate the influence of the total intensity. Note that $\theta = 1.5$ represents quite strong inhomogeneity, compare with Fig. 2.

For each combination of the parameters, 1000 point patterns were generated using MCMC and the distribution of $\widehat{\theta}_0$ was approximated by the empirical distribution from the 1000 realisations. To reduce the edge effects in the simulation the process was generated on a bigger window $[-0.2, 1.5] \times [-0.5, 1.5]$ so that $\bigcup_{x \in [0,1]^2} b_c(x, 2\delta)$ was included in this bigger window.

Fig. 4 shows the empirical mean values for the estimator $\widehat{\theta}_0$. Since the function m defined by (13) is concave and $t(\mathbf{x} \cap W)/n(\mathbf{x} \cap W)$ was found to be approximately unbiased for $m(\theta)$, $\widehat{\theta}_0$ tends to overestimate θ. This can be seen in Fig. 4 for $\theta = 1.5$ and 1, however the relative bias is not larger than 1% and it does not depend on the interaction parameter γ. The 95% envelopes for $\widehat{\theta}_0$ are also shown in Fig. 4 and for reasonably high number of observed points (i.e. $\beta = 250$) the inhomogeneity is reliably detected by $\widehat{\theta}_0$. Notice, for example that for $\theta = 1$, 95% of the estimates $\widehat{\theta}_0$ falls into the interval $[0.75, 1.25]$ and even for $\theta = 0.25$ – an inhomogeneity often hardly recognizable from the realisations, 97.5% of the $\widehat{\theta}_0$ estimates are larger that zero.

Note that since the scaling function has been normalised as in (14), (16) implies that

$$\mathbf{E}_{\theta,\psi}\, n(X_{c_\theta} \cap W) = \lambda_{0\psi}\nu^k(W)\ ,\qquad (24)$$

i.e. the mean number of points in W does not depend on the inhomogeneity parameter θ. Since (16) does not hold exactly for the Strauss process, we investigated whether (24) holds approximately, using the simulated data. The approximation is excellent in this example, cf. Fig. 5.

Let us next study the estimation of the template parameters. The density of the Strauss process is of exponential family form with one nuisance parameter δ – the interaction radius (see Example 1). Thus $\psi = (\beta, \gamma, \delta)$ and

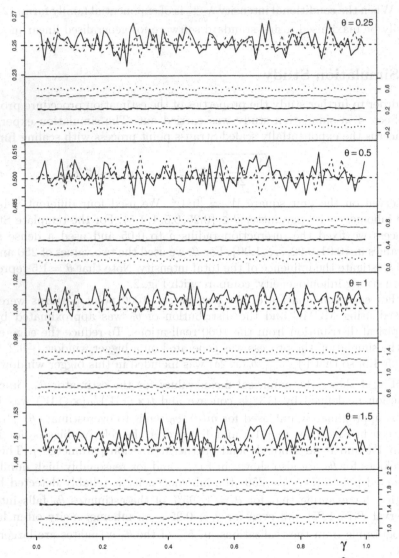

Fig. 4. Empirical mean values and 95% envelopes for the estimator $\widehat{\theta}_0$ for four different values of the inhomogeneity parameter θ (values are indicated in the plots) and for template parameter $\beta = 250$ (*full drawn lines*, resp. *dashed lines* for envelopes) and $\beta = 100$ (*dashed lines*, resp. *dotted lines* for envelopes), as a function of the template parameter γ. The central lines in the envelope plots are the empirical means again.

$$\log \lambda_j = \log \beta + s_c(u_j; \mathbf{x}) \log \gamma \,,$$

where

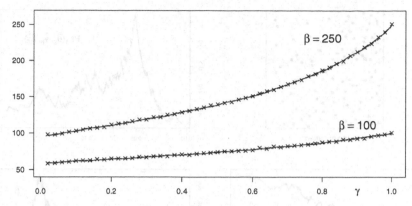

Fig. 5. Comparison of the intensities $\mathbf{E}_{\theta,\psi}\, n(X_{c_\theta} \cap W)$ (estimated by averages over 200 realisations) on the unit square W for the exponentially scaled Strauss point process X_{c_θ} with different $\theta \in \{0.25, 0.5, 1, 1.5\}$ (*dashed, chain-dotted, full and dotted curves*) and the intensity of the template processes $\lambda_{0\psi}\nu^k(W)$ (*crosses*) as a function of the template parameter γ. Results are showed for two different values of the template parameter $\beta = 100$ and 250 and the same $\delta = 0.05$ for all the processes. Since for each beta the differences between all the four curves and the crosses are hardly distinguishable we found a perfect agreement with equation (24)

$$s_c(u_j; \mathbf{x}) = \sum_{x \in \mathbf{x} \setminus \{u_j\}} 1\{\nu_c^1([u_j, x]) \leq \delta\}\,.$$

To find the estimate of ψ we have to compute and compare the profile pseudolikelihood

$$\overline{PL}_{W,1}(\delta) = \max_{\beta,\gamma} PL_{W,1}(\widehat{\theta}_0, \beta, \gamma, \delta; \mathbf{x})$$

on a grid of values of δ. We let β_δ and γ_δ be the values of β and γ at which

$$PL_{W,1}(\widehat{\theta}_0, \cdot, \cdot, \delta; \mathbf{x})$$

is maximal (the subscript δ indicates the dependence on δ). In Figs. 6 and 7 we illustrate the procedure on a simulated exponentially scaled Strauss point pattern with the scaling function (23) and parameters $\theta = 1$, $\beta = 250$, $\gamma = 0.25$, $\delta = 0.05$, $W = [0,1]^2$. The parameter θ has been fixed to the correct value and a regular grid of 100×100 dummy points was used. In the plots presented in Figs. 6 and 7, the profile pseudolikelihood and the estimates $\hat{\beta}_\delta$ and $\hat{\gamma}_\delta$ are plotted as functions of the nuisance parameter δ. The jaggedness of the plots is due to the discontinuity of the interpoint distance function s_c as a function of δ. In Fig. 6 we used no border correction (the pseudolikelihood (18) with $\mathbf{x}$ replaced by $\mathbf{x} \cap W$) while in Fig. 7 we used a border correction of $\nu_c^1 = 0.05$ (the psedolikelihood (18) with W replaced by an irregular observation window $\widetilde{W} = \{u \in W : \nu_c^1(u, \partial W) > 0.05\}$, where ∂W denotes the boundary of W).

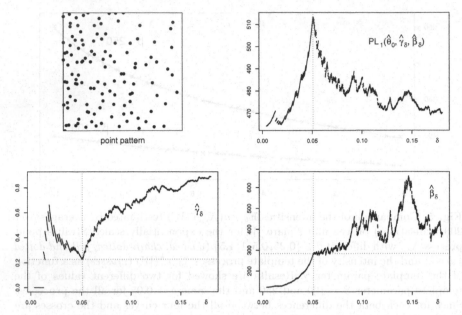

Fig. 6. Profile pseudolikelihood estimation of the template parameters β, γ, δ of a simulated exponentially scaled Strauss process on $[0,1]^2$ without any border correction. The first picture shows the data **x**. In the 3 graphs, the profile pseudolikelihood and the corresponding estimates $\hat{\beta}_\delta$ and $\hat{\gamma}_\delta$ are plotted as functions of δ. The final estimates $\hat{\delta} = 0.0508$, $\hat{\beta} = 283$, $\hat{\gamma} = 0.23$ are indicated by the dotted lines. The true values are $\delta = 0.05$, $\beta = 250$, $\gamma = 0.25$.

The obtained estimates of ψ are in good agreement with the true values, especially the estimate of the interaction radius is very precise. It is also important that the estimates with and without border correction do not differ substantially (which is probably caused by the sufficiently large number of observed points in W).

The results concerning pseudolikelihood estimation were confirmed in repeated simulation experiments.

6 Two Step Inference Where Scaling Function Is Estimated Using Other (non ML) Methods

Going one step further, one could also estimate c in some other way from the local intensity

$$\lambda_c(u) = \mathbf{E}\,\lambda_c(u \mid X_c)\,,$$

$u \in \mathcal{X}$, of the locally scaled process X_c, using the approximate relation

$$\lambda_c(u) \approx c(u)^{-k}\lambda_0\,, \tag{25}$$

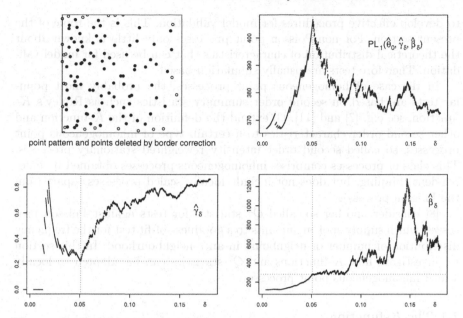

Fig. 7. Profile pseudolikelihood estimation of the template parameters β, γ, δ of a simulated exponentially scaled Strauss process on $W = [0, 1]^2$ with border correction $\nu_c^1 = 0.05$. The first picture shows the same data used in Fig. 6. The full circles are the data points used for the estimation. The 3 graphs are constructed as in Fig. 6. The obtained estimates are $\hat{\delta} = 0.0508$, $\hat{\beta} = 282$, $\hat{\gamma} = 0.22$. The true values are $\delta = 0.05$, $\beta = 250$, $\gamma = 0.25$.

where λ_0 is the intensity of the template process. In order to estimate the scaling function, we could use an estimate $\hat{\lambda}_c(u)$ of the local intensity, and set

$$\hat{c}(u) = \left[C \hat{\lambda}_c(u) \right]^{-1/k},$$

where $C = 1/\lambda_0$ is some constant that can be arbitrarily fixed. For convenience, one may choose $C = 1$.

If, in a parametric setting, $\hat{\lambda}_c(u)$ is the maximum likelihood estimator of the intensity of an inhomogeneous Poisson process, then $\hat{c}(u)$ is the same partial MLE as the one based on L_0. On the other hand, $\lambda_c(u)$ can also be estimated non parametrically, for example, using kernel methods or Voronoi tessellations. Or parametrically by other methods than maximum likelihood, e.g. regression methods.

7 Model Validation

Since the two-step estimation procedure, suggested in the present paper, can only be justified theoretically in special cases, it is particularly important to develop effective procedures for model validation. This is the topic of the present section. For non-Poisson point processes only little is known about the theoretical distribution of characteristics that can be used for model validation. Therefore tests are usually simulation based.

In the case of homogeneous point processes, the probably most popular tests are based on second order summary statistics such as Ripley's K-function, see e.g. [7] and [11]. [3] extend the definition of the K-function and other second order characteristics to a certain type of inhomogeneous point processes, so called second order intensity reweighted stationary processes. This class of processes comprises inhomogeneous processes obtained by independent thinning, but does not include locally scaled processes (apart from the Poisson process).

[8] consider another so-called Q^2 statistic for tests against Poisson processes, which simply spoken amounts to a goodness-of-fit test for the frequency distribution of number of neighbours in an r-neighbourhood. In this section we investigate how K-functions and Q^2-statistics can be adapted to locally scaled inhomogeneous point processes.

7.1 The K-function

Let λ_0 and K_0 denote the intensity and K-function of the template process, and let $\mathfrak{K}_0 := \lambda_0 K_0$. A ratio-unbiased estimator of $\mathfrak{K}_0(r)$ is given by

$$\widehat{\mathfrak{K}}_0(W, r) = \frac{1}{n(\mathbf{x} \cap W)} \sum_{x \in \mathbf{x} \cap W} \sum_{y \in \mathbf{x} \setminus \{x\}} \mathbf{1}\{\nu^1([x, y]) \leq r\},$$

where $\mathbf{x}$ is an observed point pattern from the stationary point process X. In order to obtain an estimate of $K_0(r)$, $\widehat{\mathfrak{K}}_0$ is combined with an estimate of λ_0. If instead a locally scaled point pattern $\mathbf{x}$ is observed, we suggest to use a locally scaled analogue of $\widehat{\mathfrak{K}}_0(W, r)$, viz.

$$\widetilde{\mathfrak{K}}_0(W, r) = \frac{1}{n(\mathbf{x} \cap W)} \sum_{x \in \mathbf{x} \cap W} \sum_{y \in \mathbf{x} \setminus \{x\}} \mathbf{1}\{\nu_c^1([x, y]) \leq r\}. \tag{26}$$

Note that $\widetilde{\mathfrak{K}}_0(W, r)$ is ratio-unbiased for $\mathfrak{K}_0(r)$ if c is constant. Furthermore, $\widetilde{\mathfrak{K}}_0(W, r)$ is ratio-unbiased for general scaling functions and distance-interaction point processes defined on an interval I of $\mathbb{R}^1$. To see this, we use that in $\mathbb{R}^1$ a locally scaled distance-interaction process X_c has the same distribution as $h(X)$ where h is a 1-1 differentiable transformation of I onto I with $(h^{-1})' = c^{-1}$. (A proof of this result can be found in the Appendix.) Therefore, we have

$$\mathbf{E}(\sum_{x \in X_c \cap W} \sum_{y \in X_c \setminus \{x\}} \mathbf{1}\{\nu_c^1([x,y]) \leq r\})$$

$$= \mathbf{E}(\sum_{x \in h(X) \cap W} \sum_{y \in h(X) \setminus \{x\}} \mathbf{1}\{\nu^1([h^{-1}(x), h^{-1}(y)] \leq r\})$$

$$= \mathbf{E}(\sum_{x \in X \cap h^{-1}(W)} \sum_{y \in X \setminus \{x\}} \mathbf{1}\{\nu^1([x,y]) \leq r\}) .$$

Accordingly, the ratio-unbiasedness of $\hat{\mathfrak{K}}_0$ follows from the ratio-unbiasedness of $\hat{\mathfrak{K}}_0$. Generally, $\tilde{\mathfrak{K}}_0$ is expected to be (approximately) ratio-unbiased if r is small such that c varies little in a scaled neighbourhood. In any case, one should use simulations of the scaled null hypothesis model, not only of the template, for model validation.

A further simplification is accomplished by applying $\nu_c^1([x,y]) \approx 2/(c(x) + c(y)) \times \nu^1([x,y])$, which was introduced for distance-interaction processes in [9]. The corresponding statistic

$$\check{\mathfrak{K}}_0(W,r) = \frac{1}{n(\mathbf{x} \cap W)} \sum_{x \in \mathbf{x} \cap W} \sum_{y \in \mathbf{x} \setminus \{x\}} \mathbf{1}\{\nu^1([x,y]) \leq \tfrac{1}{2}(c(x) + c(y))r\} , \quad (27)$$

is particularly useful if c is estimated nonparametrically, because it requires evaluation of c only in the data points.

In practical situations, both λ_0 and c have to be estimated from the data. As discussed in the preceding sections, the estimation of c cannot be separated from the estimation of λ_0. Since the template is unique only up to a constant scale factor which determines λ_0, the scaling function c is unique only up to a constant as well. We suggest to normalize c such that $\nu_c^k(W) = \nu^k(W)$, see (14). Thus, we set $\hat{\lambda}_0 := n(\mathbf{x} \cap W)/\nu^k(W)$ since $\mathbf{E}n(X_c \cap W) = \int_W \lambda_c(x) \nu^k(dx) \approx \int_W \lambda_0 c(x)^{-k} \nu^k(dx) = \lambda_0 \nu^k(W)$. In what follows, we use the notation $\tilde{K}_0(W,r)$ for $\tilde{\mathfrak{K}}_0(W,r)/\hat{\lambda}_0$.

7.2 The Q^2 Statistic

The Q^2-statistic proposed by [8] is (in the simplest case) based on the numbers $M_\ell(W,r)$ of points in W with ℓ r-close neighbours, $\ell = 0, 1, \ldots, q$. For a homogeneous Poisson point process, the expectation μ and the covariance matrix Σ of the vector $M = (M_0, M_1, \ldots, M_q)^\top$ can easily be calculated. A finite range dependency argument is used to show that the statistic

$$Q^2 = (M - \mu)^\top \Sigma^{-1}(M - \mu)$$

(squared Mahalanobis distance) is asymptotically χ^2-distributed for increasing size of the observation window W. By simulation experiments, [8] showed that Q^2 discriminates well between patterns from a mixed cluster and regular point process and the Poisson process.

Since μ and Σ can also be calculated for an inhomogeneous Poisson point process, it would be possible to use the same Q^2-statistic also for tests of inhomogeneous Poisson processes. However, the expected number of neighbours in a ball of radius r around a point x would depend on the local intensity $\lambda(x)$. Thence, inhomogeneity introduces much extra variation to M which would largely cut down the diagnostic value of Q^2.

This effect can be avoided by adjusting r to the local intensity. We propose to replace the Euclidean neighbour distance by the locally scaled neighbour distance. In an inhomogeneous Poisson point process with intensity $\lambda_c(x) = c(x)^{-k}\lambda_0$, the number of r-scaled-close neighbours of a point x is Poisson distributed with parameter $\lambda_0 \nu_c^k(b_c(x,r))$. Since

$$\frac{\nu_c^k(b_c(x,r))}{\nu^k(b(x,r))} \to 1 \text{ as } r \to 0 \,,$$

the distribution of r-scaled-close neighbours does hardly depend on the location for small r, and is close to the distribution of r-close neighbour number in the homogeneous case.

The local scaling analogue of M_ℓ is

$$M_{\ell\text{inhom}}(W,r) = \sum_{x \in \mathbf{x} \cap W} \mathbf{1}\{n(b_c(x,r) \cap \mathbf{x} \setminus \{x\}) = \ell\} \,. \tag{28}$$

Since calculation of μ and Σ is feasible only for the Poisson point process with slowly varying scaling function, we suggest to do simulation tests. This would allow to test any hypothesis. While any distance between observed and expected neighbour number distribution can be used, we still recommend to use the statistic Q^2, however to replace μ and Σ with estimates obtained by simulation. Note that the simulations for estimating μ and Σ are not to be reused for the test.

8 Data Analysis

The map shown in Fig. 1 was recorded in the Australian heath. This vegetation is subjected to regular fires, the study area having been last burnt ten years before the collection of data [1]. The species under study, *Scholtzia* aff. *involucrata*, is a long lived shrub that regenerates from root stock after fire yielding daughter plants that stand close together. Furthermore, seed germinates after fire, with young plants coming up within a distance of at most 2m of the parent plant. However, only very few seedlings survive the dry summer (Paul Armstrong, personal communication). These facts explain the slight clustering observable in the point pattern of plant locations. The heterogeneous intensity is likely to be due to soil mosaic, affecting mostly the seedlings that are very sensitive to shortage of water.

For the Scholtzia data set, a test on complete spatial randomness and homogeneity based on quadrat counts was highly significant. The point pattern appears inhomogeneous, particularly in the y-direction, and clustered. We therefore need to model attractive interaction between the plants. The exponentially scaled area-interaction model appears to be a good candidate because the area of a location dependent neighbourhood around each plant enters explicitly into the model density.

We used the two-step fitting procedure. For convenience we rescaled the data to the unit square $W = [0,1]^2$. As the pattern exhibits obvious inhomogeneity in the vertical direction but appears quite homogeneous in the horizontal direction we used an exponential scaling function of the form

$$c(u) = \sqrt{\frac{1 - e^{-2\theta}}{2\theta}}\, e^{\theta u_2}\,, \quad u = (u_1, u_2) \in \mathbb{R}^2\,. \tag{29}$$

Based on $L_0(\theta; \mathbf{x} \cap W)$ we obtained the following estimate of θ

$$\widehat{\theta}_0 = 1.0839\,,$$

with $\alpha(\widehat{\theta}_0) = 0.6391$, see (15).

Secondly, we maximised the pseudolikelihood $PL_{W,1}(\widehat{\theta}_0, \psi; \mathbf{x})$ with $\widehat{\theta}_0$ fixed. The density of the area-interaction process is of an exponential family form with one nuisance parameter δ – the interaction radius. As for the Strauss process, $\psi = (\beta, \gamma, \delta)$ and for the estimation we use the same weights as in (20) and

$$\log \lambda_j = \log \beta - \nu_c^2(U_{c,\delta}(u_j; \mathbf{x})) \log \gamma\,,$$
$$U_{c,\delta}(u_j; \mathbf{x}) = \{y \in W : \nu_c^1([y, u_j]) \le \delta, \nu_c^1([y, \mathbf{x} \backslash \{u_j\}]) > \delta\}\,.$$

We used a grid of 100×100 dummy points which were equidistant in the horizontal direction and $\nu_{c_{\widehat{\theta}_0}}^1$ –equidistant in the vertical direction (actually this means that the dummy points were $\nu_{c_{\widehat{\theta}_0}}^1$ –equidistant in both directions – compare with (29)).

We maximised the profile pseudolikelihood on a grid of δ-values. The main problem is the computation of the scaled volumes $\nu_c^2(U_{c,\delta}(u_j; \mathbf{x}))$ for all the points $u_j, j = 1, \ldots, m$. This can be done only approximately. To approximate these scaled volumes with a reasonable precision it is necessary to compute the scaled distances from the points $\{u_j, j = 1, \ldots, m\}$ to each point in a very fine grid of points in W. This job is computationally quite demanding.

The approximate profile pseudolikelihood $PL_{W,1}(\widehat{\theta}_0, \psi; \mathbf{x})$ was computed with border correction $\nu_c^1 = 0.05$. This degree of border correction was chosen as a compromise between minimizing the bias caused by missing unobserved points and not excluding too many observed points from the estimation (with the chosen border correction one forth of the points was not used in the estimation). The profile pseudolikelihood and estimates of the parameters as

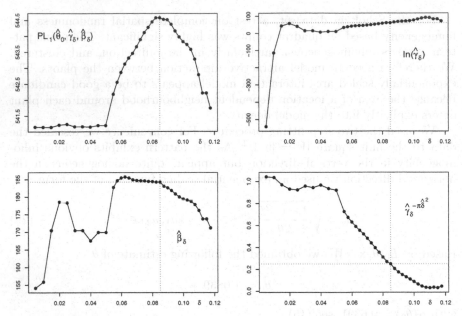

Fig. 8. Pseudolikelihood estimation of the template parameters β, γ, δ of the exponentially scaled area-interaction process for the plant data from Fig. 1 rescaled to $[0,1]^2$ with border correction $\nu_c^1 = 0.05$. The graphs show the profile pseudolikelihood and the corresponding estimates $\hat{\beta}_\delta$ and $\hat{\gamma}_\delta$ as functions of δ. The last graph of $\hat{\gamma}^{-\pi\hat{\delta}^2}$ shows the strength of the attractive interaction. The resulting estimates are $\hat{\delta} = 0.085$, $\hat{\beta} = 184$, $\hat{\gamma} = 3.99 \cdot 10^{26}$.

functions of δ are plotted in Fig. 8. Note that the curves are smoother than in the case of the Strauss process because now the interaction function is continuous as a function of δ. We obtained the following values

$$\hat{\delta} = 0.085 , \quad \hat{\beta} = 184 , \quad \hat{\gamma} = 3.99 \cdot 10^{26} , \quad \hat{\gamma}^{-\pi\hat{\delta}^2} = 0.25 . \qquad (30)$$

The value of $\hat{\gamma}^{-\pi\hat{\delta}^2}$ is included because it gives a better impression of the strength of the interaction, as this is actually the term which appears in the template density. The fit indicates a slightly clustered point pattern as we expected.

For model validation we used the $\widetilde{K}_0(\widetilde{W}, r)$ and Q^2 statistics from Sect. 7. Fig. 9 shows the locally scaled estimate $\widetilde{K}_0(\widetilde{W}, r)$ with $\widetilde{W} = \{u \in W : \nu_c^1(u, \partial W) > 0.05\}$ (full-drawn line) together with the empirical mean and 95% envelopes for $\widetilde{K}_0(\widetilde{W}, r)$ calculated from 399 simulations under the fitted exponentially scaled area-interaction model (dashed lines). The locally scaled estimate $\widetilde{K}_0(\widetilde{W}, r)$ for the plant data lies inside the envelopes of the fitted area-interaction model.

Next we tested the locally scaled Poisson hypothesis on the plant data. We used the

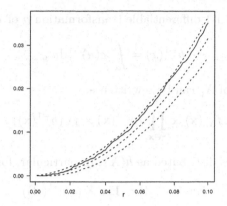

Fig. 9. The estimate $\widetilde{K}_0$ for the plant data (*full drawn line*) and mean and 95% envelopes for $\widetilde{K}_0$ for the exponentially scaled area-interaction model (*dashed lines*)

$$Q_P^2 = (M_{\text{inhom}} - \mu_P)^\top \Sigma_P^{-1} (M_{\text{inhom}} - \mu_P)$$

statistic with $r = 0.05$ and $M_{\text{inhom}} = (M_{0\text{inhom}}, \ldots, M_{6\text{inhom}})$ defined by (28). The subscript P indicates that in the formula for Q^2 we use as μ and Σ the mean μ_P and the covariance matrix Σ_P of M_{inhom} for the fitted locally scaled Poisson model with $\theta = \widehat{\theta}_0$. The values of μ_P and Σ_P were estimated from 8000 simulated realisations of the fitted locally scaled Poisson model.

The simulation test (using 399 realisations of the hypothesis locally scaled Poisson model with $\theta = \widehat{\theta}_0$) gives the p-value of 0.05. Thus the plant data is not very well described by the Poisson model.

Then we used the Q_A^2 statistic (i.e. the mean value μ_A and covariance matrix Σ_A of M_{inhom} are computed for the fitted exponentially scaled area-interaction model) for testing of the fitted locally scaled area-interaction model. The test gave the p-value of 0.106.

Appendix: Proof of Equation (16) for Distance-Interaction Processes in $\mathbb{R}^1$

Let us suppose that X is a distance-interaction process on an interval $I = [a, b]$ of $\mathbb{R}^1$ with density

$$f_X(\mathbf{x}) \propto \beta^{n(\mathbf{x})} \prod_{\mathbf{y} \subseteq_2 \mathbf{x}} \varphi(\{\nu^1([u, v]) : \{u, v\} \subseteq \mathbf{y}, u \neq v\}),$$

where $\subseteq_2$ indicates that $\mathbf{y}$ should have at least two elements. The density of X_c is then

$$f_{X_c}(\mathbf{x}) \propto \prod_{x \in \mathbf{x}} c(x)^{-1} \times \beta^{n(\mathbf{x})} \prod_{\mathbf{y} \subseteq_2 \mathbf{x}} \varphi(\{\nu_c^1([u, v]) : \{u, v\} \subseteq \mathbf{y}, u \neq v\}).$$

Let us consider the 1-1 differentiable transformation h of I onto I defined by

$$h^{-1}(x) = \int_a^x c(u)^{-1}\mathrm{d}u \ .$$

Then, the density of X_c can be rewritten as

$$f_{X_c}(\mathbf{x}) \propto \prod_{x \in \mathbf{x}} Jh^{-1}(\mathbf{x}) \times f_X(h^{-1}(\mathbf{x})) \ .$$

It follows that X_c is distributed as $h(X)$. In particular, for $A \in \mathcal{B}(I)$,

$$\begin{aligned}
\mathbf{E}n(X_c \cap A) &= \mathbf{E}n(X \cap h^{-1}(A)) \\
&= \int_{h^{-1}(A)} \lambda_0 \mathrm{d}x \\
&= \int_A c(u)^{-1}\lambda_0 \mathrm{d}u \ ,
\end{aligned}$$

or

$$\lambda_c(u) = c(u)^{-1}\lambda_0 \ .$$

Acknowledgements

The authors wish to thank Dr Paul Armstrong for kindly providing the data set and explaining its biological background, and an anonymous referee for valuable suggestions to improve the paper. MP was supported by a Marie Curie Fellowship of the European Community Programme Improving the Human Research Potential and the Socio-economic Knowledge Base under contract number HPMT-CT-2001-00364 and by the grant GAČR 201/03/0946.

References

[1] P. Armstrong. *Species patterning in heath vegetation of the Northern Sandplain*. Honours Thesis, Murdoch University, Western Australia, 1991.

[2] A.J. Baddeley, M. Hazelton, J. Møller and R. Turner. Residuals and diagnostics for spatial point processes. Abstract, 54 ISI Session Berlin, 13-20 August, 2003.

[3] A.J. Baddeley, J. Møller and R.P. Waagepetersen. Non- and semiparametric estimation of interaction in inhomogeneous point patterns. *Statistica Neerlandica*, 54:329–350, 2000.

[4] A.J. Baddeley and R. Turner. Practical maximum pseudolikelihood for spatial point processes. *Australian and New Zealand Journal of Statistics*, 42:283–322, 2000.

[5] A.J. Baddeley and M.N.M. van Lieshout. Area-interaction point processes. *Annals of the Institute of Statistical Mathematics*, 47:601–619, 1995.

[6] O.E. Barndorff-Nielsen. *Information and Exponential Families in Statistical Theory*. Wiley, Chichester, 1978.

[7] P.J. Diggle. *Statistical Analysis of Spatial Point Patterns*. Edward Arnold, London, 2003.

[8] P. Grabarnik and S.N. Chiu. Goodness-of-fit test for complete spatial randomness against mixtures of regular and clustered spatial point processes. *Biometrika*, 89:411–421, 2002.

[9] U. Hahn, E.B.V. Jensen, M.N.M. van Lieshout and L.S. Nielsen. Inhomogeneous spatial point processes by location-dependent scaling. *Advances in Applied Probability*, 35:319–336, 2003.

[10] M.N.M. van Lieshout. *Markov Point Processes and their Applications*. London: Imperial College Press, 2000.

[11] J. Møller, and R.P. Waagepetersen. *Statistical Inference and Simulation for Spatial Point Processes*. Chapman and Hall/CRC, Boca Raton, 2003.

[12] X.X. Nguyen and H. Zessin. Punktprozesse mit Wechselwirkung. *Zeitschrift für Wahrscheinlichkeitstheorie und verwandte Gebiete*, 37:91–126, 1976.

[13] L.S. Nielsen. *Point process models allowing for interaction and inhomogeneity*. PhD Thesis, Department of Mathematical Sciences, University od Aarhus, 2001.

[14] L.S. Nielsen and E.B.V. Jensen. Statistical inference for transformation inhomogeneous point processes. *Scandinavian Journal of Statistics*, 31:131–142, 2004.

[15] D.J. Strauss. A model for clustering. *Biometrika*, 63:467–475, 1975.

Nonparametric Testing of Distribution Functions in Germ-grain Models

Zbyněk Pawlas[1,2] and Lothar Heinrich[3]

[1] Charles University, Faculty of Mathematics and Physics, Department of Probability and Mathematical Statistics. Sokolovská 83, 186 75 Praha 8, Czech Republic, `pawlas@karlin.mff.cuni.cz`

[2] Institute of Information Theory and Automation, Academy of Sciences of the Czech Republic. Pod Vodárenskou věží 4, 182 08 Praha 8, Czech Republic.

[3] University of Augsburg, Institute of Mathematics. Universitätsstr. 14, D-86135 Augsburg, Germany, `lothar.heinrich@math.uni-augsburg.de`

Summary. Germ-grain models are random closed sets in the d-dimensional Euclidean space $\mathbb{R}^d$ which admit a representation as union of random compact sets (called grains) shifted by the atoms (called germs) of a point process. In this note we consider the distribution function F of an m-dimensional random vector describing shape and size parameters of the typical grain of a stationary germ-grain model. We suggest a ratio-unbiased weighted (Horvitz-Thompson type) empirical distribution function $\hat{F}_n$ to estimate F, based on the corresponding data vectors of those shifted grains which lie completely within the sampling window $W_n \subseteq \mathbb{R}^d$. Since, as W_n increases, the empirical process $\hat{F}_n(t) - F(t)$ (after scaling) converges weakly to an m-parameter Brownian bridge process, it is possible for the particular case where $m = 1$, to examine the the goodness-of-fit of observed data to a hypothesised continuous distribution function F, analogous to the Kolmogorov-Smirnov test.

Key words: Germ-grain model, Horvitz-Thompson-type estimator, Kolmogorov-Smirnov test, Multivariate empirical process, Weak convergence

1 Introduction

Let $\mathcal{K}'$ be the family of non-empty compact subsets of the d-dimensional Euclidean space $\mathbb{R}^d$ endowed with the Hausdorff metric making $\mathcal{K}'$ to a Polish space with Borel σ-algebra $\mathcal{B}(\mathcal{K}')$. A point process on the metric space $\mathcal{K}'$ is called a *particle process*, see [7], Chap. 4, for details.

Further let $\mathcal{K}'_0 = \{K \in \mathcal{K}' : c(K) = 0\}$, where $c(K) \in \mathbb{R}^d$ denotes a reference point assigned to each $K \in \mathcal{K}'$ such that the mapping $K \mapsto c(K)$ is $(\mathcal{B}(\mathcal{K}'), \mathcal{B}(\mathbb{R}^d))$-measurable and equivariant under translations (i.e. $c(K + x) = c(K) + x$ for all $K \in \mathcal{K}'$ and $x \in \mathbb{R}^d$). The most frequent choices of $c(K)$ are lexicographical minimum of the set K or a centroid of K.

Now, a particle process can be defined by means of a marked point process

$$\Psi_m = \sum_{i \geq 1} \delta_{[X_i, \Xi_i]} \tag{1}$$

in $\mathbb{R}^d$ with the mark space $\mathcal{K}'_0$. A *germ-grain model* Ξ is then defined to be the union set

$$\Xi = \bigcup_{i \geq 1} (X_i + \Xi_i). \tag{2}$$

The points $\{X_i\}_{i \geq 1}$ are called *germs and* $\{\Xi_i\}_{i \geq 1}$ *are the* grains of the germ-grain model (2).

Throughout this note we work under the following basic assumptions:

(A1) The unmarked point process $\Psi(\cdot) := \Psi_m(\cdot \times \mathcal{K}'_0) = \sum_{i \geq 1} \delta_{X_i}(\cdot)$ is simple and weakly (second-order) stationary with finite and positive intensity λ.

(A2) The grains $\{\Xi_i\}_{i \geq 1}$ are independent copies of Ξ_0 and independent of Ψ. The random compact set Ξ_0 (called the typical grain) has distribution Q on $\mathcal{K}'_0$.

(A3) There exists a number $q \geq d$ such that

$$\mathsf{E}\|\Xi_0\|^q < \infty, \quad \text{where } \|K\| = \sup\{\|x\| : x \in K\} \text{ for } K \in \mathcal{K}'_0. \tag{3}$$

(A4) The convex compact sampling window $W_n \subseteq \mathbb{R}^d$ expands without bounds in all directions such that for $q \geq d$ from (3)

$$\frac{\mathcal{H}^{d-1}(\partial W_n)}{|W_n|^{1-1/q}} \leq c_0 < \infty \quad \text{and} \quad \rho(W_n) \xrightarrow[n \to \infty]{} \infty,$$

where $\mathcal{H}^k$ denotes the k-dimensional Hausdorff measure on $\mathbb{R}^d$, $|\cdot| = \mathcal{H}^d(\cdot)$ is the Lebesgue measure on $\mathbb{R}^d$, $b(x, r)$ designates the ball of radius $r > 0$ centred at $x \in \mathbb{R}^d$ and $\rho(W_n) = \sup\{r > 0 : b(x, r) \subseteq W_n, x \in W_n\}$ is the inball radius of W_n.

(A5) The reduced covariance measure $\gamma_{red}^{(2)}$ (see [2], Chap. 10.4) of Ψ is of bounded total variation.

Let $f(\Xi_0) = (f_1(\Xi_0), \ldots, f_m(\Xi_0))$ be an m-dimensional random vector with distribution function

$$F(t) = \mathsf{P}(f(\Xi_0) \leq t) = \mathsf{P}(f_1(\Xi_0) \leq t_1, \ldots, f_m(\Xi_0) \leq t_m), \tag{4}$$

$t = (t_1, \ldots, t_m) \in \mathbb{R}^m$. The vector $f(\Xi_0)$ can describe various shape and size parameters of the typical grain.

We assume that only a single realisation of the germ-grain models (2) can be observed in W_n and that the data vectors $f(\Xi_i)$ are available for all shifted grains $X_i + \Xi_i$ lying completely in W_n. In other words, the germ-grain model

(2) consists either of isolated grains or the overlapping effects do not prevent the exact measurement of the vectors $f(\Xi_i)$ as, for example, in the case of fibre or manifold processes.

Our aim is to construct an estimation of distribution function (4) with good asymptotic properties based on the data vectors $f(\Xi_i)$ of grains satisfying $X_i + \Xi_i \subseteq W_n$. This sampling procedure (minus-sampling) leads to a weighted estimator of the form

$$\widehat{\lambda F(t)} = \sum_{i \geq 1} \frac{\mathbf{1}_{\{X_i + \Xi_i \subseteq W_n\}}}{|W_n \ominus \check{\Xi}_i|} \mathbf{1}_{(-\infty, t]}(f(\Xi_i)), \quad t = (t_1, \ldots, t_m) \in \mathbb{R}^m, \quad (5)$$

where $(-\infty, t] = \prod_{i=1}^{m}(-\infty, t_i]$, $\mathbf{1}_B$ is the indicator function of a set or event B and $A \ominus \check{B} = \{x : x + B \subseteq A\}$ is the erosion of A by B. From Campbell's theorem for stationary marked point processes (see Chap. 4.2 in [8] or Chap. 10.5 in [2]) it follows that

$$\mathsf{E}\widehat{\lambda F(t)} = \lambda \int_{\mathcal{K}_0'} \int_{\mathbb{R}^d} \frac{\mathbf{1}_{W_n \ominus \check{K}}(x)}{|W_n \ominus \check{K}|} \mathbf{1}_{(-\infty, t]}(f(K)) \, \mathrm{d}x \, Q(\mathrm{d}K) = \lambda F(t),$$

which shows that (5) is an unbiased estimator of $\lambda F(t)$. In Sect. 2 we define the empirical distribution function $\hat{F}_n(t)$ as the ratio of $\widehat{\lambda F(t)}$ and $\widehat{\lambda F(\infty)}$. As an immediate consequence of the weak convergence of the corresponding m-variate empirical process (7) stated in Sect. 3 (Theorem 1), a Kolmogorov-Smirnov test can be established at least for $m = 1$ and continuous $F(\cdot)$. In Sect. 4 this result will be applied to real data taken from some porous ceramic material. The empirical distribution functions of the volume and of some shape characteristic of the typical pore are compared with corresponding hypothesised distribution functions.

2 Empirical Distribution Functions

A quite natural empirical counterpart of the distribution function (4) is given by

$$\tilde{F}_n(t) = \frac{1}{N_n} \sum_{i \geq 1} \mathbf{1}_{\{X_i + \Xi_i \subseteq W_n\}} \mathbf{1}_{(-\infty, t]}(f(\Xi_i)), \quad t = (t_1, \ldots, t_m) \in \mathbb{R}^m,$$

where $N_n = \sum_{i \geq 1} \mathbf{1}_{\{X_i + \Xi_i \subseteq W_n\}}$ is the number of completely observable grains in W_n. However, it turns out that the empirical distribution function $\tilde{F}_n(t)$ is not close enough to $F(t)$ in order to provide a zero mean weak limit of $\sqrt{|W_n|} \left(\tilde{F}_n(t) - F(t) \right)$. It is intuitively clear that smaller particles are more likely to lie completely in W_n than larger ones. Therefore, we need to weight the event $\{X_i + \Xi_i \subseteq W_n\}$ by an appropriate factor. We define the empirical distribution function based on (5) by

$$\hat{F}_n(t) = \frac{1}{\hat{\lambda}_n} \sum_{i \geq 1} \frac{\mathbf{1}_{\{X_i + \Xi_i \subseteq W_n\}}}{|W_n \ominus \breve{\Xi}_i|} \mathbf{1}_{(-\infty, t]}(f(\Xi_i)), \quad t = (t_1, \ldots, t_m) \in \mathbb{R}^m, \quad (6)$$

where

$$\hat{\lambda}_n = \sum_{i \geq 1} \frac{\mathbf{1}_{\{X_i + \Xi_i \subseteq W_n\}}}{|W_n \ominus \breve{\Xi}_i|}$$

is an unbiased estimator of the intensity λ. Therefore, $\hat{F}_n(\cdot)$ is a so-called ratio-unbiased estimator of $F(\cdot)$.

3 Weak Convergence of an Empirical Multiparameter Process

To construct asymptotic goodness-of-fit tests for the distribution function $F(\cdot)$ we use the weak convergence of the m-parameter empirical process

$$Y_n(t) = \sqrt{\Psi(W_n)} \left(\hat{F}_n(t) - F(t) \right), \quad t \in \mathbb{R}^m. \quad (7)$$

The process $(Y_n(t), t \in \mathbb{R}^m)$ has random jumps depending on the weights $|W_n \ominus \breve{\Xi}_i|^{-1}$. Its trajectories belong to the Skorohod space $D(\mathbb{R}^m)$, see [1] for $m = 1$ and [4] for $m \geq 2$, of right continuous real functions on $\mathbb{R}^m$ with finite left limits existing everywhere. For a precise definition of the limits, see [4, 6].

Theorem 1. *Under the assumptions (A1)–(A5), the sequence $(Y_n(t), t \in \mathbb{R}^m)_{n \geq 1}$ defined by (7) converges weakly (as $n \to \infty$) in $D(\mathbb{R}^m)$ to a zero mean Gaussian process $(Y(t), t \in \mathbb{R}^m)$ with covariance function $\mathsf{E} Y(s) Y(t) = F(s \wedge t) - F(s) F(t)$, where $s \wedge t = (\min(s_1, t_1), \ldots, \min(s_m, t_m))$.*

The proof and further details can be found in [3]. In the case $m = 1$ the Gaussian limit process $Y(\cdot)$ has the same finite-dimensional distributions as $W^\circ(F(\cdot))$, where $W^\circ(\cdot)$ denotes the Brownian bridge process being a zero mean Gaussian process on $[0, 1]$ with covariance $\mathsf{E} W^\circ(s) W^\circ(t) = s \wedge t - st$, see [1]. This case is of special interest for testing the goodness-of-fit of a continuous distribution function $F(t) = \mathsf{P}(f(\Xi_0) \leq t)$, $t \in \mathbb{R}$. Using the continuous mapping theorem (see [1], Theorem 2.7) we get the following

Corollary 1. *Under the assumptions of Theorem 1 for $m = 1$, we have*

$$\mathsf{P}\left(\sup_{t \in \mathbb{R}} |Y_n(t)| \leq x \right) \xrightarrow[n \to \infty]{} \mathsf{P}\left(\sup_{t \in \mathbb{R}} |W^\circ(F(t))| \leq x \right) \quad \text{for} \quad x > 0.$$

Furthermore, if $F(\cdot)$ has no jumps then the limit $\mathsf{P}(\sup_{t \in [0,1]} |W^\circ(t)| \leq x)$ does not depend on $F(\cdot)$ and coincides with the Kolmogorov distribution function

$$K(x) = 1 + 2 \sum_{k=1}^{\infty} (-1)^k e^{-2k^2 x^2} \quad \text{for} \quad x > 0.$$

4 Centres of Pores in a Ceramic Coating

Corollary 1 enables us to perform a Kolmogorov-Smirnov test in analogy to the classical case of i.i.d. samples drawn from an unknown source with one or several hypothetical distribution functions. The Kolmogorov-Smirnov test requires the continuity of the hypothesised distribution of the data implying that the critical values of the test statistic do not depend on this distribution.

In what follows we present a practical application to the microstructure of ceramic plasma-sprayed coatings. A specimen has been prepared in the Institute of Plasma Physics, Academy of Sciences of the Czech Republic, Prague. The data analysed here (kindly provided by Dr. Pavel Ctibor) consist of approximately convex pores in a three-dimensional sampling window W_n. For further details about the data set, see [5], where the spatial distribution of particles has been investigated.

Fig. 1. The 3D specimen of plasma-sprayed ceramic coating

The specimen is a rectangular block with dimensions $450 \times 350 \times 240\,\mu\text{m}$, it was subsequently sliced to obtain serial sections of small distance perpendicular to the grinding direction (see Fig. 1). In the two-dimensional image of a section (Fig. 2) both cracks and approximately convex pores can be observed. Using image analysis technique the parameters of only approximately convex pores were measured. The centroid coordinates, size and shape of individual pores were determined. In particular, for each observable pore we get the volume, the maximal diameter and the minimal diameter. The sample of centres of pores is shown in Fig. 3. The number of shifted grains $X_i + \Xi_i$ (pores) lying completely in W_n is $N_n = 1976$ and the number of reference points X_i (centroids) in W_n is $\Psi(W_n) = 2085$. Note that the shape of W_n entails that each of the eroded windows $W_n \ominus \check{\Xi}_i$ is a rectangular block with dimension depending on the widths of Ξ_i measured parallel to the edges of W_n. This fact facilitates considerably the computation of the weighted estimator (6).

At first we consider the distribution of volume of the typical grain, i.e. we put $f(\Xi_0) = |\Xi_0|$. Since very small particles could not be detected and so are omitted in the study, there exists a lower threshold $a > 0$ such that $F(a) = 0$

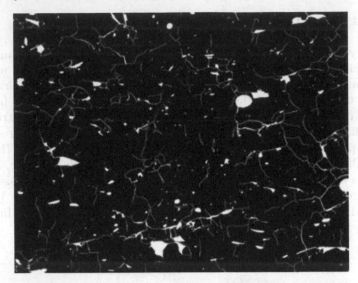

Fig. 2. The binary image of the 2D microstructure in a section including pores and cracks

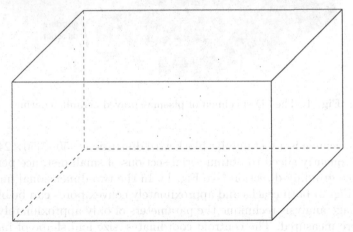

Fig. 3. The sample of the point process Ψ observed in the rectangular block W_n

and $F(t) > 0$ for $t > a$. Similarly, the absence of large pores indicates the existence of an upper bound $b > 0$. At the first glance and supported by the experience of the material scientists, the empirical distribution function $\hat{F}_n(t)$ seems to be approximately Pareto distributed. Thus, we will examine the null hypothesis that the distribution function $\mathsf{P}(f(\Xi_0) \leq t)$ coincides with a truncated Pareto distribution function $F_0(\cdot)$ given by

$$F_0(t) = 1 - \left(\frac{a}{t}\right)^c \times \frac{b^c - t^c}{b^c - a^c}, \quad a \leq t \leq b.$$

Correcting sampling bias effects by the weights $|W_n \ominus \breve{\Xi}_i|$ we modify the maximum likelihood method leading to the following estimates of the location parameters $a > 0$, $b > 0$ and the shape parameter $c > 0$, respectively:

$$\hat{a} = 31.462\,\mu\text{m}^3,$$
$$\hat{b} = 6152.056\,\mu\text{m}^3,$$
$$\hat{c} = 1.012.$$

Here, it should be noted that the weak convergence of the empirical process (7) stated in Theorem 1 does not hold in general when parameters in $F(\cdot)$ are replaced by corresponding (maximum likelihood) estimators. For this reason we are not allowed to plug in the parameter estimates $\hat{a}$, $\hat{b}$ and $\hat{c}$ in the hypothesised distribution function $F_0(\cdot)$. This fact is already well-known from the classical i.i.d. case and can be interpreted as higher sensitivity of the Kolmogorov-Smirnov test against the null hypothesis.

However, the above estimates give us at least a hint about where the true parameters could be located. We have chosen a small discrete grid around the parameters estimated from the data and calculated the maximal deviations. The best fit was obtained for $a = 30.9$, $b = 6500$ and $c = 1.005$. We perform the test for these values.

In Fig. 4 the plot of the empirical distribution function $\hat{F}_n(\cdot)$ defined by (6) is compared with the hypothesised distribution function $F_0(\cdot)$, where the curves are plotted in log-scale.

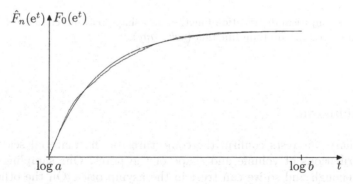

Fig. 4. The empirical distribution function of volume (*dashed line*) and fitted truncated Pareto distribution function (*solid line*)

The maximal deviation of $\hat{F}_n(\cdot)$ from $F_0(\cdot)$ is $\sup_{t \in \mathbb{R}} |\hat{F}_n(t) - F_0(t)| = 0.025$. This means that $\sup_{t \in \mathbb{R}} |Y_n(t)| = 1.141$ is not greater than 1.358 ($= 95\%$-quantile of the Kolmogorov distribution function) and so the null hypothesis is not rejected at the 5%-level.

We have also computed the maximal deviation of $\tilde{F}_n(\cdot)$ from $F_0(\cdot)$ which is slightly higher than the value for $\hat{F}_n(\cdot)$: $\sup_{t\in\mathbb{R}} |\tilde{F}_n(t) - F_0(t)| = 0.030$.

As a second example we choose the shape parameter $f(\Xi_0)$ being equal to the natural logarithm of the ratio of the maximal diameter and the minimal diameter of the typical grain. Suggested by material scientists we check the null hypothesis whether the corresponding shape parameter of the pores in our specimen is Weibull distributed with parameters $\alpha = 1.04$ and $\beta = 0.31$

$$F_0(t) = 1 - e^{-(t/\beta)^\alpha}, \quad t \geq 0.$$

The functions $\hat{F}_n(t)$ and $F_0(t)$ are compared in Fig. 5, where again the log-scale is used. The maximal deviation is $\sup_{t\in\mathbb{R}} |\hat{F}_n(t) - F_0(t)| = 0.020$ implying that $\sup_{t\in\mathbb{R}} |Y_n(t)| = 0.934$. Hence, the null hypothesis is again not rejected at the 5%-level.

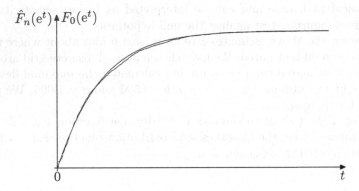

Fig. 5. The empirical distribution function of a shape parameter (*dashed line*) and the hypothesised distribution function (*solid line*)

5 Conclusion

In summary our tests confirm the conjectures of the material scientists on the distributions of volume and shape of the pores. Our sampling window is large enough and so we can trust in the asymptotics. On the other hand, since the data are based on sections, the knowledge of volume and shape of individual pores is not precise. We should also be aware that Theorem 1 and our goodness-of-fit test rely essentially on the independence assumption (A2). There are grounds for the assumption that more or less weak dependencies between neighbouring pores exist. Mathematically spoken, the system of pores modelled by (2) is driven by a stationary marked point process (1) involving dependencies between different grains as well as between grains and germs. Under certain mixing and regularity conditions an analogue to Theorem 1 seems provable but the covariance structure of the corresponding limit process would be more complicated than that of $Y(\cdot)$. A detailed study of such weakly dependent structures with applications to testing the goodness-of-fit of certain marginal distributions of the typical grain should be a meaningful subject of future research.

Acknowledgements

This research was supported by grant IAA 1075201 "Asymptotics of spatial random processes" of the Grant Agency of the Academy of Sciences of the Czech Republic and by grant MSM 0021620839 "Methods of Modern Mathematics and Their Applications" of the Czech Ministry of Education. The authors are grateful to Pavel Ctibor (Institute of Plasma Physics, Academy of Sciences of the Czech Republic) for providing the data and Radka Jůzková for technical assistance.

References

[1] P. Billingsley. *Convergence of Probability Measures*. 2nd Edition, Wiley & Sons, New York, 1999.

[2] D.J. Daley and D. Vere-Jones. *An Introduction to the Theory of Point Processes*. Springer-Verlag, New York, 1988.

[3] L. Heinrich and Z. Pawlas. Weak and strong convergence of empirical distribution functions in germ-grain models. Submitted, 2004.

[4] B.G. Ivanoff. The function space $D([0, \infty)^q, E)$. *Canadian Journal of Statistics*, 8:173–186, 1980.

[5] R. Júzková, P. Ctibor and V. Beneš. Analysis of porous structure in plasma-sprayed coating. *Image Analysis and Stereology*, 23:45–52, 2004.

[6] G. Neuhaus. On the weak convergence of stochastic processes with multidimensional time parameter. *Annals of Mathematical Statistics*, 42:1285–1295, 1971.

[7] R. Schneider and W. Weil. *Stochastische Geometrie*. B.G. Teubner, Stuttgart, Leipzig, 2000.

[8] D. Stoyan, W.S. Kendall and J. Mecke. *Stochastic Geometry and its Applications*. 2nd Edition, Wiley & Sons, Chichester, 1995.

Principal Component Analysis for Spatial Point Processes – Assessing the Appropriateness of the Approach in an Ecological Context

Janine Illian, Erica Benson, John Crawford and Harry Staines

SIMBIOS, University of Abertay Dundee, Scotland, UK,
j.illian@abertay.ac.uk

Summary. There is a need to characterise spatial point patterns of ecological plant communities in which a very large number of points exist for many different plant species. We further investigate principal component analysis for spatial point patterns using functional data analysis tools on second-order summary statistics as introduced in [10, 11]. The approach is used to detect different types of point patterns in a multi-type pattern to classify the species by their spatial arrangement. The developed method is evaluated in a detailed feasibility study, giving rise to a number of recommendations including the choice of the appropriate summary statistic. In addition, we investigate the performance of the method under noisy conditions simulating a number of settings typically occurring in an ecological context. Overall, the method produces stable results, the best results being achieved when the pair-correlation function is used. In all settings the level of noise needs to be very high to invalidate the results.

Key words: Ecological plant communities, Functional data analysis, Functional principal component analysis, Multi-type spatial point patterns

1 Introduction

The study of patterns in ecosystem diversity and functioning is driven by the need to understand the processes that organise ecological communities. The link between ecosystem function, observed at the level of the community, and ecosystem diversity, manifest at the scale of the individual plant, has long been a subject of debate in ecology, and arises as a consequence of the spatial mixing of individuals over time. Understanding the mechanisms that promote and sustain biodiversity and allow a large number of species to coexist is therefore a key interest within community ecology [14, 16]. Species coexistence in turn is directly linked to local inter- and intraspecific competition structures in a community [6], and spatial processes play a fundamental role within ecology and particularly in plant ecology, since plants are non-motile organisms. Each plant has a dependence on local growing conditions in its respective location [20]. Plants interact mainly with their immediate neighbours so an approach that models the exact location of a plant will support the elicitation of the underlying processes which have caused the observed pattern and thus yield an understanding of community dynamics and consequently the mechanisms involved in the functioning of ecosystems [8, 9]. Spatial point processes would appear to be an ideal tool to investigate the interaction structure between individual plants, since they are statistical models describing the exact locations of objects in space (see, e.g. [4, 5, 15]). Model parameters reflect overall properties of the observed pattern that are of ecological interest, for instance strength and direction of interaction among individuals [13].

However, the pattern of a typical plant community, may be very complex, due to the large number of individuals and species and the resulting extremely large number of potential inter- and intra-species interactions. [11] and [10] introduce a methodology that may be used to reduce the dimensionality of a spatial point pattern dataset. Functional data analysis tools [17, 18] are applied to the second-order statistics of multi-type point processes, in particular to L-functions and pair-correlation functions, to derive a PCA method for spatial point pattern data.

This paper assesses in detail the feasibility of the approach through a simulation study in order to make recommendations as to which second-order summary statistic is the most appropriate. The recommendations are derived from simulated spatial point patterns with different characteristics, including random, regular and clustered patterns.

An important consideration with respect to ecological data is that recorded data are susceptible to a certain amount of noise, unlike simulated patterns. This paper therefore also investigates the performance of the method under noisy conditions simulating a number of settings typically occurring in ecological data set construction. This is done in order to provide information to the applied researcher as to when the method fails and how detailed the data recording has to be for the method to classify patterns properly.

Section 2 briefly summarizes the approach introduced in [11] and [10]. In Sect. 3 we describe an extended simulation study used to assess the performance of the method under the assumption of homogeneity. Section 4 examines the performance of the methods with added noise typical of data collection in the context of ecological studies. Section 5 briefly describes the results from an application of the method to a data set.

2 Methods

2.1 Functional Data Analysis

For a detailed introduction to functional data analysis see [17] and [18]. In functional data analysis observations are functions and these are interpreted as single entities rather than as consecutive measurements. Generally speaking, the record of a functional observation x consists of n pairs (t_j, y_j), where y_j is an observation of $x(t_j)$ at time t_j. Since the functions are usually observed at only a finite number of values of t, interpolation or smoothing techniques have to be applied to yield a functional representation of the data.

For functional PCA (see [12, 17, 18]) consider function values $x_i(t)$ and define $f_i(w) = \int w(t)x_i(t)\, dt$, where $w(t)$ is a weight function. Maximise $N^{-1}\sum_i^N f_i^2(w)$ under the constraint $\|w\|^2 = \int w(t)^2\, dt = 1$ and get an eigenequation

$$\int v(t, s)w(s)\, ds = \lambda w(t) \tag{1}$$

with variance covariance function $v(t, s) = N^{-1}\sum_{i=1}^N x_i(t)x_i(s)$. The solution to this eigenequation with the largest eigenvalue solves the maximisation problem and will be denoted by w_1 with corresponding scores $f_{i1} = f_i(w_1)$. The second largest eigenvalue with eigenfunction w_2 yields the second principal component with scores f_{i2} etc. Further analysis will mainly examine the scores f_{ik} for each of the curves on the first p principal components, where typically $p \ll N$.

2.2 Functional Principal Component Analysis of Second-Order Summary Statistics

Let Z be a spatial point process on $\mathbb{R}^2$. Let X be a multitype point process $X = \{(\zeta, m_\zeta) : \zeta \in Z\}$ with $m_\zeta \in \mathcal{M}$ and $\mathcal{M} = \{1, \ldots, k\}$ a set, where no other marks are available, and subprocesses $X_i \subset X$ with $X_i = \{(\zeta, m_\zeta) : \zeta \in Z \text{ and } m_\zeta = i\}$ and $i = 1, \ldots, k$. Consider a realisation x of X and use second-order summary statistics to characterise the spatial behaviour of the individual subpatterns x_i. Apply a functional principal component analysis to the smoothed L-functions or pair-correlation functions to group the point patterns by their spatial behaviour.

We estimate the L-functions using the following estimator for the K-function [19]:

$$\hat{K} = n^{-2}|A|\sum_{\zeta \neq \xi}\sum w_{\zeta,\xi}^{-1}I_d(d_{\zeta,\xi}),$$

where n is the number of points in region $A \in \mathbb{R}$ with area $|A|$, $d_{\zeta,\xi}$ is the distance between point ζ and ξ and $w_{\zeta,\xi}$ is an edge correction factor – the proportion of the circle with centre ζ passing through ξ which lies in A. Clearly, $\hat{L}(r) = \sqrt{\hat{K}/\pi}$. Note that we plot $\hat{L}(r) - r$ when visualising results.

For the pair-correlation function, we apply spline smoothing to $Z(r) = \frac{\hat{K}(r)}{\pi r^2}$, constrain $Z(0) = 1$ and estimate its derivative.

We smooth the estimated second-order summary statistics using B-splines (see [7]), i.e. splines with compact support, as they are capable of picking up local features. We subsequently perform a functional PCA on the smoothed functions. Through this, the subprocesses may be grouped on the basis of their scores on the principal components. We use hierarchical cluster analysis on these scores, in particular Ward's methods [2], to detect clusters of similar second-order summary statistics and hence groups of point processes with similar spatial behaviour. The result of the cluster analysis is plotted in a dendrogram and, together with a plot of the first p principal components, will reveal groups of points processes with different spatial behaviour. In addition, the finer structure of the dendrogram will display similarities between individual patterns. Thus the dendrogram summarises the most distinctive features in the population as well as the position of the individual species within the structure.

See Fig. 1 for an example using simulated data of 20 clustered and 20 random patterns (for more details on the simulations see Sect. 3). Figure 1 a) shows the smoothed L-functions, 1 b) the first two principal components, 1 c) a plot of the scores of all patterns on the first principal components and d) 1 a dendrogram of the scores.

Also note, that $L(r)$-values for regular patterns tend to lie in $[0, r]$ whereas $L(r)$-values for clustered patterns are usually larger than r with no upper bound. Hence, if we want to distinguish between clustered, random and regular patterns, the difference between the L-function for a clustered pattern and a random pattern tends to be larger than the difference between the L-function for a regular pattern and a random pattern. In analogy to the approach taken in a standard PCA context when variables have been measured on different scales, we perform a FPCA on the correlation matrix rather than on the covariance matrix. I.e. equation (1) now becomes

$$\int v^*(t, s) w(s) \, \mathrm{d}s = \lambda w(t),$$

where v^* is the correlation function $v(t, s) = N^{-1} \sum_{i=1}^{N} x_i^*(t) x_i^*(s)$, i.e. the co-variance function of a standardised data matrix x^*. A similar situation occurs when the pair-correlation function is being used – regular and random pattern appear more similar than clustered and random patterns so the correlation matrix will be used instead. The simulation study in Sect. 3 has investigated this aspect and compares the performance of the two statistics in this context.

3 Feasibility Study

In a detailed simulation study we examine the capability of the approach to separate groups of simulated point patterns with different spatial behaviour.

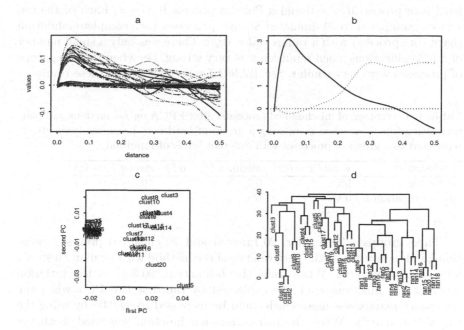

Fig. 1. Results from a functional principal component analysis on L-functions for 20 clustered and 20 random patterns, (**a**) smoothed L-functions; (**b**) first two principal components; (**c**) a plot of the scores of all patterns on the first principal components; (**d**) dendrogram of the scores

The aim is to validate the methodology in general and also to identify which second-order summary statistic should be used with real data.

In the simplest case, a set of 20 homogeneous Poisson cluster processes [5] were simulated on the unit square with a parent process of intensity $c = 10$ and daughter processes with radius $rad = 0.025$. In addition a set of 20 binomial process, i.e. Poisson processes with a fixed number of points, were generated. L-functions as well as pair-correlation functions were calculated for each of the patterns. These were smoothed with 10 cubic B-splines and then submitted to a functional principal component analysis followed by a cluster analysis of their scores in the first two principal components. The method was capable of distinguishing the two groups of patterns perfectly: there was no misclassification, either with L-functions or with pair-correlation functions. A similar result was achieved with two sets of 20 random and 20 regular (hard core) processes, respectively [4].

In order to assess the sensitivity of the method when the groups of pattern become increasingly similar, Strauss processes [21] with different levels of regularity were compared to hard core processes. The interaction parameter was chosen as $\gamma \in \{0.1, 0.2, 0.3, 0.4, 0.5, 0.6, 0.7, 0.8, 0.9, 1\}$, where we have a hard core process if $\gamma = 0$ and a Poisson process if $\gamma = 1$. Each of the ten sets was compared to 20 simulated Strauss processes with complete inhibition (hard core process) with a radius $rad = 0.05$. There was only a small number of misclassifications when repulsion was very strong, i.e. when the two groups of processes were very similar. See Table 1 for an overview of these results.

Table 1. Percentage of misclassified processes after FPCA on L-functions and pair-correlation functions when comparing a group of hard core processes (very strong repulsion) to a group of processes with different levels of repulsion

repulsion	weak ($\gamma = 0.9$)	medium ($\gamma = 0.5$)	strong ($\gamma = 0.1$)
L-function	0%	0%	10 %
pair-correlation f.	0%	0%	5%

Three groups – 20 regular, 20 random and 20 clustered process – were simulated to investigate the performance of the method with regular, clustered and random patterns. When using the L-function, 90.36% of the variation amongst the functions could be explained by the first two PC's when the covariance matrix was used which could be increased to 94.94% by using the correlation matrix. When the pair-correlation function was used, both the approach using the covariance matrix and the approach using the correlation matrix, could account for 99.5% of the variation. Figures 2 and 3 show that for both summary statistics, the three clusters were clearly identified by the method, and in both cases the regular patterns seemed more similar to the random ones, as predicted. The three clusters appear more distinct when the pair-correlation is used as a summary statistic.

4 Erroneous Data

The performance of the method when used with noisy data was assessed, since data used in ecological studies are susceptible to error, in particular when data collection is challenging, as for example in [1]. Indeed, identifying the exact location of a plant and distinguishing individuals may be a complex task in itself [20].

In order to investigate the robustness of the method, three different types of error common to ecological applications were considered:

a) The location was inaccurately recorded due to human error or technical problems.
b) The location was recorded on a grid. This grid is fine enough that the probability for any resulting cell to contain more than one data point is very small. Nevertheless, strictly speaking the recorded location does not reflect the exact location of the individual plants.

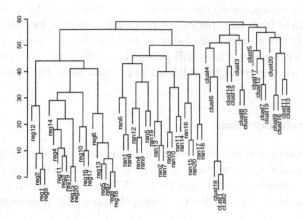

Fig. 2. Clustered scores on the first two principal components for 20 clustered, 20 regular and 20 random patterns using FPCA on *L*-functions

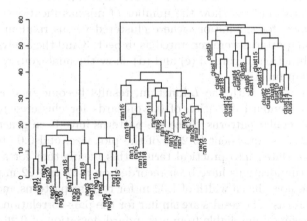

Fig. 3. Clustered scores on the first two principal components for 20 clustered, 20 regular and 20 random patterns using FPCA on pair-correlation functions

c) The wrong species was recorded, i.e. the marks were accidentally confounded.

The simulation study mimicked these three cases by generating point patterns of different spatial behaviour and subsequently increasing the noise in the data. For a) an increasing degree of random noise is deliberately added to the original locations; for b) the data are discretised; and for c) an increasing number of marks from one species are randomly replaced by marks of another species. In all three cases, the focus is to identify the degree of distortion that may be introduced whilst still preserving sufficiently accurate results, in order to provide the ecologists with an indication as to the degree of quality required.

4.1 Inaccurate Location

In order to mimic erroneously recorded locations, sets of clustered, random and regular point patterns were generated as described in Sect. 3. Subsequently, values from a normal distribution $N(0, \sigma^2)$ were generated and added to the original x- and y-coordinates. Here, the strength of noise is reflected in the size of σ^2.

Note that after this procedure the pattern generated from homogeneous Poisson processes will still show complete spatial randomness whereas the clustered and regular processes will become increasingly similar to the Poisson processes. The simulation study was undertaken in order to reveal exactly what degree of inaccuracy would result in the procedure failing and thus to advise ecologists on the accuracy needed for data collection.

Figures 4 (a) and (b) show the number of misclassifications for different degrees of noise, for simulations where clustered versus random and regular versus random patterns were generated as in Sect. 3 and the analysis was done using the L-function. Figures 4 (c) and (d) show the analogous results for the pair-correlation function.

For the analyses with the L-function, results become increasingly unreliable from a standard deviation of 0.2 onwards for clustered patterns and from 0.03 for regular patterns. Since the patterns have been generated on the unit-square, 0.2 is equivalent to 20% of the plot size, and 0.03 to 3% of the plot size. Translated into practical terms, this means that for a data set as in [1] where the locations have been recorded on a 22 m ×22 m plot, the 2σ region for the noise has a width of 1.32 m for regular patterns and 8.80 m for clustered patterns. The results are similar for the pair-correlation with results getting increasingly unreliable from a standard deviation of 0.26 onwards for clustered patterns and from 0.05 for regular patterns.

4.2 Data Collected on a Grid

Spatial point process models assume that the location of objects has been recorded on a continuous scale. In practice, however, this is not the case since data is typically sampled from a discretised space due to real world constraints. Ecologists typically record a plant's location on a grid, as in for example [1]. This grid has a very fine resolution such that it is very unlikely for any two points to appear in the same grid cell but this is not impossible. Here, we

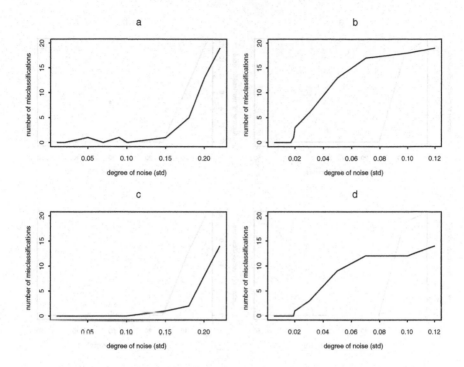

Fig. 4. Number of misclassifications as a function of the strength of noise. (**a**) clustered patterns, L-function; (**b**) regular patterns, L-function; (**c**) clustered patterns, pair-correlation function; (**d**) regular patterns, pair-correlation function

generate patterns as in 3 and then modify the location to an increasingly coarse grid by rounding the coordinates.

Figure 5 shows the number of misclassifications resulting from an increasingly coarse grid. Figures 5 (a) and (b) show the results when the L-function was used for clustered and regular patterns, respectively. Figure 5 c and d show the analogous results when the pair-correlation function was used.

The results are very similar, for both regular and clustered patterns as well as for the two summary statistics. Misclassifications only occur when the grid becomes as coarse as consisting of $4 \times 4 = 16$ cells. Hence, the method only fails when the grid has been extremely coarse-grained, so coarse that this choice of grid would be highly impractical in application. In order to measure the exact location of 20 sets of approximately 100 points each, a grid of this coarseness would not be chosen. Also, note that, theoretically, spatial point processes are defined as simple processes where the probability of two points occurring at the same location is 0 [15]. However, the finer structure is lost leading to less detailed results and changing the relative differences between

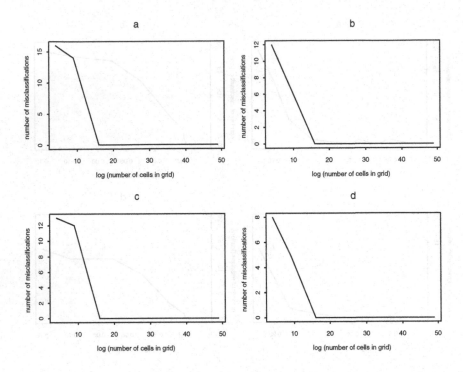

Fig. 5. Number of misclassifications as a function of the number of gridcells used. (a) clustered patterns, L-function; (b) regular patterns, L-function; (c) clustered patterns, pair-correlation function; (d) regular patterns, pair-correlation function

species. Thus, the relative position of the individual patterns in the hierarchy is obscured and changed.

4.3 Wrong Species Recorded

In practice, human error might lead to the data collector recording the wrong species. This can be the result of a number of mistakes, such as misrecorded species and other technical problems. Most of these problems are not systematic and will not be considered here. We consider a situation where individual plants from one specific species have been erroneously recorded as being another specific species with a different spatial behaviour. We assume it is unlikely that it is an arbitrary misclassification, but with one with similar phenotypic characteristics since this is the most likely error that can happen.

In order to mimic this situation, sets of regular and random as well as clustered and random patterns were simulated as described above (see 3). Subsequently, individuals from one of the clustered or regular patterns, respectively, were randomly labelled with the label of one of the random patterns with increasing probability. For two point patterns x and y let $d(x, y)$ be the Euclidean distance between the score vectors of the corresponding first principal components. For two sets of point patterns $\mathbf{x} = \{x_1, \ldots, x_n\}$ and $\mathbf{y} = \{y_1, \ldots, y_m\}$ we can now calculate the average of these distances for pattern x_i to the other patterns in its group relative to the average distance of pattern x_i to all other patterns, i.e. we use the relative distance defined by

$$rd(x_i, \mathbf{x}, \mathbf{y}) = \frac{(n-1)^{-1} \sum_{j=1}^{n} d(x_i, x_j)}{(n-1)^{-1} \sum_{j=1}^{n} d(x_i, x_j) + m^{-1} \sum_{k=1}^{m} d(x_i, y_k)}$$

and assess the performance of the method on rd, assuming that a pattern will be classified into the group that it is closer to. If the distance to the original group is larger than the distance to the wrong group, i.e. when $rd(x_i, \mathbf{x}, \mathbf{y}) \geq 0.5$, the modified pattern is more likely to be wrongly classified.

Figure 6 shows the results from 100 simulations for regular patterns with complete inhibition versus random patterns, using the L-function. $rd(x_i, \mathbf{x}, \mathbf{y})$ is plotted as a function of the probability of an individual point in a subpattern being re-labelled.

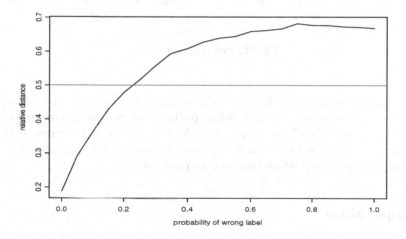

Fig. 6. $rd(x_i, \mathbf{x}, \mathbf{y})$ as a function of the probability of an individual point in a subpattern being re-labelled for regular with complete inhibition and random patterns using the L-function

Figure 7 shows a plot of how an individual subpattern slowly "moves" from it's own group of regular patterns into the other group of random patterns with increasing probability of an individual point in a subpattern being re-labelled (see circle).

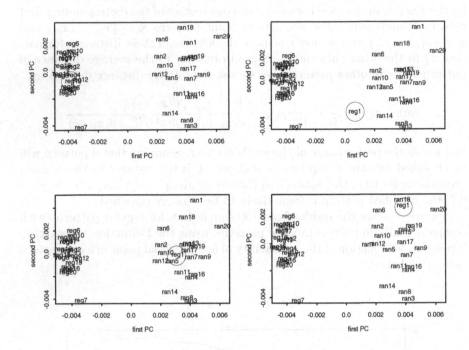

Fig. 7. Relative distance

From a probability of re-labelling of more than 0.24 onwards $rd(x_i, \mathbf{x}, \mathbf{y}) \geq$ 0.5 in the case of regular versus random patterns when L-functions are being used. It increases to 0.27 for the pair-correlation function in the same setting, and to 0.32 and 0.36 in the case clustered versus random patterns, for L-function and pair-correlation function, respectively.

5 Application

In [11] the methods investigated here are applied to a data set as described in [1]. This data set is a multi-type spatial point pattern formed by a natural plant community in the heathlands of Western Australia, consisting of the exact locations of 6378 plants from 67 species on a 22m by 22m plot. The data have been recorded on a fine grid; there is only one instance of two plants appearing in the same grid-cell. Since both the simulations described above revealed that the pair correlation function yields more reliable results in this context, these were estimated for all species with an abundance larger than 20. The functions were smoothed using 10 cubic b-splines and a FPCA was performed on these functions.

The first principal component represents clustering at close distances and the second clustering at larger distances (for details see [11]). Figure 8 shows the result of the cluster analysis of the scores on the first two principal components. Four different groups of patterns can be identified. The first group

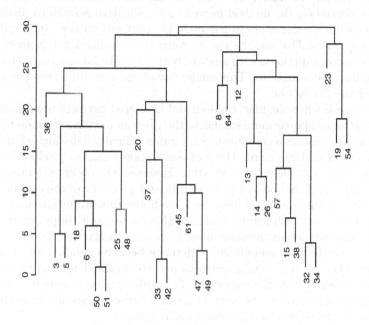

Fig. 8. Dendrogram for the Western Australian data set after cluster analysis (Ward's method) of the scores on the first two principal components

scores low on the first principal component. This indicates that the patterns in this group show slight repulsion at close distances i.e. they are patterns with a hard core. The second group scores low on both PCs, i.e. some repulsion to randomness at close and further distances. The third group scores high on the first PC so shows clustering at closer distances, i.e. no hard core. The fourth group scores even higher on the first PC so shows stronger clustering at closer distances and no hardcore again.

6 Discussion

We derive principal component analysis for spatial point patterns by applying functional principal component analysis to the second-order summary statistics of multi-type spatial point patterns. This yields a classification of the subpatterns into groups of similar spatial arrangement.

A feasibility study revealed that the method was capable of distinguishing clearly between both clustered versus random, and regular versus random patterns. Furthermore, the method proved to be sensitive enough to distinguish between rather similar types of patterns, i.e. patterns with a very similar degree of regularity. Overall, the results were very similar for L-functions and pair-correlation functions, with slightly better results for the pair-correlation function in some situations. This might have been a result of the cumulative nature of the L-function.

In a setting where regular, random and clustered patterns are present, the regular patterns appear more similar to the random patterns than do the clustered patterns, and as a consequence are more difficult to distinguish from the random patterns. Here again, the pair-correlation function produced slightly better results with a clearer classification. However, the observed similarity between regular and random arises as an inherent property of the classifications of the patterns themselves. Whilst there is a finite limit to the spacings and inhibition associated with regular patterns, there is no limit (in principle) to the spacings and attractions associated with clustered patterns. Therefore there may be a "tending to unbounded" difference between random and clustered patterns, whereas the regular patterns are constrained to complete inhibition. Consequently, when interpreting the results from a principal component analysis this will have to be kept in mind when comparing the strength of clustering with respect to the strength of regularity.

A detailed simulation study investigated the performance of the method in the presence of noise typical of ecological data. When random noise was added to the location of individuals, the method proved to be very stable. Only strong degrees of noise lead to serious misclassifications. Similarly, when data were discretised into a grid of increasing coarseness, only a very coarse grid prohibited the overall classification of the subpatterns into the largest groups. However, the finer similarity structure becomes lost as a result of less precise data. Finally, the probability of an individual species being classified into a group with different spatial behaviour was only high when the probability of the species to be misidentified was at least 0.24.

In all three cases, again, the pair-correlation function yields slightly better results. Overall, the results from the study enable us to inform the applied researcher about the degree of noise which will invalidate the analysis. Due to the ongoing technological development, larger numbers of similar data sets, as described in [3], will become available. Hence, there is much room for an extension of the approach, for example to incorporate interspecies interactions based on two-dimensional L-functions or pair-correlation functions, or to incorporate marks into the analysis using a mixed (i.e. functional and non-functional) principal component analysis.

References

[1] P. Armstrong. Species patterning in the heath vegetation of the Northern Sandplain. Honours thesis, University of Western Australia, 1991.

[2] B. Everitt, S. Landau and M. Leese. *Cluster Analysis*. Arnold, London, 2001.

[3] D.F.R.P. Burslem, N.C. Garwood and S.C. Thomas. Tropical forest diversity – the plot thickems. *Science*, 291:606–607, 2001.

[4] N.A.C. Cressie. *Statistics for Spatial Data*. New York: Wiley, 1991.

[5] P.J. Diggle. *Statistical Analysis of Spatial Point Patterns, 2nd ed*. Oxford University Press, 2003.

[6] R. Durrett and S. Levin. Spatial aspects of interspecific competition. *Theoretical Population Biology*, 53:30–43, 1998.

[7] P.J. Green and B.W. Silverman. *Nonparametric regression and generalized linear models: a roughness penalty approach*. London: Chapman and Hall, 1994.

[8] P. Greig Smith. *Quantitative Plant Ecology*. Oxford: Blackwell, 1983.

[9] T. Herben, H. During and R. Law. Spatio-temporal patterns in grassland communities. In Dieckmann et al., editor, *The Geometry of Ecological Interactions: Simplifying Spatial Complexity*, pages 11–27. Cambridge University Press, 2000.

[10] J.B. Illian, E. Benson, J. Crawford and H.J. Staines. Multivariate methods for spatial point process – a simulation study. In A.J. Baddeley et al., editor, *Spatial point process modelling and its applications*, pages 125–130. Castelló de la Plana: Publicacions de la Universitat Jaume I, 2004.

[11] J.B. Illian. Multivariate methods for spatial point processes with applications in plant community ecology. *(in preparation)*, 2005.

[12] I.T. Jolliffe. *Principal component analysis*. Springer, 2002.

[13] R. Law, T. Herben and U. Dieckmann. Non-manipulative estimates of competition coefficients in grassland communities. *Ecology*, 85:505–517, 1997.

[14] M. Loreau, S. Naeem, P. Inchausti, J. Bengtsson, J.P. Grime, A. Hector, D.U. Hooper, M.A. Huston, D. Raffaelli, B. Schmid, D. Tilman and D.A. Wardle. Biodiversity and ecosystem functioning: current knowledge and future challenges. *Science*, 294:804–808, 2001.

[15] J. Møller and R.P. Waagepetersen. *Statistical Inference and Simulation for Spatial Point Processes*. Chapman and Hall/CRC, 2003.

[16] D.J. Murrell, D.W. Purves and R. Law. Uniting pattern and process in plant ecology. *Trends in Ecology and Evolution*, 16:529–530, 2001.

[17] J.O. Ramsay and B.W. Silverman. *Functional data analysis*. Springer, 1997.

[18] J.O. Ramsay and B.W. Silverman. *Applied functional data analysis*. Springer, 2002.

[19] B.D. Ripley. The second-order analysis of stationary point processes. *Journal of Applied Probability*, 13:255–266, 1976.

[20] P. Stoll and J. Weiner. A neighbourhood view of interactions among individual plants. In Dieckmann et al., editor, *The Geometry of Ecological Interactions: Simplifying Spatial Complexity*, pages 11–27. Cambridge University Press, 2000.

[21] D. Strauss. A model for clustering. *Biometrika*, 63:467–475, 1975.

Practical Applications of Spatial Point
Processes

Part III

On Modelling of Refractory Castables
by Marked Gibbs and Gibbsian-like Processes

Felix Ballani

Institut für Stochastik, Technische Universität Bergakademie Freiberg,
Prüferstr. 9, 09596 Freiberg, Germany, ballani@math.tu-freiberg.de

Summary. The modelling of self-flowing refractory castables, a special kind of
concrete, is discussed. It consists of two phases: a system of randomly distributed
spherical hard grains and a cement matrix. The focus is on marked Gibbs and
Gibbsian-like processes but also some other models are discussed. It turns out that a
particular canonical marked Gibbsian-like process is useful for modelling the samples
and more plausible than the classical stationary marked Gibbs process.

Key words: Canonical Gibbsian-like process, Marked Gibbs point processes, Random-
shift model, Refractory castables

1 Introduction

Concrete is an important building material, widely used for example in dams,
highways and buildings. Its versatile use derives from the numerous types of
its composition. Concrete can often be considered to be a two-phase material,
consisting of a system of randomly distributed hard grains embedded in a
matrix, which is the binding system and constitutes the second phase.

Important subjects of engineering research for concrete are its mechanical
properties, such as cold pressure strength and fracture behaviour, and percolation properties, see e. g. [5]. These properties are closely related to the inner
geometrical microstructure. They are not merely determined by the volume
fraction of the particle phase. Therefore, statistical analysis of this structure
and its modelling are very important.

Useful models of concrete belong to the class of models of randomly hard
particles, and of hard spheres, respectively, if the grains are assumed to be
spherical as in the present contribution. Such models are also of interest in
many other fields of engineering and materials research, for example in the
context of sinter metals or granular matter. Most notably the models developed by engineers and physicists are very successful [5, 15, 21, 30], but they
are mathematically intractable. Even simulation is often not straightforward.

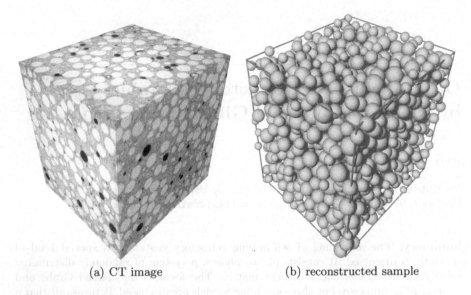

(a) CT image   (b) reconstructed sample

Fig. 1. Visualization of a sample of concrete. The black spheres in (a) are pores.

There are indeed simpler models based on Poisson processes like the Matérn model, the Stienen model, the lilypond model and the dead leaves model. But the small volume fraction of space occupied by the spheres makes these models unsuitable for a successful application in the present contribution.

As described in Sect. 2, the samples which are discussed here consist of spherical grains with varying sizes. Some of the corresponding classical models from spatial statistics and statistical physics make use of a proposal size distribution which differs from an observable resulting size distribution. This is discussed e.g. in [17] and [25] and leads in the case of modelling a sample to the problem of determining the right proposal distribution, see [28]. Therefore, after a discussion of the stationary marked Gibbs process in Sect. 3, a particular canonical Gibbsian-like process will be introduced in Sect. 4 which does without a proposal size distribution. For these and some other models it is investigated how far they can serve as models for the concrete samples.

2 The Data

The examined material is a special concrete, which ranks among the class of self-flowing refractory castables. In the following a sketch of data extraction is given. Afterwards some statistics is presented.

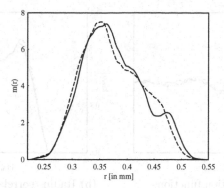

Fig. 2. Estimated probability density function of radii of the concrete samples: Sample A (—), sample B (- - -)

2.1 Production and Data Extraction

With the goal of varying the cold crushing strength under strong pressure, particles (called refractory aggregates) are added to a compound consisting of cement, water and superplasticiser. The three main steps of the production are dry-mixing, wet-mixing and firing at 1000°C. The result is a concrete which essentially consists of two phases, namely of grains and the matrix (the binding system). Usually the grains have an irregular shape, with some similarity to convex polyhedra. Especially for research purposes, castables with spherical grains consisting of corundum were manufactured in Freiberg, as described in [11]. Samples of size $10 \times 10 \times 10$ mm^3 were cut out of larger cylindrical samples of the castables and investigated by computerised tomography (CT). For the following treatment of the obtained three-dimensional images it was helpful to have spherical grains. The data were processed by methods of Bayesian image analysis (cf. e.g. [14],[29]). That means that configurations of non-overlapping ideal spheres which matched the CT images best were searched for. The applied optimization technique was simulated annealing as described in [29]. Existing air pores were ignored and added to the matrix phase. The result of this reconstruction, samples of the sphere centres $\mathbf{x}_i$ and radii r_i, are considered as data from a marked point process in which the radii are the *marks* associated with the *points* $\mathbf{x}_i$. Figure 1 shows one of the samples geometrically, both as a CT image and reconstructed as a sample of ideal spheres.

The following statistical analysis is based on the reconstructed point process data of two samples A and B and uses methods commonly applied to point processes and random sets (cf. [24]).

2.2 Statistics

The numbers of spheres are 1454 in sample A and 1862 in sample B, and the volume fractions are 0.368 and 0.423, respectively. Figure 2 shows the

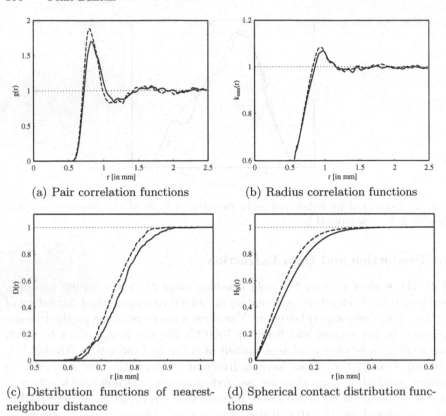

(a) Pair correlation functions

(b) Radius correlation functions

(c) Distribution functions of nearest-neighbour distance

(d) Spherical contact distribution functions

Fig. 3. Estimated characteristics of the concrete samples: Sample A (—), sample B (- - -)

empirical distributions of radii of the spheres for the two samples. The particles of both samples come from the same material and thus differ only statistically. The mean radius for sample A is 0.376 mm and 0.373 mm for sample B, both with standard deviation 0.055 mm.

Figure 3(a) shows the empirical pair correlation functions of the point patterns of sphere centres, obtained by the method described in [20] and [26]. Here, some form of short-range order is observable, which appears to extend to about 1.5 mm.

The mark correlation functions $k_{mm}(r)$ for the radius marks (see [26]) given in Fig. 3(b) show that there is some form of spatial correlation between the spheres. The curves indicate a tendency connected with the property of the spheres to be "hard": the price two spheres have to pay for being close together is to have both a radius which is smaller than the mean radius. This geometrical repulsion property is true for distances smaller than 0.8 mm.

Figure 3(c) shows the distribution function of the nearest-neighbour distance, a further characteristic of the point pattern of sphere centres.

Finally, the set-theoretic summary statistic "spherical contact distribution function" $H_s(r)$ [24] shown in Fig. 3(d) characterizes the structure in another way. It is the distribution function of the random distance from a randomly chosen point in the matrix to the closest point on any sphere surface.

The following sections introduce several models for random systems of hard spheres and discuss whether they fit the samples of concrete in a suitable way. The choice of this kind of models is in accordance with the prior information of non-overlapping ideal spheres used for the reconstruction in Sect. 2.1, and no further restriction of the investigated models is necessary.

3 The Stationary Marked Gibbs Proccss

The concept of marked Gibbs processes is based on ideas from statistical physics. The special case of identical hard spheres is a frequently used model investigated in many papers and books, see [15] and the references therein. The case of spheres with random radii is discussed more rarely [17, 21], but for this case there are also papers which study statistical problems, both for the canonical and grand canonical (even stationary) case (see [7, 28]). Analogous problems in the planar case are studied in [1, 9, 10, 12, 13] and [22].

Firstly, the stationary case of a Gibbs process of polydisperse hard spheres is investigated. For exact definitions the reader is referred to [17, 21] and [26]. Heuristically, one can think of a marked Gibbs process as an independently marked Poisson process under the condition that the spheres do not intersect, where the radii are the marks.

The stationary marked Poisson process is completely characterised by an intensity λ and a density function $m(r)$ of the radius distribution, which could be called "proposal" radius distribution (cf. [25]). The radii in accepted samples tend to be smaller than the radii corresponding to the mark distribution of the initial Poisson process. This can best be explained with a simulation method for the stationary marked Gibbs process, namely the Metropolis-Hastings algorithm, see [20]. If one tries to add a new sphere with radius according to the proposal radius distribution to the current configuration, it is more likely that a smaller sphere is accepted because of the non-overlapping condition. Therefore, there is a further radius distribution, the radius distribution of the Gibbs process, which is called "resulting" radius distribution, and its density function is denoted by $m^*(r)$. λ and $m(r)$ are quantities which cannot be observed directly in samples if one supposes that they are Gibbsian, whereas $m^*(r)$ can be observed. It is known that proposal and resulting quantities are related by

$$\lambda^* m^*(r) = \lambda m(r)(1 - V_V)(1 - H_s(r)), \tag{1}$$

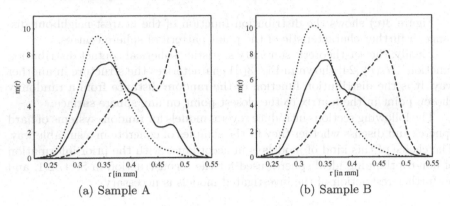

(a) Sample A   (b) Sample B

Fig. 4. Radius probability density functions: empirical distribution (—), proposal distribution (- - -), resulting distribution (· · ·)

as shown in [17], where λ^* is the intensity of the Gibbs process, $H_s(r)$ is the spherical contact distribution of the random set resulting from the union of all spheres of the Gibbs process, and V_V is the corresponding volume fraction.

For fitting the marked Gibbs process to the concrete samples, relation (1) can be applied for estimating the proposal intensity and radius distribution: λ^*, the resulting radius distribution m^* and the spherical contact distribution function H_s can be estimated from the samples, and λm can then be determined. Figure 4 shows the estimated density function of the proposal distribution using relation (1) and the corresponding resulting distribution, which was obtained by simulating the Gibbs process via the Metropolis–Hastings algorithm. The resulting density function (· · ·) widely differs from the empirical radius probability density function (—); when the model is correct, both functions should be close together. For testing the statistical procedure, the same method was used again, but now starting from the resulting distribution shown in Fig. 4 and the corresponding spherical contact distribution. In this case the method worked well: the proposal distribution was reproduced exactly, because one is now working in the world of Gibbs distributions. This suggests to conclude that the model of a stationary marked Gibbs process is not appropriate for the concrete data. This statement should be independent of the choice of the used particular simulation method.

A similar result is obtained if one tries to construct a proposal intensity and radius distribution which definitely reproduces the right resulting radius distribution and volume fraction. To achieve this aim, in [28] a statistical method is described in another context. This method is complicated and laborious. For this reason the proposal quantities are determined instead via the so-called Percus-Yevick approximation [15]. It is a well-established tool in statistical physics for solving problems, first investigated for the stationary Gibbs process of monodisperse hard spheres. This approximation can also be derived for the case of polydisperse hard spheres, see e.g. [28], and leads to the following relation between $\lambda^* m^*(r)$ and $\lambda m(r)$. (A comparison with (1) shows that the $\exp(\cdot)$-term is in fact the Percus-Yevick approximation of $1 - H_s(r)$.)

$$\lambda^* m^*(r) = \lambda m(r)(1 - \zeta_3) \exp\left(-\frac{8r^3\zeta_0 + 12r^2\zeta_1 + 6r\zeta_2}{1 - \zeta_3}\right. \\ \left. -\frac{24r^3\zeta_1\zeta_2 + 18r^2\zeta_2^2}{(1 - \zeta_3)^2} - \frac{24\zeta_2^3}{(1 - \zeta_3)^3}\right), \tag{2}$$

where $\zeta_k = \lambda^* 2^k \pi R_k^*/6$, and R_k^* is the kth moment of $m^*(r)$. Hence, for the estimation of $\lambda m(r)$ one only needs to know λ^* and $m^*(r)$. Applying this to the concrete data, the Metropolis–Hastings algorithm does reproduce the resulting radius distribution and the volume fraction but now the spherical contact distribution differs considerably.

Besides the question whether this model is appropriate for the concrete samples and the difficulties of determining the right proposal quantities a further aspect should be mentioned. In the beginning of the production process of the castables, the size distribution of the used corundum balls is fixed once and for all since after that no particle is sorted out. Therefore, the observed radius distribution is the only one, and there is no interpretation for the "proposal" radius distribution of the marked Gibbs process. Sections 4 and 5 discuss models which try to overcome this problem.

4 The Random-Shift Model

In the following a modified process is introduced which has some similarities to marked Gibbs processes. It is called *random-shift model*. The aim is to avoid the use of a proposal size distribution as is needed for the model in Sect. 3. This constraint to one size distribution seems to be feasible only as a finite process of hard spheres but not as a stationary one.

4.1 Definition

Let the system of hard spheres live in a (bounded) cuboid region $W = [0, a_1] \times [0, a_2] \times [0, a_3]$ where periodic boundary conditions are assumed. Let the number of spheres be n and the radii of these n spheres be fixed.

A configuration of spheres is denoted by $(\mathbf{x}_1, r_1), \ldots, (\mathbf{x}_n, r_n)$, where the $\mathbf{x}_i$ are their centres and the r_i their radii, ordered in decreasing size. Of course, the $\mathbf{x}_i$ are elements of W. The energy of such a configuration is

$$E_{r_1,\ldots,r_n}(\mathbf{x}_1, \ldots, \mathbf{x}_n) = \sum_{i<j} \phi_{ij}(\mathbf{x}_i, \mathbf{x}_j),$$

where the radii-dependent hard-sphere pair potential ϕ_{ij} is given by

$$\phi_{ij}(\mathbf{x}_i, \mathbf{x}_j) = \begin{cases} \infty & \text{if } \|\mathbf{x}_i - \mathbf{x}_j\| \le r_i + r_j, \\ 0 & \text{otherwise}, \end{cases}$$

and $\| \cdot \|$ is the distance due to the periodic boundary conditions, given by

$$\|\mathbf{x}\| = \sqrt{(\min(x_1, a_1 - x_1))^2 + (\min(x_2, a_2 - x_2))^2 + (\min(x_3, a_3 - x_3))^2}$$

for $\mathbf{x} = (x_1, x_2, x_3) \in W$. Any configuration with vanishing energy consists of nonoverlapping spheres and is therefore called *admissible*. Only the centres of the spheres are variable; the radii are fixed. Hence, admissibility depends on the $\mathbf{x}_i$.

A distribution similar to a Gibbs distribution can be constructed as follows. For Borel subsets $B_1, \ldots, B_n$ of W set

$$\Pi^*_{r_1, \ldots, r_n}(B_1 \times \cdots \times B_n)$$
$$= \frac{1}{Z_{r_1, \ldots, r_n}} \int_{B_1} \cdots \int_{B_n} \exp\left(-E_{r_1, \ldots, r_n}(\mathbf{x}_1, \ldots, \mathbf{x}_n)\right) \, \mathrm{d}\mathbf{x}_1 \ldots \mathrm{d}\mathbf{x}_n$$

where

$$Z_{r_1, \ldots, r_n} = \int_W \cdots \int_W \exp\left(-E_{r_1, \ldots, r_n}(\mathbf{x}_1, \ldots, \mathbf{x}_n)\right) \, \mathrm{d}\mathbf{x}_1 \ldots \mathrm{d}\mathbf{x}_n$$

is the radii-dependent partition function. The distribution $\Pi^*_{r_1, \ldots, r_n}$ is not symmetric. A permutation of the sets B_i yields different values because the $\{r_i\}$ are ordered and in general not pairwise equal. In order to obtain a distribution not distinguishing points, $\Pi^*_{r_1, \ldots, r_n}$ is symmetrised by permutation as follows:

$$\Pi_{r_1, \ldots, r_n}(B_1 \times \cdots \times B_n)$$
$$= \frac{1}{n! Z_{r_1, \ldots, r_n}} \sum_{\substack{(i_1, \ldots, i_n) \\ \in \mathrm{Perm}(n)}} \int_{B_{i_1}} \cdots \int_{B_{i_n}} \exp\left(-E_{r_1, \ldots, r_n}(\mathbf{x}_1, \ldots, \mathbf{x}_n)\right) \, \mathrm{d}\mathbf{x}_1 \ldots \mathrm{d}\mathbf{x}_n.$$

The integrand $\exp(\cdot)$ is in fact $\{0, 1\}$-valued. $\Pi_{r_1, \ldots, r_n}$ has therefore a density with respect to a binomial point process with n i.i.d. points in W with values $k/(n! Z_{r_1, \ldots, r_n})$, $k \in \{0, 1, \ldots, n!\}$. Because of the sum in $\Pi_{r_1, \ldots, r_n}$ the random-shift model is Gibbsian-like but not of pure Gibbsian type.

4.2 Simulation

Like many other models of spatial statistics and statistical physics the random-shift model is mathematically intractable. Therefore simulation is the main tool for its investigation. To simulate the random-shift model various methods are possible. The method which is recommended and used for this contribution

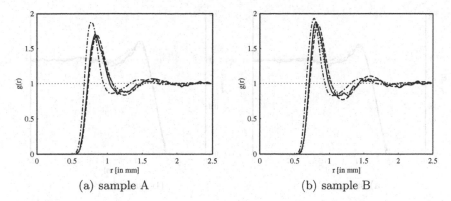

(a) sample A   (b) sample B

Fig. 5. Pair correlation function of the concrete sample (—), of the random-shift model (− −), of the ordered RSA model (· · ·) and of the dense random packing (− · −)

is practically the same as the original method of [19] and explains the name *random-shift model*.

One starts with some configuration of non-overlapping spheres with the prescribed radii. This can be obtained by simulations as e. g. described in Sect. 5, or one simply takes the respective concrete sample itself. Then the spheres are shifted randomly, avoiding overlappings. This method works well for the volume fractions observed in the concrete samples. For higher volume fractions more sophisticated simulation methods are necessary, e. g. molecular dynamics [27] and simulated tempering [6, 18].

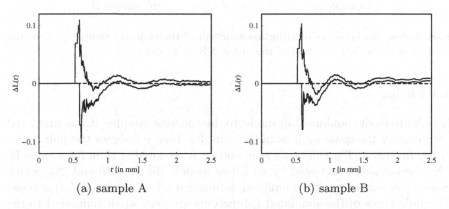

(a) sample A   (b) sample B

Fig. 6. 99% envelopes of the L-function (deviations from the L-function of the concrete sample) for the random-shift model (—)

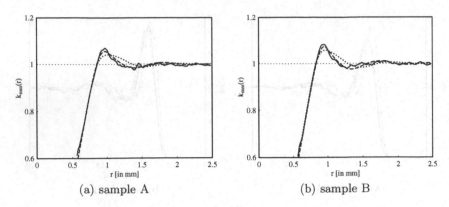

(a) sample A   (b) sample B

Fig. 7. Mark correlation function of the concrete sample (—), of the random-shift model (− −) and of the ordered RSA model (· · ·)

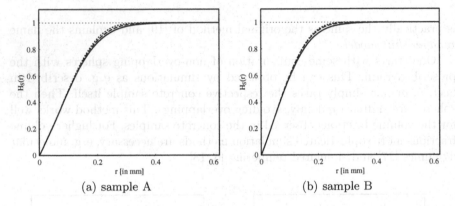

(a) sample A   (b) sample B

Fig. 8. Spherical contact distribution function of the concrete sample (—), of the random-shift model (− −) and of the ordered RSA model (· · ·)

4.3 Fitting

In order to fit the random-shift model to the concrete samples, it was simulated with exactly the same radii as in the samples. Figure 5 shows the pair correlation functions. The differences for sample A are smaller than for sample B. The goodness-of-fit is tested by an L-test as described in [20] and [26], which consists in comparing the empirical L-function with simulated L-functions. The differences of the simulated L-functions are very small compared to the values of $L(r) - r$ for r less than the smallest possible sphere centre distance because $L(r) = 0$ in that range. Therefore, instead of plotting $L(r) - r$ as proposed in [20], Fig. 6 shows the difference between the empirical L-function and the 99% envelopes calculated from 999 simulations of the random-shift model. Hence, the r-axis corresponds to the empirical L-function. This formal

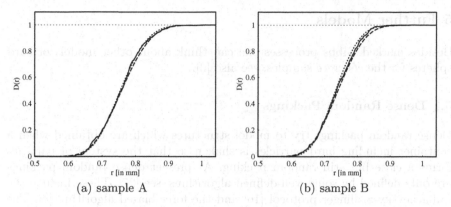

Fig. 9. Distribution functions of nearest-neighbour distance of the concrete sample
(—), of the random-shift model (− −) and of the ordered RSA model (· · ·)

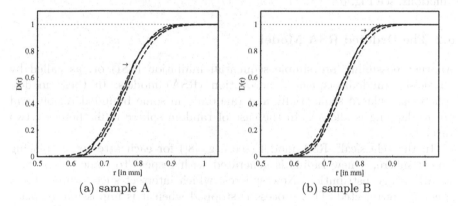

Fig. 10. 99% envelopes of the D-function for the random-shift model (—)

test suggests rejection of the random-shift model, for sample B clearer than
for sample A.

A similar behaviour can be observed with the distribution functions of the
nearest-neighbour distance. The curves in Fig. 9 are close together but the
respective test with the 99% envelopes for the random-shift model suggests
again rejection.

A further check shows the radius correlation functions in Fig. 7 and the
spherical contact distribution functions in Fig. 8. For both characteristics it
seems that the random-shift model is at least better than the ordered RSA
model (cf. Sect. 5).

5 Further Models

Besides marked Gibbs processes one can think about other models of hard spheres for the concrete samples, see also [3].

5.1 Dense Random Packings

Dense random packings try to model structures which are obtained when a container including hard particles is shaken so that the system of particles forms a close but still random packing. At present dense random packings are only defined by some well-defined algorithms, see [27]. These include the Lubachovsky-Stillinger protocol [16] and the force-biased algorithm [4]. The latter was applied to the samples of concrete and seems not to be a suitable model because of considerable differences of the respective pair correlation functions, see Fig. 5.

5.2 The Ordered RSA Model

Another possibility are simple sequential inhibition (SSI) or, as called by physicists, random sequential adsorption (RSA) models. In these models spheres are placed sequentially and randomly in some bounded region, and no overlapping is allowed. In the case of random sphere radii there are two variants.

In the "classical" RSA model (see e.g. [8]) for each attempt at placing a new sphere, a new radius is generated with respect to some given (proposal) radius distribution. New spheres, which intersect successfully placed spheres, are rejected. This process is stopped when it is impossible to place any new sphere. During this placement procedure there is the tendency to accept smaller spheres more often than larger spheres because the available space reduces. Therefore, in this variant of the RSA model one has also to discriminate between a proposal and a resulting radius distribution as in the stationary marked Gibbs process in Sect. 3; for a discussion see [25]. But the actual problem is the maximum reachable volume fraction of spheres, which is too low for an application in the case of the concrete samples.

This problem can be overcome by a variation of the RSA model in the sense of [5]: A fixed number of spheres is placed where all radii are generated before the placing procedure. In order to assure that the larger spheres can be placed, the spheres are placed from largest to smallest. If a placement trial is not successful, then it is repeated with the same radius but with a new random centre. Therefore, any radius generated at the beginning is kept. Only those attempts are accepted where all spheres can be placed. This variation leads to higher possible volume fractions of spheres and also seems to be a passable model for the samples of concrete, see Figs. 5, 7, 8 and 9.

6 Conclusions

As already set out in Sect. 3, the well-known model of a stationary marked Gibbs process does not seem to be appropriate for the concrete samples. Instead the newly developed model of Sect. 4, the random-shift model, is better, though not entirely satisfactory, because it is formally rejected by some tests. The extent of the differences seems to depend on the volume fraction; for sample A the fit by the random-shift model is better than for sample B. One reason for these differences may be indeed a different compound of spheres. This is e. g. indicated by a smallish increase of the first maximum of the pair correlation functions of the samples in comparison to the random-shift model (see Fig. 5) which may reflect that in the samples spheres may sometimes be relatively close together. Besides that, rejection can also be a statistical problem because the sample size of order 1000 is quite large compared with the sample size of order 100 that until now have usually been used in point process statistics. This large sample size leads to very small fluctuations in the estimated characteristics of the random-shift model.

Compared to other possible models like the dense random packing model and the ordered RSA model, the random-shift model seems to be better not only because of the comparison of several statistics (Figs. 5–10). Firstly, the author is convinced that the grains in the refractory castables are wrapped up by a thin cement film which therefore are not in direct contact. This is well-reflected by the random-shift model. Furthermore, the random-shift model might be arguably better because the simulation method described in Sect. 4 resembles the process of producing concrete that includes such operations as mixing, stirring and grains pushing away other grains. Nevertheless, because of the pair correlation functions (Fig. 5) one may think about a model intermediate between the random-shift model and the random packing model; even for higher volume fractions of spheres the pure packing model may be preferable.

Acknowledgements

The author would like to thank D.J. Daley and D. Stoyan for their cooperation and inspiring discussions, J. Hubálková for her work with the concrete samples, and I. Kadashevich for performing the dense packing simulations. He wants to express his gratitude to J. Ohser for his help with the CT analysis of the samples and the permit of using the CT data.

References

[1] A.J. Baddeley and J. Møller. Nearest-neighbour Markov point processes and random sets. *International Statistical Review*, 57:89–121, 1989.

[2] A.J. Baddeley and R. Turner. Practical maximum pseudolikelihood for spatial point patterns. *Australian and New Zealand Journal of Statistics*, 42:283–322, 2000.

[3] F. Ballani, D.J. Daley and D. Stoyan. Modelling the microstructure of concrete with spherical grains. To appear in *Computational Materials Science*, 2005.

[4] M. Bargieł and J. Moscinski. C language program for irregular packing of hard spheres. *Computer Physics Communications*, 64:183–192, 1991.

[5] D.P. Bentz, E.J. Garboczi and K.A. Snyder. A hard core/soft shell microstructural model for studying percolation and transport in three-dimensional composite media. Building and Fire Research Laboratory, National Institute of Standards and Technology, Gaithersburg, Maryland, http://ciks.cbt.nist.gov/~garbocz/hcss/useguide.html, 1999.

[6] G. Döge, K.R. Mecke, J. Møller, D. Stoyan and R.P. Waagepetersen. Grand canonical simulations of hard-disk systems by simulated tempering. *International Journal of Modern Physics C*, 15:129–147, 2004.

[7] G. Döge and D. Stoyan. Statistics for non-sparse homogeneous Gibbs point processes. In: K.R. Mecke, D. Stoyan (eds) *Morphology of Condensed Matter. Physics and Geometry of Spatially Complex Systems.* Lecture Notes in Physics, No. 600, 418–427, Springer, Berlin Heidelberg, 2002.

[8] J.W. Evans. Random and cooperative sequential adsorption. *Review of Modern Physics*, 65:1281–1304, 1993.

[9] A. Frey and V. Schmidt. Marked point processes in the plane I – a survey with applications to spatial modelling of communication networks. *Adv. Perf. Anal.*, 1:65–110, 1998.

[10] M. Goulard, A. Särkkä and P. Grabarnik. Parameter estimation for marked Gibbs processes through the maximum pseudolikeliood method. *Scandinavian Journal of Statistics*, 23:365–379, 1996.

[11] J. Hubálková and D. Stoyan. On a quantitative relationship between degree of inhomogeneity and cold crushing strength of refractory castables. *Cement Concrete Research*, 33:747–753, 2003.

[12] J.L. Jensen and J. Møller. Pseudolikelihood for exponential family models of spatial point processes. *Annals of Applied Probability*, 1:445–461, 1991.

[13] T. Kokkila, A. Mäkelä and E. Nikinmaa. A method for generating stand structures using Gibbs marked point process. *Silva Fennica*, 36:265–277, 2002.

[14] M.N.M. van Lieshout. *Stochastic Geometry Models in Image Analysis and Spatial Statistics.* CWI Tract, 108, Amsterdam, 1995.

[15] H. Löwen. Fun with hard spheres. In: K.R. Mecke, D. Stoyan, (eds) *Statistical Physics and Spatial Statistics. The Art of Analyzing Spatial Structures and Pattern Formation.* Lecture Notes in Physics, No. 554, 295–331, Springer, Berlin Heidelberg, 2000.

[16] B.D. Lubachevsky and F.H. Stillinger. Geometric properties of random disk packings. *Journal of Statistical Physics*, 60:561–583, 1990.

[17] S. Mase. On the possible form of size distributions for Gibbsian processes of mutually non-intersecting discs. *Journal of Applied Probability*, 23:646–659, 1986.

[18] S. Mase, J. Møller, D. Stoyan, R.P. Waagepetersen and G. Döge. Packing densities and simulated tempering for hard-core Gibbs point processes. *Annals of the Institute of Statistical Mathematics*, 53:661–680, 1999.

[19] N. Metropolis, A.W. Rosenbluth, M.N. Rosenbluth, H. Teller and E. Teller. Equation of state calculations by fast computing machines. *Journal of Chemical Physics*, 21:1087–1092, 1953.

[20] J. Møller and R.P. Waagepetersen. *Statistical Inference and Simulation for Spatial Point Processes*. Chapman and Hall/CRC, Boca Raton, 2003.

[21] P. Sollich. Predicting phase equilibria in polydisperse systems. *Journal of Physics: Condensed Matter*, 14:79–117, 2002.

[22] D. Stoyan. Statistical inference for a Gibbs point process of mutually non-intersecting discs. *Biometrical Journal*, 51:153–161, 1989.

[23] D. Stoyan. Random systems of hard particles: models and statistics. *Chinese Journal of Stereology and Image Analysis*, 7:1–14, 2002.

[24] D. Stoyan, W.S. Kendall and J. Mecke. *Stochastic Geometry and its Applications*. John Wiley & Sons, Chichester, 1995.

[25] D. Stoyan and M. Schlather. Random sequential adsorption: Relationship to dead leaves and characterization of variability. *Journal of Statistical Physics*, 100:969–979, 2000.

[26] D. Stoyan and H. Stoyan. *Fractals, Random Shapes and Point Fields*. John Wiley & Sons, Chichester, 1994.

[27] S. Torquato. *Random Heterogeneous Materials: Microstructures and Macroscopic Properties*. Springer, New York, 2002.

[28] N.B. Wilding and P. Sollich. Grand canonical ensemble simulation studies of polydisperse fluids. *Journal of Chemical Physics*, 116:7116–7126, 2002.

[29] G. Winkler. *Image Analysis, Random Fields and Markov Chain Monte Carlo Methods*. Springer, Berlin, 2003.

[30] D.N. Winslow, M.D. Cohen, D.P. Bentz, K.A. Snyder and E.J. Garboczi. Percolation and pore structure in mortars and concrete. *Cement Concrete Research*, 24:25–37, 1994.

Source Detection in an Outbreak of Legionnaire's Disease

Miguel A. Martínez-Beneito[1], Juan J. Abellán[2], Antonio López-Quílez[3], Hermelinda Vanaclocha[1], Óscar Zurriaga[1], Guillermo Jorques[4] and José Fenollar[4]

[1] Dirección General de Salud Pública, Generalitat Valenciana, Spain,
    martinez_mig@gva.es, vanaclocha_her@gva.es, zurriaga_osc@gva.es
[2] Small Area Health Statistics Unit, Imperial College of London, UK,
    j.abellan@imperial.ac.uk
[3] Departament d'Estadística i Investigació Operativa, Universitat de València,
    Spain, Antonio.Lopez@uv.es
[4] Centro de Salud Pública de Alcoi, Generalitat Valenciana, Spain,
    jorques_gui@gva.es, fenollar_jos@gva.es

Summary. Spatial statistics have broadly been applied, developed and demanded from the field of epidemiology. The point process theory is an appropriate framework to analyse the spatial variation of risk of disease from information at individual level.

We illustrate an application of point pattern tools to study a few legionnaire's disease outbreaks. Specifically, these techniques are applied to explore the geographical distribution of cases resulting from three legionnaire's disease outbreaks that occurred successively in Alcoi, a city placed in the East of Spain.

Key words: Epidemiology, Geographic Information Systems (GIS), Legionnaire's disease outbreaks, Spatial point processes

1 Introduction

Spatial statistics have broadly been applied, developed and demanded from the field of epidemiology. This discipline covers the study of the distribution and determinants of health-related states or events in specified populations, and the applications of this study to control health problems [14]. The distribution of health events can be studied depending on population groups, risk factors, time or space, and that last factor is the reason why spatial statistics is so important in epidemiology. There are several monographs [1, 12, 15, 16] devoted to such applications and a wide bibliography regarding this topic. Interestingly, the problem of analyzing the geographical variation of risk of disease in a given region, so common in epidemiology, has led to important developments in spatial statistics and continues offering new challenges to it. This fact motivates that developments of both topics are intimately related.

The statistical methods to analyse the spatial distribution of a disease depend mainly on the type of available data. Not too long ago, the information mostly consisted of area-level counts of cases, the areas usually being administrative units that constitute a partition of the region of interest. Lattice methods, based on the comparison of observed to expected (according to a certain standard) counts, are the most appropriate and flexible tool to analyse this type of data. Nevertheless, in the last years medical databases have experienced a great improvement and individual location of cases are starting to be collected. Moreover, Geographic Information Systems (GIS) are becoming a daily tool in epidemiology and they easily allow to geocode the spatial component of the information stored in those databases. All these advances make possible to work with individual data instead of aggregated ones. Therefore, the use of techniques capable to deal with this kind of information is becoming more popular and demanded [7, 8, 10]. The point process theory is an appropriate framework to analyse the spatial variation of risk of disease from information at individual level.

The present work illustrates an application of point pattern tools to study a few legionnaire's disease outbreaks. Specifically, these techniques are applied to explore the geographical distribution of cases resulting from three legionnaire's disease outbreaks that occurred successively in Alcoi, a city placed in the East of Spain.

The present contribution is structured in five sections, this introduction being the first one. In Sect. 2 it is described the case study that motivates the present work. Section 3 presents the methodology used to study the configuration of cases resulting from the scenario described in the previous section. Results derived from such application are presented in Sect. 4. The last section is devoted to discuss the results and findings of this work.

2 Case Study Description

In 1976 the American Legion, an ex-military association, organised a meeting in Philadelphia (USA) to commemorate the signature of the United States Declaration of Independence. More than 180 delegates got ill during the meeting and 29 of them finally died. These people suffered legionnaire's disease, a respiratory illness unknown until that time. The origin of this disease is *Legionella pneumophila*, a bacterium that develops in humid and warm environments if appropriate nutrients are present. Bacterium transmission occurs by inhalation from contaminated aerosols present in the environment, whereas its ingestion is inoffensive. So all devices able to produce aerosols are a threat for public health since they are potential emitters of the bacteria in the case that they were colonised by them. In urban areas there are plenty of places with the appropriate conditions to develop legionella's colonies and spread them. Examples of this kind of devices, among others, are cooling towers associated to air conditioning systems and industrial processes, evaporative coolers, showers and fountains. Therefore a special surveillance should be done to this kind of installations to prevent bacterial growth, which in turn could produce a legionnaire's disease outbreak when spread out on the atmosphere.

Alcoi is an industrial city placed at the oriental side of Spain with about 60000 inhabitants in 2003. The textile industry is the main economic sector in this town and a great proportion of its population works for this industry. The textile production has a long tradition in Alcoi and as a consequence there is a high number of factories placed in the urban area, with all the problems and threats for health that it entails. In particular, the industrial cooling towers of these factories spread aerosols uninterruptedly into the atmosphere so they are a potential source of transmission of legionella. The presence of cooling towers, the peculiar orographic conditions and the soil properties of this region provokes very favorable conditions to legionnaire's disease outbreaks occur. All these particular facts promote that this disease can be considered nearly endemic in this city unless a prevention effort is carried out.

Strikingly, from September 20th, 1999 to December 1st, 2000 three consecutive outbreaks of legionnaire's disease occurred in Alcoi. The first of them lasted from September 20th, 1999 to February 27th, 2000 with 36 people affected, the second outbreak extended from April 9th, 2000 to July 30th, that year with a total of 11 persons ill and the third and biggest one began the September 9th and lasted to December 1st, 2000 affecting to 97 people. It was suspected that the source of the outbreaks could be one or several of the cooling towers (colonised by the bacteria) placed inside the urban area of the city. A key epidemiological issue that arose from the successive outbreaks was to assess whether or not the geographical distribution of the cases could be considered "random" (according to the population at risk in Alcoi). If the answer is negative, it would be a symptom of an extra aggregation present in the spatial distribution of the cases, as a consequence of one or more local sources of emission of the bacteria. In that case it would also be very interesting to determine the areas in the city with higher risk of disease, to focus on those locations the search of the installations that could be involved in the spreading of the outbreak.

Opposite to applications in other fields, the hypothesis of complete spatial randomness usually lacks of sense in epidemiological studies. This is due to the fact that the intensity of the cases depends directly on that of the population at risk across the region of interest. Moreover, the incidence of the disease can be also influenced by other personal factors, for example the age, gender or immune deficiency conditions. Therefore, the fact that the cases do not present the same distribution that general population should not be surprising. Thus, the main question in this kind of studies is if the spatial distribution of the observed cases can be considered a random sample of one population with similar characteristics to those presented in the cases. If not, we can conclude that there exist an exogenous factor that is influencing the geographical distribution of the cases over the city. To contrast this fact it has been collected a matched population sample. All the individuals collected had an admission to hospital for other reason different to legionnaire's disease in the same period than the cases in the first outbreak, the same gender and they shared approximately the same age. No more samples were collected for the other two outbreaks because risk factors and populational characteristics of the cases were very similar to those in the first one. Thus the cases in all three outbreaks were compared against the same sample, that is intended as a collection of people with similar features as the cases but that have not contracted the disease. The individuals in this sample are known in epidemiology as controls and 65 of them were included in the study. So the main interest of the study will rely on the comparison of the distribution of the available cases and controls. Fig. 1 shows the spatial distribution of cases for the three outbreaks and controls.

To our knowledge little has been studied about the distribution of legionnaire's disease cases in urban areas. [2] shows that sporadic cases of this disease over the Scottish city of Glasgow presented aggregations in its spatial distribution, although the time between the dates pointed no common source for the cases. There are other works, though, where it is first established the hypothetical location of the sources responsible for the outbreak and then it is tested if the proximity to those places increases the risk of observing a case. In this sense, [4] states that risk decreases in a 20% for each 0.1 mile (0.16 km) increase in distance from cooling tower. [3] show that the relative risk to get the disease for those people living less than that distance from a cooling tower is 3 times bigger than the risk for those people living more than 1 km away. They also established that the radius of influence of a cooling tower in an outbreak was about 500 m.

3 Methods

Four point patterns are considered, taking as events the residence locations of the cases in the three consecutive outbreaks and in the control set. Home addresses of all cases and controls were geocoded in a georeferenced map of the city (scale 1:2000).

A proper description of a point pattern can be obtained by a trend measure, as the density of events across the region, and a dispersion measure. Both characteristics are related with moment measures of the point process that is assumed to have generated the pattern, the intensity function and the Ripley's K-function [17], respectively.

The estimation of the intensity function can be performed by kernel smoothing [6], that is, the estimation of the density in a generic point (x, y) of the study region is given by

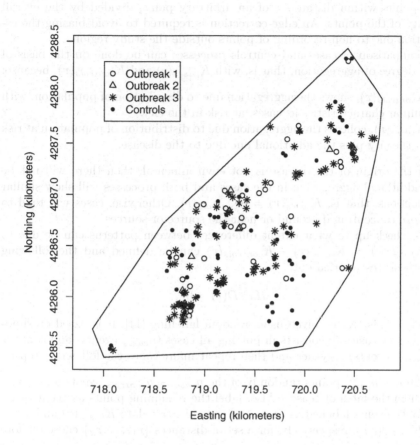

Fig. 1. Point patterns of cases and controls

$$\hat{\lambda}_h(x,y) = \sum_{i=1}^{n} f_h[(x,y) - (x_i, y_i)]$$

where n is the number of events in the point process and $f_h(x,y)$ is the kernel function, a bivariate and symmetric density function which depends on parameter h, called bandwidth, that accounts for local variation and controls the degree of smoothing. The larger is h the smoother is the resulting surface. A wise choice of this parameter is crucial, even more important than the choice of the kernel function [19].

The kernel function employed was the quartic kernel whose expression is

$$f_h(a,b) = \begin{cases} \frac{3}{\pi h^2}(1 - \frac{a^2+b^2}{h^2})^2 & \text{if } a^2 + b^2 \leq h^2 \\ 0 & \text{if } a^2 + b^2 > h^2 \end{cases}$$

The K-function can be estimated by means of the average number of further points within distance r of an arbitrary point, divided by the overall density of the points. An edge-correction is required to avoid biasing the estimation due to non-recording of points outside the study region.

Comparison of cases and controls processes can be done on the basis of their degree of aggregation, that is, with $K_{cases}(r)$ and $K_{controls}(r)$, because:

- $K_{controls}(r)$, shows the aggregation due to distribution of population, with similar characteristics as cases, at risk in the city.
- $K_{cases}(r)$, collects the aggregation due to distribution of population at risk in the city plus the additional one due to the disease.

If the origin of the disease is not environmental, then there will not be any additional aggregation in the cases and both processes will show similar K-functions, that is, $K_{cases}(r) = K_{controls}(r)$. Otherwise, cases will tend to be more concentrated around or near the source or sources.

To check up to what extent differences between patterns can be due to chance, $D(r) = K_{cases}(r) - K_{controls}(r)$ can be defined and the following hypothesis test carried out:

$$H_0 : D(r) = 0,$$

for all r. This test is solved using random labelling [11]. It is based on simulations and basically consists in putting all cases (n_{cases} events) and controls ($n_{controls}$ events) together and then repeat many times the following steps:

1. Choose n_{cases} points randomly of the $n_{cases} + n_{controls}$ events and assign them the label of "case". Also, label the remaining points as "control".
2. With events labelled as cases and controls, calculate $K_{cases}(r)$ and $K_{controls}(r)$, respectively, for a set of distances $\{r_1, \ldots, r_k\}$ chosen before hand.
3. Calculate $D(r_i) = K_{cases}(r_i) - K_{controls}(r_i), \forall i = 1, \ldots, k$.

Once the simulation is done, we have got a large number of values of $D(r)$ under the null hypothesis. Percentiles 2.5% and 97.5% can be obtained from the simulated values for each distance r_i providing approximate limits of the 95% confidence band for $D(r)$. The observed value of $D(r)$ with original cases and controls can be represented beside this confidence band. If the original $D(r)$ lies within the band, it means that there is no evidence against null hypothesis. If it goes out above the upper limit, it means that for that specific distance there is greater aggregation of cases than controls. Conversely, an observed $D(r)$ under the lower limit indicates inhibition between cases, showing less aggregation than controls for that distance.

A plot of $D(r_i), \forall i = 1, \ldots, k$, and their 95% confidence limits would help to quickly visualise the distance values for which cases exhibit more aggregation than controls, but it should not be interpreted as a formal hypothesis test, since it would imply multiple testing (a statistical test for each distance) and, therefore, the significance level would be much higher than the nominal 0.05. To avoid that problem, Diggle and Chetwynd [9] propose the statistic $D = \sum_{i=1}^{k} \frac{D(r_i)}{\sqrt{Var[D(r_i)]}}$ to test the global difference of aggregations between cases and controls patterns. The value of $Var[D(r_i)]$ under the null hypothesis can also be computed from random labelling simulations.

Apart from the test to know if there is greater aggregation in cases than controls, another interesting task is to obtain a risk surface and to determine the zones of the city where disease had more impact and that are likely to contain or be near to the cooling towers that flow out the bacteria.

A possible risk measure is given by the difference between the observed probability to be case and the expected one. From intensities of both point patterns, $\lambda_{cases}(x,y)$ and $\lambda_{controls}(x,y)$, the probability for an event at point (x,y) to be a case can be estimated as:

$$p(x,y) = \frac{\lambda_{cases}(x,y)}{\lambda_{cases}(x,y) + \lambda_{controls}(x,y)}$$

whereas expected probability to be case for the whole city can be computed:

$$p_E = \frac{n_{cases}}{n_{cases} + n_{controls}}$$

so a measure of risk $R(x,y)$ could be given by the difference between the observed probability and the expected one, that is:

$$R(x,y) = p(x,y) - p_E$$

The respective intensity functions can be approached with kernel estimation, and the observed probability $p(x,y)$ expressed using the following variable Z associated to events:

$$Z_i = \begin{cases} 1 & \text{if the } i\text{-th event is a case} \\ 0 & \text{if the } i\text{-th event is a control} \end{cases}$$

So the kernel estimator of the observed probability $p(x,y)$ is:

$$\hat{p}_h(x,y) = \frac{\sum_{i=1}^{n} f_h[(x,y) - (x_i,y_i)]z_i}{\sum_{i=1}^{n} f_h[(x,y) - (x_i,y_i)]}$$

where $n = n_{cases} + n_{controls}$ is the number of events (either cases or controls) and h is the parameter that tunes smoothing of the intensity.

The bandwidth h can be chosen using the maximum likelihood cross-validation method proposed by Kelsall and Diggle [13]. It is based on the likelihood calculated for the binary regression framework of the risk measure.

In addition to the value of the risk function, it is interesting to assess its statistical significance. This can be done using the simulations provided by the random labelling. In each iteration, not only $K_{cases}(r)$, $K_{controls}(r)$ and

$D(r)$ are calculated, but also $R(x, y)$. The rank of the observed risk among the simulated ones is computed afterwords, to obtain a Monte Carlo p-value representing the statistical significance of the risk value.

All computations were made using R language and the free distribution library Splancs [18].

4 Results

Spatial distributions of cases in the three outbreaks and controls were compared by means of second order properties. Estimated K-functions are shown in Fig. 2A. Disease patterns present greater values of K than population pattern at any distance between 0 and 400 meters, suggesting that cases are much more crowded than controls.

The differences between $K_{cases}(r)$ and $K_{controls}(r)$ are gathered in $D(r)$ and tested by random labelling. Results for the three outbreaks can be seen in Figs. 2B, 2C and 2D. Since $D(r)$ is above the upper bands, it seems clear that $D(r) > 0$, that is, cases show patterns more aggregated than that of controls.

D statistic was calculated for the comparison of each outbreak pattern and control pattern, and also for the comparison among the three outbreaks. Results are given in Table 1.

Table 1. D statistic comparing K-functions, with p-value of random labelling test between brackets

	Controls	Outbreak 1	Outbreak 2
Outbreak 1	101.54 (0.046)		
Outbreak 2	140.07 (0.026)	69.66 (0.249)	
Outbreak 3	285.79 (0.000)	121.70 (0.047)	8.26 (0.923)

Results of D statistic confirm an evidence of higher aggregation of cases than of controls. This reinforces the environmental origin of the disease.

Risk surfaces, as well as their statistical significance, were estimated for each outbreak. In the kernel estimation method, the bandwidth h was selected attending to maximum likelihood criterion simultaneously for all three outbreaks, obtaining an optimum value of 550 meters. With this bandwidth, smoothed images of risk and contour lines for the statistically significant values, are produced and shown in Figs. 3, 4 and 5.

In these risk maps, the darker zones correspond to those points where $R(x, y)$ is higher, that is, where the probability of an event to be case is upper than expected. Similar areas are highlighted as regions with statistically significant risk, one in the north-east and another one in the south-west of the city. Though the upper zone is more evident in the first and second outbreaks, and lower zone is clearly marked in the third outbreak.

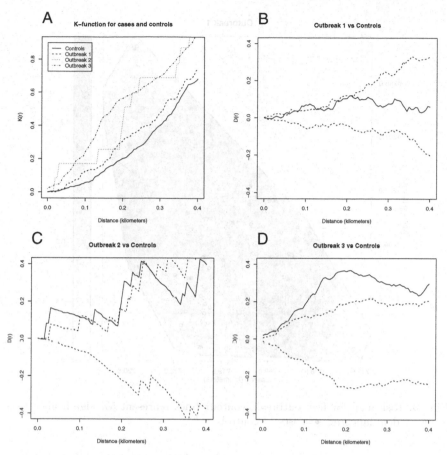

Fig. 2. Comparison of K-functions: Estimation of K-function for cases in the three outbreaks and controls (**A**); D function for cases in each outbreak versus controls (**B, C** and **D**), dashed lines represent 95% confidence bands using random labelling

5 Discussion

Point pattern analysis allows to study disease outbreaks in urban areas. In this context, aggregation of cases according to administrative divisions entails an important loose of information that may be not feasible to tackle the analysis. So data analysts of epidemiological problems should be aware of this collection of techniques that profit individual geographic information whenever it is available. Moreover, point pattern methodology provides more accurate conclusions than the lattice one because the former provides its results in a continuous domain while the latter one does not provide any information under the administrative division.

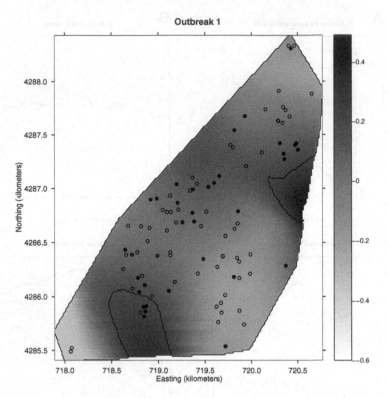

Fig. 3. Risk map for first outbreak, contour lines represent 5% significance risks using random labelling, ●=cases ○=controls.

One of the limitations of the analysis is that addresses of residence were geocoded assuming implicitly that home was the place of exposure for all cases, which might not be too realistic. Also, many cases were retired people who did not work but used to go for a walk. It turned out that there were a couple of routes quite common in the city for those people, and of course the bacterium could have been inhaled during those walks, even if the walk lasted one hour or even less. Nevertheless it is a fact that people spend a big amount of time in their homes and so it is expected that the pattern observed in the analysis could reflect the variation of risk along the city. Moreover, in the third outbreak, three cases did not get out home during the latency period of the disease (it ranges from 2 to 10 days) and their apartments were placed in the main cluster of cases. This fact supports the results of the study.

Another issue of discussion concerns about cases and controls matching. [5] shows an adaptation of point pattern methodology to that kind of designs. The controls for the present work have been sampled in a 1 case - 2 control basis, but only for the first outbreak. The other ones do not dispose of matched controls. It was neither possible to get such control for every case in the

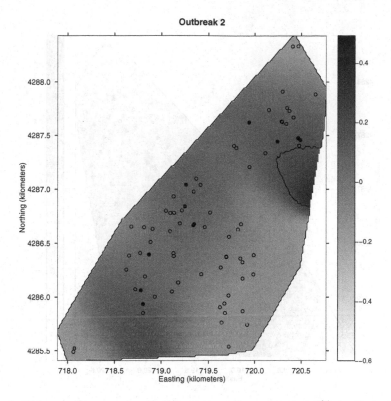

Fig. 4. Risk map for second outbreak, contour lines represent 5% significance risks using random labelling, •=cases ○=controls

first outbreak, in fact there were only 65 controls available for 36 cases. It has to be also taken into account that the hospital where cases and controls were compiled from covers the whole population, so it is not expected to find a strong spatial association between the matching pairs. This impression is confirmed by the fact that the mean distance between cases' homes and their matched controls ones is 1.65 km while that distance for cases and not matched controls is 1.59 km, even lesser than that for the matched ones. So it has not been considered as necessary to use the matched methodology for this work.

Conclusions from the present job have a great value from an epidemiological view. The first conclusion that can be extracted is that cases show a more aggregated pattern than controls in all three outbreaks. This fact suggests the presence of a mechanism that concentrates the incidence of the disease around some specific locations, maybe due to the effect of one or more risk facilities that could have been colonised by the bacteria. But the most valuable conclusion of such aggregation is that it can be discarded that the origin of the outbreak comes from the main water supply of Alcoi which provides water for

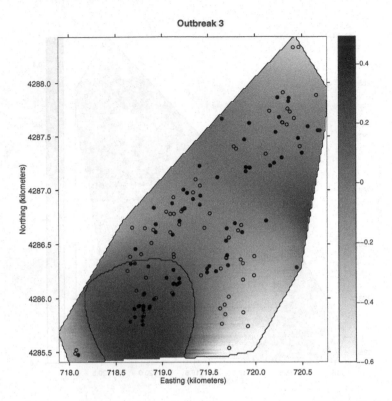

Fig. 5. Risk map for third outbreak, contour lines represent 5% significance risks using random labelling, ●=cases ○=controls

the whole population. If it would have been that way, the distribution of cases and controls would be more similar because the exposition to the bacteria in the whole city would be the same for all the population and it would not be expected to observe the patterns that have occurred. So the study supports the environmental origin of the outbreaks as was thought previously. As a consequence the efforts to stop the outbreaks should be focus on the search of risk facilities within the zones with higher risk in the city, that could be involved in the spread of the bacteria.

Comparison of K-functions between different outbreaks reveals that these do not have exactly the same behavior in relation to aggregation of patterns. This fact does not allow to perform a joint analysis of all the three outbreaks against the controls. Moreover the results of the analysis show that the first outbreak has a less aggregated pattern than the third one and it is not believable that those differences can be attributed to chance.

Estimation of risk and p-value surfaces mean a valuable tool from an epidemiological point of view. These representations provide a useful plot to guide the efforts of health intervention policy in an outbreak detection problem. In our particular study we can appreciate that risk estimation points towards a multifocal origin of the outbreaks in all three cases. But we can see that although the three outbreaks point more or less the same areas as regions of risk, there are some features that distinguish them. For example, the proportion of people involved in each cluster or the spreading of cases in the different outbreaks.

Acknowledgements

This research was funded partially by the Dirección General de Salud Pública de la Generalitat Valenciana and grant MTM2004-03290 from the Ministerio de Educación y Ciencia.

References

[1] F.E. Alexander and P. Boyle (Editors). *Methods for Investigating Localized Clustering of Disease.* International Agency for Cancer Research, Lyon, 1996.

[2] R.S. Bhopal, P.J. Diggle and B. Rowlingson. Pinpointing clusters of apparently sporadic cases of legionnaires' disease. *British Medical Journal,* 304:1022–1027, 1992.

[3] R.S. Bhopal, R.J. Fallon, E.C. Buist, R.J. Black and J.D. Urquhart. Proximity of the home to a cooling tower and risk of non-outbreak Legionnaire's Disease. *British Medical Journal,* 302:378–383, 1991.

[4] C.M. Brown, P.J. Nuorti, R.F. Breiman, A.L. Hathcock, B.S. Fields, H.B. Lipman, G.C. Llewellyn, J. Hoffman and M. Cetron. A community outbreak of legionnaire's disease linked to hospital cooling towers: an epidemiological methos to calculate dose of exposure. *International Journal of epidemiology,* 28:353–359, 1999.

[5] A.G. Chetwynd, P.J. Diggle, A. Marshall and R. Parslow. Investigations of spatial clustering from individually matched case-control studies. *Biostatistics,* 2:277–293, 2001.

[6] P.J. Diggle. A kernel method for smoothing point process data. *Journal of the Royal Statistical Society C,* 34:138–147, 1985.

[7] P.J. Diggle. A point process modelling approach to raised incidence of a rare phenomenon in the vicinity of a prespecified point. *Journal of the Royal Statistical Society A,* 153:349–362, 1990.

[8] P.J. Diggle. Point process modelling in environmental epidemiology. In: V. Barnett and K. Turkman (eds.) *Statistics for the Environment.* Wiley, New York, 1993.

[9] P.J. Diggle and A. Chetwynd. Second-order analysis of spatial clustering for inhomogeneous populations. *Biometrics,* 47:1155–1163, 1991.

[10] P.J. Diggle and P. Elliot. Disease risk near point sources: statistical issues in the analysis of disease risk near point sources using individual or spatially aggregated data. *Journal of Epidemiology and Community Health*, 49:20–27, 1995.

[11] P.J. Diggle and B. Rowlingson. A conditional approach to point process modelling of elevated risk. *Journal of the Royal Statistical Society A*, 157:433–440, 1994.

[12] P. Elliot, J.C. Wakefield, N.G. Best and D.J. Briggs (Editors). *Spatial Epidemiology: Methods and Applications*. Oxford University Press, Oxford, 2000.

[13] J.E. Kelsall and P.J. Diggle. Spatial variation in risk of disease: a nonparametric binary regression approach. *Applied Statistics*, 47:559–573, 1998.

[14] J.M. Last. *A Dictionary of Epidemiology*. Oxford University Press, New York, 1995.

[15] A.B. Lawson. *Statistical Methods in Spatial Epidemiology*. Wiley, Chichester, 2001.

[16] A. Lawson, A. Biggeri, D. Böhning, E. Lesaffre, J.F. Viel and R. Bertollini (Editors). *Disease Mapping and Risk Assessment for Public Health*. Wiley, Chichester, 1999.

[17] B.D. Ripley. *Spatial Statistics*. Wiley, New York, 1981.

[18] B. Rowlingson and P.J. Diggle. Splancs: Spatial Point Pattern Analysis Code in S-Plus. *Technical Report*, Lancaster University, Lacanster, U.K., 1993.

[19] B.W. Silverman. *Density Estimation for Statistics and Data Analysis*. Chapman and Hall, London, 1986.

Doctors' Prescribing Patterns in the Midi-Pyrénées rRegion of France: Point-process Aggregation

Noel A.C. Cressie[1], Olivier Perrin[2] and Christine Thomas-Agnan[3]

[1] Department of Statistics, The Ohio State University, Columbus, OH 43210, USA, ncressie@stat.ohio-state.edu
[2] GREMAQ – Université Toulouse 1 (and LERNA-INRA), 21 allée de Brienne, 31000 Toulouse, France
[3] GREMAQ – Université Toulouse 1 (and LSP – Université Toulouse 3), 21 allée de Brienne, 31000 Toulouse, France

Summary. People can be located according to their residence, their place of work, their doctor's office, their pharmacy, and so forth. It is sometimes of interest to look for patterns in people's locations in relation to their behaviours.

In this article, we are particularly interested in the cost of patients' prescriptions per doctor consultation. In particular, we consider the common problem brought about by aggregation when only the less-precise locational information and the associated variable 'average prescription amount per consultation' are available.

We build a spatial regression model for the spatially aggregated data depending on covariates. We fit initially a non-spatial version of the model to the doctor-prescribing data. We then consider spatial dependence in the data after the large-scale variation has been accounted for, and propose a final model that explains doctors' prescribing patterns.

Key words: EDA, ESDA, Region of Midi-Pyrenees (France), Spatial analysis of doctors' prescribing patterns, Spatial regression

1 Introduction

People can be located according to their residence, their place of work, their doctor's office, their pharmacy, and so forth. It is sometimes of interest to look for patterns in people's locations in relation to their behaviors. In this article, we are particularly interested in the cost of patients' prescriptions per doctor consultation, during the period January 1, 1999-December 31, 1999, in the region of southwest France known as the Midi-Pyrénées.

Because of confidentiality requirements or the manner in which data are recorded, the precise locations of residences or doctors' offices are unknown. As a consequence, the locational data are less precise, but this does not prevent a spatial analysis from being carried out. For example, in the Midi-Pyrénées region, there are 268 cantons with at least one doctor, and the data we have are organised by these cantons; that is, individual prescription amounts are aggregated to an average amount per consultation, where the average is taken over all consultations in a given canton in 1999.

Consider the i-th canton. Point-level spatial data are $(\mathbf{s}_{ij}, Y(\mathbf{s}_{ij}))$, where $\mathbf{s}_{ij}$ is the location (residence or pharmacy or ...) of the j-th prescription amount $Y(\mathbf{s}_{ij})$ in canton i, $j = 1, \ldots, E_i$, and E_i is the number of consultations in canton i in 1999. Aggregated spatial data are $(i, \sum_{j=1}^{E_i} Y(\mathbf{s}_{ij}) / \sum_{j=1}^{E_i} 1)$, where, in general, there are $i = 1, \ldots, n$ cantons of interest. Aggregated data come from the Union Régionale des Caisses d'Assurance Maladie (URCAM), which is interested in relations between the many variables they collect. In this paper, we shall focus on the activity of all the general-practitioner doctors in the $n = 268$ cantons, after elimination of the cantons that have no doctors; in particular, we consider the dependent variable,

$$Y_i \equiv \text{average prescription amount (in FF) per consultation}, \qquad (1)$$

where FF denotes French Francs and $i = 1, \ldots, 268$.

To get (1), there is $\{(\mathbf{s}_{ij}, Y(\mathbf{s}_{ij}))\}$ an underlying (marked) spatial point pattern, where $\{\mathbf{s}_{ij} : j = 1, \ldots, E_i\}$ are well defined locations (e.g., patients' residences) of all prescriptions written in the i-th canton in 1999, and $Y(\cdot)$ is the "mark" variable given by the prescription amount. Then the point pattern over the whole Midi-Pyrénées region is

$$Z \equiv \bigcup_{i=1}^{n} \{\mathbf{s}_{ij} : j = 1, \ldots, E_i\}. \qquad (2)$$

An analysis of the marked point pattern (Z, Y) could proceed in ways described in [2, Ch. 8], or in [4, Ch. 15], *provided* the point pattern (2) and the mark variable $Y(\cdot)$ are available. This article considers the common problem brought about by aggregation when only the less-precise locational information (e.g., canton neighbourhood structure) and the associated variable (1) are available. Importantly, the aggregation has consequences for the statistical analysis, which we shall demonstrate in the sections that follow.

In Sect. 2, we build a spatial regression model for the spatially aggregated $\{Y_i\}$ depending on p covariates $\{x_{ki}\}$; $k = 1, \ldots, p$. In Sect. 3, we fit initially a non-spatial version of the model in Sect. 2 to the doctor-prescribing data $\{Y_i\}$ using two covariates: percentage of patients 70 or older and per-capita income. The fit is done in a series of exploratory-data-analysis (EDA) steps that recognize that aggregation demands a careful *weighted* analysis. Section 4 is concerned with spatial dependence in the data after the large-scale variation (i.e., covariates and heteroskedasticity) has been accounted for; through ESDA (exploratory *spatial* data analysis) we look for spatial dependence and propose a final model that explains doctors' prescribing patterns. Section 5 contains discussion and conclusions.

2 Spatial Regression Model

Because of the spatial nature of locations $\{\mathbf{s}_{ij}\}$, we expect that the data aggregated into cantons show spatial dependence. To model this dependence, we use a Markov random field known as the conditional autoregressive (CAR) model, which has a joint Gaussian distribution. With sufficient aggregation, the central limit theorem tells us that the variable,

$$Y_i \equiv \sum_{j=1}^{E_i} Y(\mathbf{s}_{ij}) / \sum_{j=1}^{E_i} 1 ; \quad i = 1, \ldots, n , \tag{3}$$

is approximately Gaussian. That is, $\mathbf{Y} \equiv (Y_1, \ldots, Y_n)'$ is modelled as:

$$\mathbf{Y} \sim \mathrm{Gau}(\boldsymbol{\mu}, \Sigma) , \tag{4}$$

a multivariate Gaussian distribution with mean $\boldsymbol{\mu} \equiv (\mu_1, \ldots, \mu_n)'$ and $(n \times n)$ variance-covariance matrix Σ. What makes (4) a CAR model is that Σ takes on a special form:

$$\Sigma = (I - C)^{-1} M , \tag{5}$$

where $C \equiv (c_{ij})$, $c_{ii} = 0$; $i = 1, \ldots, n$, and $M \equiv \mathrm{diag}(\tau_1^2, \ldots, \tau_n^2)$ are parameters in the conditional distributions,

$$Y_i | \mathbf{Y}_{-i} \sim \mathrm{Gau}(\mu_i + \sum_{j=1}^{n} c_{ij}(Y_j - \mu_j), \tau_i^2) , \tag{6}$$

for $\mathbf{Y}_{-i} \equiv (Y_1, \ldots, Y_{i-1}, Y_{i+1}, \ldots, Y_n)'$ and $i = 1, \ldots, n$. The $\{c_{ij}\}$ are spatial-dependence parameters and the $\{\tau_i^2\}$ are heteroskedasticity parameters that together satisfy:

$$\begin{aligned} M^{-1}C & \text{ is symmetric (symm.)} ; \\ M^{-1}(I - C) & \text{ is positive-definite (p.d.)} ; \end{aligned} \tag{7}$$

see [1].

We assume that covariates are included in the model linearly through $\boldsymbol{\mu} = X\boldsymbol{\beta}$, where $\boldsymbol{\beta}$ is a $(p \times 1)$ vector of regression parameters; $p < n$. Also, we assume that the conditional variances $\tau_1^2, \ldots, \tau_n^2$ are known up to a normalising constant (as is the case when the data are rates); that is, $M = \Phi\tau^2$, where $\Phi \equiv \mathrm{diag}(\phi_1, \ldots, \phi_n)$ is a known $(n \times n)$ diagonal matrix. Finally, we assume that C is a function of a $(q \times 1)$ vector of spatial-dependence parameters $\boldsymbol{\gamma}$, which we write as $C(\boldsymbol{\gamma})$.

To summarize, the CAR model we shall consider in this article is,

$$\mathbf{Y} \sim \mathrm{Gau}(X\boldsymbol{\beta}, (I - C(\boldsymbol{\gamma}))^{-1}\boldsymbol{\Phi}\tau^2), \tag{8}$$

which is clearly a special case of a general linear model. We shall make inference on parameters $\boldsymbol{\beta}$, τ^2, and $\boldsymbol{\gamma}$ through maximum likelihood estimation. The joint Gaussian form of the CAR model means that its normalising constant is known analytically and can be evaluated straightforwardly when n is moderate in size. When n is large, [3] shows that maximum likelihood estimation can be achieved through Monte-Carlo-based algorithms.

Under the CAR model (8), the parameter space is

$$\mathcal{P} \equiv \{\boldsymbol{\beta}, \tau^2, \boldsymbol{\gamma} : \boldsymbol{\beta} \in \mathbb{R}^p; \ \tau^2 > 0; \ \boldsymbol{\gamma} \in \mathbb{R}^q; \ p + q < n; $$
$$\boldsymbol{\Phi}^{-1}(I - C(\boldsymbol{\gamma})) \text{ is symm., p.d.}\}. \tag{9}$$

To estimate the parameters, we use maximum likelihood estimation. The likelihood is

$$\ell_{\mathbf{Y}}(\boldsymbol{\beta}, \tau^2, \boldsymbol{\gamma}) \equiv \{(2\pi\tau^2)^{-n/2}|\boldsymbol{\Phi}|^{-1/2}/k(\boldsymbol{\gamma})\} \times$$
$$\exp\{-(1/2)(\mathbf{Y} - X\boldsymbol{\beta})'\boldsymbol{\Phi}^{-1}(I - C(\boldsymbol{\gamma}))(\mathbf{Y} - X\boldsymbol{\beta})/\tau^2\},$$

where the normalising constant is $k(\boldsymbol{\gamma}) \equiv |I - C(\boldsymbol{\gamma})|^{-1/2}$.

We now show that a simple transformation reduces the problem to one where conditional variances are equal (conditional homoskedasticity). Write

$$\widetilde{\mathbf{Y}} \equiv \boldsymbol{\Phi}^{-1/2}\mathbf{Y}. \tag{10}$$

Then

$$\widetilde{\mathbf{Y}} \sim \mathrm{Gau}(\widetilde{X}\boldsymbol{\beta}, (I - \widetilde{C}(\boldsymbol{\gamma}))^{-1}\tau^2) \equiv \mathrm{Gau}(\widetilde{X}\boldsymbol{\beta}, \widetilde{\Sigma}), \tag{11}$$

where $\widetilde{X} \equiv \boldsymbol{\Phi}^{-1/2}X$ and $\widetilde{C}(\boldsymbol{\gamma}) \equiv \boldsymbol{\Phi}^{-1/2}C(\boldsymbol{\gamma})\boldsymbol{\Phi}^{1/2}$. Since the information content of $\widetilde{\mathbf{Y}}$ and $\mathbf{Y}$ is identical, inference on $\boldsymbol{\beta}$, τ^2, and $\boldsymbol{\gamma}$ can be based equivalently on the likelihood of $\widetilde{\mathbf{Y}}$,

$$\ell_{\widetilde{\mathbf{Y}}}(\boldsymbol{\beta}, \tau^2, \boldsymbol{\gamma}) \equiv \{(2\pi\tau^2)^{-n/2}/k(\boldsymbol{\gamma})\} \tag{12}$$
$$\times \exp\{-(1/2)(\widetilde{\mathbf{Y}} - \widetilde{X}\boldsymbol{\beta})'(I - \widetilde{C}(\boldsymbol{\gamma}))(\widetilde{\mathbf{Y}} - \widetilde{X}\boldsymbol{\beta})/\tau^2\}.$$

Notice that with regard to the normalising constant, it is immaterial whether $C(\boldsymbol{\gamma})$ or $\widetilde{C}(\boldsymbol{\gamma})$ is used in its evaluation:

$$k(\boldsymbol{\gamma}) = |I - C(\boldsymbol{\gamma})|^{-1/2} = |\boldsymbol{\Phi}^{1/2}(I - \widetilde{C}(\boldsymbol{\gamma}))\boldsymbol{\Phi}^{-1/2}|^{-1/2}$$
$$= |I - \widetilde{C}(\boldsymbol{\gamma})|^{-1/2}.$$

The negative loglikelihood is,

$$L_{\widetilde{\mathbf{Y}}}(\boldsymbol{\beta}, \tau^2, \boldsymbol{\gamma}) \equiv -\log \ell_{\widetilde{\mathbf{Y}}}(\boldsymbol{\beta}, \tau^2, \boldsymbol{\gamma}). \tag{13}$$

Consider $\gamma \in \mathcal{P}$, fixed for the moment; then minimising $L_{\widetilde{\mathbf{Y}}}$ with respect to $\beta(\in \mathbb{R}^p)$ and $\tau^2(> 0)$ is easily seen to yield estimates:

$$\widehat{\beta}(\gamma) = (\widetilde{X}'(I - \widetilde{C}(\gamma))\widetilde{X})^{-1}\widetilde{X}'(I - \widetilde{C}(\gamma))\widetilde{\mathbf{Y}} \tag{14}$$

$$\widehat{\tau}^2(\gamma) = (\widetilde{\mathbf{Y}} - \widetilde{X}\widehat{\beta}(\gamma))'(I - \widetilde{C}(\gamma))(\widetilde{\mathbf{Y}} - \widetilde{X}\widehat{\beta}(\gamma))/n. \tag{15}$$

Substituting these estimates back into $L_{\widetilde{\mathbf{Y}}}$ given by (13), we obtain a negative log *profile* likelihood:

$$\mathcal{L}_{\widetilde{\mathbf{Y}}}(\gamma) \equiv L_{\widetilde{\mathbf{Y}}}(\widehat{\beta}(\gamma), \widehat{\tau}^2(\gamma), \gamma) \tag{16}$$
$$= (n/2)(\log(2\pi) + 1) + \log k(\gamma) + (n/2)$$
$$\times \log[\widetilde{\mathbf{Y}}'(I - \widetilde{C}(\gamma))\{I - \widetilde{X}(\widetilde{X}'(I - \widetilde{C}(\gamma))\widetilde{X})^{-1}\widetilde{X}'(I - \widetilde{C}(\gamma))\}\widetilde{\mathbf{Y}}/n].$$

Minimising this with respect to $\gamma \in \mathcal{P}$ yields the maximum likelihood estimator, which we denote as $\widehat{\gamma}$. Then the maximum likelihood estimators of the regression parameters, the variance parameter, and the spatial-conditional-autoregressive coefficients are, respectively, $\widehat{\beta}(\widehat{\gamma})$, $\widehat{\tau}^2(\widehat{\gamma})$, and $C(\widehat{\gamma})$.

For the rest of this article, we model the spatial dependence in $C(\gamma)$ through just one real parameter γ; specifically, we assume that

$$C(\gamma) - \gamma H, \tag{17}$$

where $H \equiv (h_{ij})$ is a known $(n \times n)$ matrix whose diagonal elements are zero and such that

$$\widetilde{H} \equiv \Phi^{-1/2}H\Phi^{1/2}, \tag{18}$$

is symmetric. Finally, (11) becomes

$$\widetilde{\mathbf{Y}} \sim \text{Gau}(\widetilde{X}\beta, (I - \gamma\widetilde{H})^{-1}\tau^2), \tag{19}$$

where $\widetilde{X} \equiv \Phi^{-1/2}X$, and $\widetilde{H} \equiv \Phi^{-1/2}H\Phi^{1/2}$ is symmetric.

A simple analysis with no spatial dependence gives

$$\text{var}(Y_i) = \sum_{j=1}^{E_i} \text{var}(Y(\mathbf{s}_{ij}))/E_i^2 = \sigma_Y^2/E_i,$$

assuming that $\text{var}(Y(\mathbf{s}_{ij})) \equiv \sigma_Y^2$, a constant. In the presence of spatial dependence, we model the *conditional* variances to be proportional to $\{1/E_i\}$, respectively:

$$\text{var}(Y_i|\mathbf{Y}_{-i}) = \tau^2/E_i; \quad i = 1, \ldots, n. \tag{20}$$

That is, in terms of (8), $\Phi = \text{diag}(E_1^{-1}, \ldots, E_n^{-1})$, and recall from (17) that $C(\gamma) \equiv \gamma H$. Reference [3] shows that in order for γ to be interpretable as a unitless correlation parameter, one should use the *spatial-rates CAR model*; that is,

$$h_{ij} = \begin{cases} (E_j/E_i)^{1/2} \, ; & j \in N(i) \\ 0 \, ; & \text{otherwise} \, , \end{cases}$$

where $N(i) \subset \{1, \ldots, i-1, i+1, \ldots, n\}$ is a prespecified neighbourhood set that represents the cantons that are neighbours of canton i; $i = 1, \ldots, n$. For the 268 cantons of interest in the Midi-Pyrénées, we computed the centroids $\{s_i\}$ and we defined $\{N(i)\}$ by $N(i) \equiv \{j : 0 < d(s_i, s_j) \leq 30 \, \text{km}\}$; $i = 1, \ldots, 268$, where $d(\cdot, \cdot)$ denotes Euclidean distance in the cartographic projection NTF (nouvelle triangulation de la France). The consequence of all these model specifications is:

$$Y_i | \mathbf{Y}_{-i} \sim \text{Gau}\left((X\beta)_i + \gamma \left\{ \sum_{j \in N(i)} (E_j/E_i)^{1/2}(Y_j - (X\beta)_j) \right\}, \tau^2/E_i \right),$$

where $(X\beta)_i$ denotes the i-th element of $X\beta$; $i = 1, \ldots, n$. Equivalently,

$$\mathbf{Y} \sim \text{Gau}(X\beta, (I - \gamma H)^{-1}\Phi\tau^2);$$

or equivalently,

$$\widetilde{\mathbf{Y}} \sim \text{Gau}(\widetilde{X}\beta, (I - \gamma\widetilde{H})^{-1}\tau^2), \tag{21}$$

where recall from (11) that $\widetilde{X} = \Phi^{-1/2}X$, and it it straightforward to derive that

$$\widetilde{H} = \begin{cases} 1 \, ; & j \in N(i) \\ 0 \, ; & \text{otherwise} \, , \end{cases} \tag{22}$$

which we shall refer to as the neighbourhood matrix. (The matrix $\widetilde{H}$ has 0-1 entries and simply records which cantons are neighbours of each other.)

3 Weighted Regression Analysis

In this section, we focus on a non-spatial maximum-likelihood approach to explain Y in terms of x_1, per-capita income, and x_2, percentage of patients 70 or older. Two analyses are given. The unweighted analysis does not recognize the importance of a changing denominator $\{E_i\}$ in the rates $\{Y_i\}$. The weighted analysis recognizes that a rate Y_i with a larger denominator E_i is more precise and so should get more weight in both non-spatial and spatial regressions.

From Fig. 1, we can contrast the scatterplots of the dependent variable Y against x_1 and x_2, with and without weighting. The comparison clearly shows that the correlation between Y and x_2 and between Y and x_1, is stronger after weighting has been carried out. Specifically, in the top two panels of Fig. 1 we plot $\{Y_i\}$ versus $\{x_{ki}\}$, and in the bottom two panels we plot weighted variables $\{\widetilde{Y}_i\}$ versus $\{\widetilde{x}_{ki}\}$, whose definitions are given in (10) and just below (11).

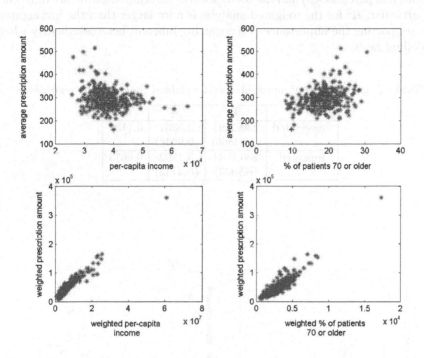

Fig. 1. Comparison of unweighted (*top*) and weighted (*bottom*) scatterplots. In the weighted scatterplots, the point beyond the main cluster is due to the canton of Toulouse

It should be remembered that there is a column of 1s in the matrix X referred to in (8). We write $\mathbf{x}_0 \equiv (1, \dots, 1)'$ and hence from (11), $\widetilde{\mathbf{x}}_0 \equiv \Phi^{-1/2}\mathbf{x}_0$. We next examine Fig. 1.2, the partial residual plots of $\widetilde{\mathbf{Y}}$ against $\widetilde{\mathbf{x}}_1$ (respectively, $\widetilde{\mathbf{x}}_2$), after accounting for the variable $\widetilde{\mathbf{x}}_0$. We see that the dependence between prescription amount and percentage of patients 70 or older, is stronger than that between prescription amount and per-capita income. This is confirmed by a t-statistic of 7.56 for the estimate of the former regression coefficient, compared to a t-statistic of -4.48 for the estimate of the latter regression coefficient. Further partial residual plots of $\widetilde{\mathbf{Y}}$ against $\widetilde{\mathbf{x}}_1$, after accounting for both $\widetilde{\mathbf{x}}_0$ and $\widetilde{\mathbf{x}}_2$ (not shown), revealed that per-capita income had no extra explanatory power in the presence of percentage of patients 70 or older. That is, in the regression, $E(\mathbf{Y}) = \beta_0\mathbf{x}_0 + \beta_1\mathbf{x}_1 + \beta_2\mathbf{x}_2$, exploratory and confirmatory analyses have led us to put $\beta_1 = 0$. Therefore, from now on, we shall keep only $\mathbf{x}_2$ (and the intercept $\mathbf{1}$) as an explanatory variable for $\mathbf{Y}$.

Comparing now the unweighted and weighted (non-spatial) regressions of $\mathbf{Y}$ against $\mathbf{x}_2$, we see in Table 1 that the maximum likelihood estimates of the regression parameters (β_0, β_2) (their corresponding standard deviations are indicated in parentheses) and the coefficients of determination R^2 are different. In particular, R^2 for the weighted analysis is a lot larger than the unweighted one, supporting the importance of recognizing inherent heteroskedascity when modelling rates.

Table 1. Comparison of unweighted and weighted (non-spatial) regressions

	β_0	β_2	R^2
unweighted	230.2491	3.5361	0.1139
	(11.4566)	(0.6048)	
weighted	236.1941	3.1292	0.9673
	(6.8033)	(0.4145)	

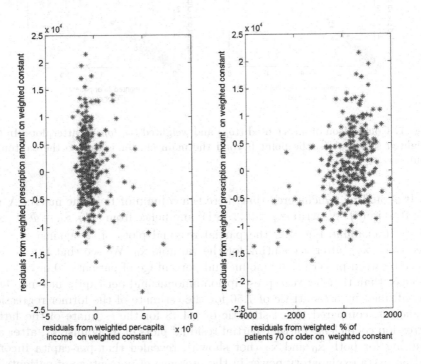

Fig. 2. Comparison of partial residual plots for weighted (non-spatial) regressions

4 Weighted Spatial Regression Analysis

We now inspect the residuals from the unweighted and weighted non-spatial regressions for presence of spatial autocorrelation. We shall use Moran's contiguity ratio (e.g., [2], p. 442) to measure this spatial autocorrelation, with the contiguity matrix as $\widetilde{H}$, where recall that $\widetilde{H}$ is the neighbourhood matrix with 0-1 entries given in (22). For the 268 cantons of interest in the Midi-Pyrénées, recall that we defined j to belong to the neighbourhood $N(i)$ if the centroids of the j-th and i-th cantons are less than or equal to 30 km apart. To present the results, we use plots from the GeoXp package[4], which in this case consists of a map of the Midi-Pyrénées region on the left and a Moran plot on the right; see Figs. 3 and 4. (For a given variable $\mathbf{Z}$ and a given contiguity matrix W, the Moran plot is a scatterplot of $W\mathbf{Z}$ against $\mathbf{Z}$.) Define

$$\widehat{\varepsilon}^{uns} \equiv \mathbf{Y} - \widehat{\beta}_0^{uns}\mathbf{x}_0 - \widehat{\beta}_2^{uns}\mathbf{x}_2$$

to be the residuals of the unweighted non-spatial model, and

$$\widehat{\varepsilon}^{wns} \equiv \widetilde{\mathbf{Y}} - \widehat{\beta}_0^{wns}\widetilde{\mathbf{x}}_0 - \widehat{\beta}_2^{wns}\widetilde{\mathbf{x}}_2$$

to be the residuals of $\widetilde{\mathbf{Y}}$ from the weighted non-spatial model, where the regression parameters (β_0, β_2) are in each case estimated by maximum likelihood, according to the model assumed.

Figures 3 and 4 present respectively the Moran plots based on $\widehat{\varepsilon}^{uns}$ and on $\widehat{\varepsilon}^{wns}$, where the matrix $W = \widetilde{H}$ was used in both cases. They both exhibit significant spatial autocorrelation; the p-value is more significant for the weighted model, with a value of 8.7935×10^{-12} (and Moran's contiguity ratio equal to 2.0772), compared with 4.2337×10^{-7} for the unweighted model (and Moran's contiguity ratio equal to 1.5067).

To take into account the spatial autocorrelation demonstrated just above, we now turn to the CAR model presented in Sect. 1.2. As in Sect. 1.3, we can fit an unweighted or a weighted model: The weighted CAR model is given by (21), and the unweighted CAR model corresponds to putting $\tau_i^2 \equiv \tau^2$; $i = 1, \ldots, n$, in (6). The unweighted CAR model we use is,

$$\mathbf{Y} \sim \text{Gau}(X\beta, (I - \gamma\widetilde{H})^{-1}\tau^2), \qquad (23)$$

where recall that $\widetilde{H}$ is the neighbourhood matrix with 0-1 entries. The results of maximum-likelihood fitting [3] are shown in Table 2. The t-statistic related to the spatial-dependence parameter γ is equal to 6.1723 for the unweighted CAR model and is equal to 11.9168 for the weighted CAR model. This confirms that there is significant spatial dependence in both models (assuming the models' respective correctness), and it appears that the spatial dependence is stronger in the weighted model.

---

[4] a Matlab toolbox downloadable at
   http://www.univ-tlse1.fr/GREMAQ/Statistique/geoxppage.htm

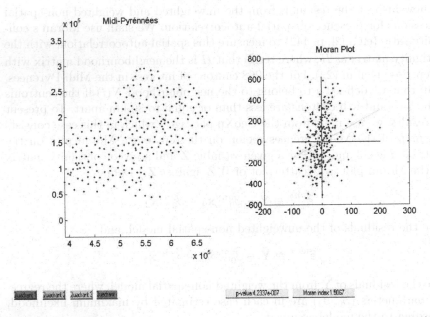

Fig. 3. Map of $n = 268$ centroids in units of km (cartographic projection NTF) and Moran plot of the residuals from the unweighted, non-spatial regression

In order to compare the four models (unweighted non-spatial, weighted non-spatial, unweighted CAR, and weighted CAR), we propose to use a cross-validation-type criterion,

$$CV \equiv \sum_{i=1}^{n} [Y_i - \widehat{E}(Y_i | \{Y_j : j \in N(i)\})]^2,$$

as the basis of the comparison. That is, CV is the sum of squared prediction errors.

The comparison based on CV is only indicative, since it depends on the dataset at hand. Based on (6), the estimated conditional expectation, $\widehat{E}(Y_i | \{Y_j : j \in N(i)\})$, in the formula for CV is computed as follows:

Unweighted non-spatial (uns):

$$\widehat{\beta}_0^{uns} x_{0,i} + \widehat{\beta}_2^{uns} x_{2,i}.$$

Weighted non-spatial (wns):

$$\widehat{\beta}_0^{wns} x_{0,i} + \widehat{\beta}_2^{wns} x_{2,i}.$$

Unweighted CAR (ucr):

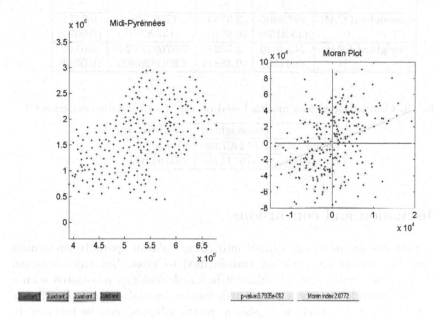

Fig. 4. Map of $n = 268$ centroids in units of km (cartographic projection NTF) and Moran plot of the residuals from the weighted, non-spatial regression

$$\widehat{\beta}_0^{ucr} x_{0,i} + \widehat{\beta}_2^{ucr} x_{2,i} + \widehat{\gamma}^{ucr} \sum_{j=1}^{n} I(j \in N(i))(Y_j - \widehat{\beta}_0^{ucr} x_{0,j} - \widehat{\beta}_2^{ucr} x_{2,j}).$$

Weighted CAR (wcr):

$$\widehat{\beta}_0^{wcr} x_{0,i} + \widehat{\beta}_2^{wcr} x_{2,i} + \widehat{\gamma}^{wcr} \sum_{j=1}^{n} (E_j/E_i)^{1/2} I(j \in N(i))(Y_j - \widehat{\beta}_0^{wcr} x_{0,j} - \widehat{\beta}_2^{wcr} x_{2,j}),$$

where $I(\cdot)$ is the indicator function and the parameters β_0, β_2, and γ are estimated by maximum likelihood for the respective models.

The values of CV for the four models are shown in Table 3. The table confirms that it is better to use a spatial model than a non-spatial model, and it is better to use a weighted regression than an unweighted regression. That is, based on this small evaluation study, weighted spatial regression offers the smallest sum of squared prediction errors. We conclude that a canton's prescription amount per consultation is positively related to its percentage of patients 70 and older, and that errors from this regression relationship are both heteroskedastic and spatially dependent. This proper modelling of the error structure leads to efficient inference on the regression parameters.

Table 2. Comparison of weighted and unweighted CAR model fits

	β_0	β_2	τ^2	γ
unweighted CAR	227.6205	3.5763	1750.5399	0.0471
	(13.3159)	(0.6515)	(152.8731)	(0.0076)
weighted CAR	245.2020	2.5324	30576927.6248	0.0515
	(8.1155)	(0.4881)	(2664826.93)	(0.0043)

Table 3. Comparison of four models based on the cross-validation criterion CV

	unweighted	weighted
non-spatial	514407.09	510833.31
spatial	476124.80	463754.04

5 Discussion and conclusions

Point patterns are often aggregated into counts within spatially contiguous regions. The counts are typically transformed to rates, but any statistical analysis should account for the inherent heteroskedasticity associated with a dependent variable that is a rate. Furthermore, spatial dependence that is perhaps due to a misspecified regression relationship, should be included in the model.

Through the use of a dataset where the rate is prescription amount per consultation, in cantons of the Midi-Pyrénées in southwest France, we illustrate the importance of doing a weighted (both non-spatial and spatial) analysis. The denominators of the rates control the weights; specifically, both response (i.e., rate) and covariates are multiplied by the square root of the corresponding denominator and an unweighted (both non-spatial and spatial) analysis is carried out. Our results show that the unweighted analysis is a much blunter tool for analyzing the data; with a weighted analysis, we are able to explain much more of the variability through the candidate covariates, we are able to account for the remaining variability with a spatially dependent error term, and the resulting weighted spatial (CAR) model has the smallest sum of squared prediction errors.

Acknowledgements

This research was carried out while Noel Cressie was Visiting Professor at the University of Toulouse 1. The authors are appreciative of the University's support of his visits.

References

[1] J.E. Besag. Spatial interactions and the statistical analysis of lattice systems. *Journal of the Royal Statistical Society B*, 36:192–225, 1974.

[2] N.A.C. Cressie. *Statistics for Spatial Data (rev. edn.)*. Wiley, 1993.

[3] N.A.C. Cressie, O. Perrin and C. Thomas-Agnan. Likelihood-based estimation for gaussian mrfs. *Statistical Methodology (forthcoming)*, 2004.

[4] D. Stoyan and H. Stoyan. *Fractals, Random Shapes and Point Fields*. Wiley, 1994.

Strain-typing Transmissible Spongiform Encephalopathies Using Replicated Spatial Data

Simon Webster[1], Peter J. Diggle[2], Helen E. Clough[1], Robert B. Green[3] and Nigel P. French[4]

[1] Department of Veterinary Clinical Sciences, University of Liverpool, UK,
   swebster@liv.ac.uk
[2] Department of Statistics, Lancaster University, UK
[3] Department of Neuropathology, Veterinary Laboratories Agency, UK
[4] Institute of Veterinary, Animal and Biomedical Sciences, Massey University,
   New Zealand

Summary. This chapter describes progress towards the development of more objective methods for discriminating between the neuropathological features produced by different TSE strains, in particular with respect to the vacuolation of brain tissue. We examine data on patterns of vacuolation in the brain tissue of mice challenged with different TSEs, and assess three separate aspects of the pattern: vacuole counts; vacuole sizes; and spatial distribution of vacuoles. The long-term goal is to develop a discriminant rule which can be used to identify a particular TSE strain should it arise in a group of animals.

Key words: Replicated spatial data, Spatial patterns of vacuolation in the brain tissue of mice, Transmissible spongiform encephalopathies (TSEs)

1 Introduction

Transmissible spongiform encephalopathies (TSEs) are a group of related fatal neurodegenerative disorders, affecting a number of animal species [7, 10]. There are several strains including Bovine Spongiform Encephalopathy (BSE) in cattle, scrapie in sheep and new-variant Creutzfeldt-Jakob Disease (vCJD) in humans. Some of these can be divided; there are numerous strains of scrapie for example, but there is only one strain of BSE in the UK. TSEs exhibit a number of distinctive neuropathological features including spongiform change, characterised by the appearance of lesions, or vacuoles (small, rounded holes), in brain tissue. Other symptoms of TSEs vary but commonly include alterations of temperament and lack of physical coordination, for example an unsteady gait or involuntary jerking.

Strain-typing, the process of distinguishing between and identifying strains, is an important tool in facilitating our understanding of the pathology, epidemiology and transmission of these diseases. TSE strains vary in a number of respects, including the length of their incubation period and the severity and distribution of the lesions they create in the brain [2]. Strains also differ in terms of their transmissibility, both within and between species. Strain-typing allows us to quantify these differences, as well as enabling exploration of the links between TSEs occurring naturally in different species. Notably, strain-typing led to the discovery that BSE and vCJD were related and that humans could have contracted vCJD by eating beef originating from BSE-affected cattle [3].

Lesion profiling was first described in [6] and remains the most commonly used method of TSE strain-typing. Its focus is the vacuolation occurring in the brain tissue of affected animals. Usually, the disease is inoculated into mice to determine infectivity or to identify the TSE strain. Severity of vacuolation is rated on a 0–5 scale in each of nine pre-defined regions of the mouse brain, located on four vertically cut two-dimensional sections of brain tissue. A severity score of 0 corresponds to no vacuolation, while 5 corresponds to very severe vacuolation. Fraser and Dickinson include a diagram in [6], illustrating the definition of the regions of interest.

A plot of region versus severity score, known as a lesion profile, helps to determine the strain because different strains have been found to show distinctive shapes. Figure 1 shows examples of typical lesion profile scores from BSE and scrapie-challenged mice. The scores shown are averages from experiments on three groups of mice, profiled at the Veterinary Laboratories Agency in Weybridge, UK. Two of the groups had been challenged with scrapie and one with BSE.

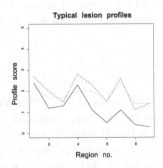

Fig. 1. Typical lesion profile scores for mice challenged with BSE (*solid line*), and representative Scrapie (*dashed and dotted lines*)

This method has several limitations. Foremost amongst these are the subjectivity and imprecision involved in assessing vacuolation severity. The 0–5 scale is based only loosely on the number of vacuoles in the region, for example a score of 3 is defined as "moderate numbers of vacuoles, evenly scattered". In addition, lesion profiling does not take into account, for example, the distribution of vacuole sizes or their spatial arrangement. Note also that the ordering of the regions along the x-axis in Fig. 1 is arbitrary.

This contribution describes progress towards the development of more objective methods for discriminating between the neuropathological features produced by different TSE strains, in particular with respect to the vacuolation of brain tissue. We examine data on patterns of vacuolation in the brain tissue of mice challenged with different TSEs, and assess three separate aspects of the pattern: vacuole counts; vacuole sizes; and spatial distribution of vacuoles. The long-term goal is to develop a discriminant rule which can be used to identify a particular TSE strain should it arise in a group of animals.

2 Data Collection

We describe the sample material under study, and the processes by which the data for our analyses are extracted.

2.1 Sample Material

Slides of brain tissue sections from mice challenged with BSE or scrapie are provided by the Veterinary Laboratories Agency (UK). The mice are all of the R_{III} line, which is a standard inbred strain of mice used in laboratory studies to eliminate genetic variability from scientific testing. We focus upon three regions, the paraterminal body, thalamus and tectum of the midbrain. Each section is strongly stained, with the vacuolation presenting as white holes. Figure 2 illustrates severe vacuolation.

Fig. 2. Severe vacuolation in the hippocampus region of a scrapie-affected mouse brain

For each mouse, the TSE strain (BSE or scrapie) is known, along with covariate information such as mouse gender and whether the mouse was challenged with sheep or cattle material. Scrapie-challenged mice have been categorised only by gender, and were all challenged using wild-type ovine material of mixed strain. Scrapie isolations from natural cases appear to be a mixture of strains upon first passage into mice. However, upon serially passaging the infection from one mouse to another, the dominant scrapie strain can be distinguished. The use of serial passage enables the original isolate to be refined until it exhibits a constant presentation, which may be identified by comparison to the presentation of other known strains. Only single-passaged material was available in the current study, but we intend to apply our methods also to multiply-passaged "pure" strains, and we would expect these to yield more consistent results. Using these pure strains would enable us to compare different strains of scrapie, as well as contrasting mixed-strain scrapie with BSE. Mice challenged with bovine BSE material, of which there is only one known strain in the UK, have been further categorised by whether the cattle were in a pre-clinical or terminal condition. Pre-clinical cattle were killed at a predetermined time-point, while terminal cattle were killed when they began to show end-stage clinical signs, to limit suffering. Mice challenged with ovine BSE material are categorised by the length of time between the sheep being challenged and its pre-determined time of death (10, 16 or 22 months).

For each combination of covariates, there are a number of replicated experiments, or groups of mice which have been challenged using an equal volume of the same inoculum. Replication offers the opportunity to quantify variability at different levels of the experiment, and to identify differences between groups of mice with respect to the neuro-anatomical features of interest.

Table 1 summarises the available material. This is limited by difficulties in obtaining usable samples. For each combination of covariates, there are between 1 and 4 replicates, each consisting of between 4 and 9 mice. There are 84 mice in total.

Some data are missing as a result of sections being inaccurately cut. Sections from seven mice have a missing paraterminal body, two have a missing thalamus and one has a missing tectum. These missing sections arise for reasons unrelated to the disease, and can therefore be treated as missing completely at random, in the sense of [9].

Table 1. Experimental design (15 groups of replicates)

	BSE Cattle Status		BSE Sheep Time post-challenge			Scrapie (sheep)
	Pre-clinical	Terminal	10 mnth	16 mnth	22 mnth	
No. males/	4/4	2/2	3/2	3/3	4/5	1/3
females in	4/3	2/3		3/2	3/5	1/3
each group	5/2	2/2				2/2
of replicates	2/2					
Total no. mice	26	13	5	11	17	12
(male/female)	(15/11)	(6/7)	(3/2)	(6/5)	(7/10)	(4/8)

We have no reason to expect differences in spatial distribution between males and females, nor are these suggested by exploratory data analysis. Hence, for the spatial analysis, we ignore sex. Severity of vacuolation is, however, believed to differ between the sexes, with females having a greater number of vacuoles on average. For the analyses of vacuole counts and vacuole sizes, therefore, sex is considered as an additional covariate.

2.2 Extraction of Data from the Sample Material

The software Histometrix (Medical Solutions plc, Nottingham, UK) allows microscopy slides to be viewed and manipulated on the computer screen. In particular, coordinates of individual vacuoles and a polygonal border for each region can be downloaded. We use a tool within the software to define the edges of each vacuole, and the edges of the regions of interest. Coordinates defining these edges are saved and downloaded into statistical software. Some of the subjectivity and imprecision of lesion profiling is removed by using the computer to generate vacuole counts, rather than making a judgement based on visual impression. The software we used can detect vacuoles automatically but is only able to record spatial coordinates for *manually* defined vacuoles, hence we retain a degree of subjectivity.

The images are first registered to a common set of coordinates by choosing one physical point in the brain tissue to be the (0, 0) spatial coordinate for all mice, as shown in Fig. 3. The orientation of the regions is fixed by choosing any two points along their bottom edge and rotating the image around the (0,0) point until these lie in a horizontal line.

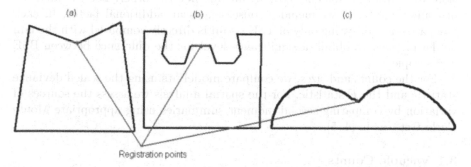

Registration points

Fig. 3. Registration points ([0,0] coordinates) for the three tissue regions: (a) paraterminal body; (b) thalamus; (c) tectum of midbrain

In order for a pair of coordinates to relate to an approximately equivalent physical location in the brains of different mice, we scale the patterns to a common area. Scaling alters vacuole sizes and distances between vacuoles, but inspection of the patterns suggests that this is desirable, since vacuoles in larger brains also tend to be larger in size. In effect, the disease in a larger brain appears as a magnification of the disease in a smaller brain. Scaling ensures that distances and relative vacuole sizes are comparable between patterns.

Once the polygonal regions have been registered and scaled, those within each group of replicates are overlaid and their intersection is used as the common region for that experiment. Any vacuoles outside this intersection are discarded.

3 Methods

Vacuole counts, vacuole sizes and spatial distributions of vacuoles are assessed in three separate analyses. In each analysis, an individual mouse contributes a single response from each of the three brain tissue regions (tectum of midbrain, thalamus and paraterminal body), and these are assessed separately.

In the study, there are 15 replicated experiments, each replicate consisting of a group of mice challenged with the same inoculum. These 15 inocula can be categorised into 6 distinct types according to the specific origin of the material, as defined by a combination of covariates, which are: source animal (cattle or sheep), TSE strain (BSE or scrapie), and stage of infection. The 6 types for which we have data, form the column headings in Table 1.

Within each analysis, we investigate whether there are significant differences between the 15 groups of replicates. If there are, we sequentially fit and compare simpler models to assess whether significant proportions of this variation are explained by differences in the origin of the infection (source animal, TSE strain and stage of infection), or more simply by the TSE strain alone. We also consider the possibility that the inoculum and/or covariate information are important for one TSE strain but not the other. For the analyses of counts and sizes, we include mouse sex as an additional factor. In each case, a model consisting only of TSE strain is directly compared with the null model in order to obtain a significance level for the difference between BSE and scrapie.

For the counts and sizes, we compare model fits using the scaled deviance statistic and the F-statistic. For the spatial analysis we assess the sources of variation by comparing second moment summaries using appropriate Monte Carlo tests, as in [4, 5].

3.1 Vacuole Counts

Vacuole counts are modelled using a quasi-Poisson log-linear model with dispersion parameter ϕ equal to the variance-to-mean ratio. The saturated model allows a separate mean count for each mouse. More parsimonious sub-models are compared using a scaled deviance criterion as follows. Let R_s and R_c be the residual deviances for the simpler and more complicated sub-models, on d_s and d_c degrees of freedom, respectively. To test the adequacy of the simpler model, we compare $D = (R_s - R_c)/(R_c/d_c)$ with critical values of χ_q^2, where $q = d_s - d_c$.

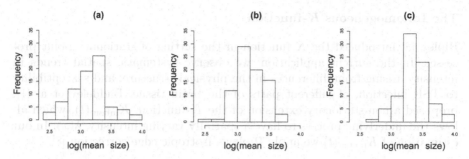

Fig. 4. Histograms of logged mean scaled vacuole sizes. (a) is the paraterminal body, (b) the thalamus and (c) the tectum

3.2 Vacuole Sizes

Each of the brain tissue regions has been scaled to a common area of $1\mu m^2$; we analyse the scaled vacuole sizes, but report sizes in units of $10^6 \mu m^2$, which has the effect of putting the units on approximately the same scale as the original. For each vacuole, the image analysis software records the lengths, r_1 and r_2, of the two principal radii, and we approximate each vacuole as an ellipse, hence area $= \pi r_1 r_2$.

The response for each mouse is the logarithm of the observed mean vacuole size. Figure 4 suggests that the variation in log-mean-size is approximately Gaussian and so we analyse these data using a Gaussian linear model. Note that by summarising the scaled vacuole sizes from each mouse by a single log-mean, we allow for the fact that individual sizes within mice are dependent.

The procedure for fitting and comparing models for the mean scaled vacuole sizes is essentially the same as for the vacuole counts, except that the scaled deviance statistic is replaced by a standard F-statistic,

$$F = \frac{(R_s - R_c)/q}{R_c/(n - d_c)}$$

where now n is the number of mice, R_s and R_c are residual sums of squares and we compare F with critical values of $F_{q,n-d_c}$.

3.3 Spatial Distribution

We use the inhomogeneous K-function [1] to summarise each spatial pattern by its degree of regularity or aggregation, allowing for a spatially varying intensity.

We compare groups of mice, with groups defined in three ways in turn: by specific inoculum; by the type of the inoculum (as defined by the different combinations of covariates shown in Table 1); and by TSE strain alone. We then suggest sources of variation in the data by comparing the attained significance level for group differences across these three analyses.

The Inhomogeneous K-function

Ripley [8] introduced the K-function in the setting of stationary point processes. In the current application, we observe systematic spatial trends in intensity arising from differences in the physical structure, and susceptibility to TSE infection, of different parts of the brain tissue. Baddeley et al [1], proposed a non-stationary extension of the K-function, $K_{\text{inhom}}(t)$, which allows the underlying process to have a spatially varying intensity, $\lambda(x)$. In our estimation of $K_{\text{inhom}}(t)$ we apply Ripley's isotropic edge correction [8].

Estimating the Underlying Intensity Function

We use a kernel smoothing procedure to estimate $\lambda(x)$. As noted in [1], it is difficult to distinguish between variation due to $\lambda(.)$ and variation due to spatial interaction, using data from a single point pattern realisation. Our approach alleviates this difficulty by exploiting the replicated nature of the data as follows.

Assume that mice within each group of replicates have proportional intensity functions, i.e. if $\lambda_{ij}(.)$ is the intensity function for the jth mouse in the ith group, then $\lambda_{ij}(x) = \alpha_{ij}\lambda_i(x)$. Inspection of the data supports this as a reasonable approximation. The K-function is invariant under scaling of $\lambda(x)$, and therefore provides a summary of the spatial distribution which is complementary to the information provided by vacuole counts. Within each group of replicates, we superimpose the points from all replicates and obtain kernel intensity estimates at the locations of each vacuole in the superimposition, using a standard Gaussian kernel. Note that the argument for the bias-correction recommended by [1] when estimating both $\lambda(x)$ and $K_{\text{inhom}}(t)$ from a single pattern does not apply when replicated data are available.

We choose the kernel smoothing parameter, h, using a cross-validation procedure. For m replicates in the ith group, let x_{jk} denote the location of the kth vacuole in the jth replicate, and $\hat{f}^{(-j)}(x;h)$ the kernel estimate with smoothing parameter h, obtained by omitting the jth replicate from the data and scaled to integrate to one over the study region. Then, the cross-validation criterion chooses h to maximise

$$L(h) = \sum_{j=1}^{m} \sum_{k=1}^{n_{ij}} \log \hat{f}^{(-j)}(x_{jk}),$$

where n_{ij} is the number of points in the ijth pattern. We scale each kernel estimate to integrate to 1, so that $L(h)$ can be interpreted as a cross-validated log-likelihood. We then estimate the constants of proportionality as $\hat{\alpha}_j = n_{ij}/n_i$, where $n_i = \sum_j n_{ij}$.

Comparing Replicated Spatial Point Patterns

Two statistics for comparing two or more groups of replicated spatial data using the K-function of [8] have been proposed [4, 5]. We modify the proposal in [5] by replacing $K(t)$ with $K_{\text{inhom}}(t)$ throughout.

Let $\hat{K}_{ij}(t)$ denote the estimate of $K_{\text{inhom}}(t)$ for the jth replicate in the ith experimental group. We then estimate group-mean K-functions by

$$\bar{K}_i(t) = n_i^{-1} \sum_{j=1}^{n_i} n_{ij} \hat{K}_{ij}(t)$$

where n_{ij} is the number of points in the jth replicate within the ith group and $n_i = \sum_j n_{ij}$. Similarly, we estimate an overall mean K-function by

$$\bar{K}(t) = n^{-1} \sum_{i=1}^{g} n_i \bar{K}_i(t)$$

where $n = \sum_i n_i$.

The statistic to test for significant differences amongst the groups is then

$$D_g = \sum_{i=1}^{g} \int_0^{t_0} w(t) n_i \left[\bar{K}_i(t) - \bar{K}(t) \right]^2 dt.$$

The choice of t_0 should capture the range of distances over which spatial interactions between vacuoles are thought to operate; in the absence of scientific guidance on this point, a sensible upper limit is one quarter the width of the region. For the weighting function $w(t)$, we use $w(t) = t^{-2}$ to reflect the asymptotic variance of $\hat{K}(t)$ for a Poisson process. The statistic D_g is loosely analogous to a residual sum of squares in a conventional one-way ANOVA.

Because the distribution of D_g is intractable, we use the following Monte Carlo significance test of the null hypothesis of no difference between the groups. We first compute a set of residual K-functions,

$$R_{ij}(t) = n_{ij}^{\frac{1}{2}} \left\{ \hat{K}_{ij}(t) - \bar{K}_i(t) \right\}$$

The $R_{ij}(t)$ are approximately exchangeable under both the null and alternative hypotheses. We now randomly permute the $R_{ij}(t)$ to give a set of permuted residual K-functions $R_{ij}^*(t)$ and calculate

$$\hat{K}_{ij}^*(t) = \bar{K}(t) + n_{ij}^{-\frac{1}{2}} R_{ij}^*(t).$$

We then use the $\hat{K}_{ij}^*(t)$ to compute a realisation D^* from the approximate sampling distribution of D under the null hypothesis. Repeated sampling of D^* leads to a Monte Carlo significance test. If D_g is the rth smallest amongst $(D_g, D_1^*, \ldots, D_N^*)$, then the attained significance level of a test of the null hypothesis of no group differences is given by $p = r/(N+1)$.

In addition to conducting a test of significance, a graphical display of estimated individual and group-mean K-functions is often informative. We plot $\hat{K}(t) - \pi t^2$, a quantity with an expected value of zero under complete spatial randomness (CSR), to facilitate visual inspection.

4 Results

4.1 Vacuole Counts

We present the results for each region in turn. Outputs from the models for counts are summarised in Table 2. For scrapie-affected mice, the mean counts are 232 in the paraterminal body, 162 in the thalamus and 88 in the tectum. For the BSE-affected mice, the mean counts are 47 in the paraterminal body, 40 in the thalamus and 37 in the tectum. There is thus a suggestion that scrapie mice demonstrate more severe vacuolation than BSE mice. In addition, there may be greater disparity between counts in the three regions for scrapie mice. We now consider the quasi-Poisson log-linear modelling approach as described in Sect. 3.1.

For the paraterminal body, extra-Poisson variation is substantial ($\hat{\phi} = 174.19$). There is a significant difference between BSE and scrapie-challenged mice ($D = 35.25, p < 0.01$). However, a model which combines all BSE mice into one group but retains the three distinct groups for scrapie mice according to specific inoculum, offers a significant improvement over a model in which scrapie mice are also defined as a single group ($D = 37.08, p < 0.01$). There is no benefit in separating groups of BSE mice according to covariate information or inoculum ($D = 13.99, p = 0.23$) and the effect of mouse sex is not significant ($D = 0.05, p = 0.83$). Quasi-Poisson modelling can therefore distinguish between BSE and scrapie mice in the paraterminal body, and there are no significant differences between mice challenged with BSE from different inocula. Variation between inocula remains important, however, for scrapie mice.

In the thalamus, extra-Poisson variation is again substantial ($\hat{\phi} = 68.71$). There is a significant difference between the counts for BSE and scrapie-challenged mice ($D = 47.58, p < 0.01$) but again, a model combining all BSE mice but retaining the distinction between mice challenged using different scrapie inocula proves significantly better than a model which also combines all scrapie mice ($D = 19.24, p < 0.01$). There is no improvement made by splitting BSE groups according to inoculum ($D = 5.31, p = 0.26$) and the inclusion of sex offers no significant improvement over the null model ($D = 0.26, p = 0.61$). BSE and scrapie can be distinguished but variation between scrapie mice challenged using different inocula is still significant.

Finally, within the tectum, extra-Poisson variation is again evident ($\hat{\phi} = 20.85$). As in the other two regions, there is a significant difference between BSE and scrapie-challenged mice ($D = 27.36, p < 0.01$) but in contrast to

Table 2. Comparisons and deviance reduction statistics for quasi-Poisson log-linear models of vacuole counts from the paraterminal body, the thalamus and the tectum of the midbrain. Inoculum indicates specific material used. Inoculum origin indicates combination of covariates. Model (a) is all mice grouped by inoculum origin, model (b) is BSE grouped by inoculum origin, scrapie by inoculum, and model (c) is BSE all combined, scrapie grouped by inoculum

Model	Deviance	df	Model comparison	df	D
Paraterminal body					
Null	7070.8	76			
Sex	7066.2	75	Null vs Sex	1	0.05
Inoculum	2049.3	62	Null vs Inoculum	14	151.94
(a)	3529.4	71	(a) vs Inoculum	9	44.78
(b)	2253.8	69	(b) vs Inoculum	7	6.19
(c)	2511.6	73	(c) vs Inoculum	11	13.99
TSE strain	3787.3	75	TSE strain vs (c)	2	37.08
			Null vs TSE strain	1	35.25
Thalamus					
Null	3336.3	81			
Sex	3325.4	80	Null vs Sex	1	0.26
Inoculum	959.8	67	Null vs Inoculum	14	165.90
(a)	1302.7	76	(a) vs Inoculum	5	23.94
(b)	1030.3	74	(b) vs Inoculum	7	4.92
(c)	1104.3	78	(c) vs (b)	4	5.31
TSE strain	1376.6	80	TSE strain vs (c)	2	19.24
			Null vs TSE strain	1	47.58
Tectum of midbrain					
Null	1360.7	82			
Sex	1298.8	81	Null vs Sex	1	3.86
Sex + inoculum	666.0	67	Sex vs Sex + inoculum	14	63.65
Sex * inoculum	498.6	53	(Sex +) vs (Sex *) inoculum	14	17.80
Sex + (a)	777.7	76	Sex + (a) vs Sex + inoculum	9	11.23
Sex + TSE strain	860.1	80	Sex + TSE strain vs Sex + (a)	4	8.12
			Null vs TSE strain	1	27.36

the other two regions, mouse sex is marginally significant ($D = 3.86, p = 0.05$). Source inoculum is significant ($D = 63.65, p < 0.01$), but interaction between sex and inoculum is non-significant ($D = 17.80, p = 0.22$). Finally, a model which includes sex and combines the groups of replicates which share covariate information, does not give a significant improvement over a model which categorises the mice only by TSE strain ($D = 8.12, p = 0.09$).

Conclusions

The preferred model for both the paraterminal body and the thalamus combines all mice challenged with BSE into one group but retains a distinction between the three groups of mice challenged with scrapie. For the tectum, the model again separates BSE and scrapie but combines all scrapie mice into one group. Sex is included in the preferred model for the tectum data only.

In the paraterminal body and thalamus, counts in this study differed significantly between groups of scrapie-challenged mice challenged using different inocula, but this was not the case for BSE-challenged mice. It may be that this variation could be reduced by using mice challenged with a pure strain of scrapie. In the tectum, groups of mice with the same covariate information (TSE strain, cattle or sheep origin and stage of infection) were similarly affected, but there were significant differences in the counts between groups whose covariate information differed. Overall, the counts differed between the three regions to a greater extent for scrapie mice than BSE mice and the results overall indicate a clear potential for count data to be useful in distinguishing TSE strains.

4.2 Vacuole Sizes

Simple analysis of the scaled vacuole size data indicates that the mean logged vacuole sizes are more variable between source groups in the paraterminal body (mean = 3.134, sd = 0.168) and the tectum (mean = 3.342, sd = 0.125) than in the thalamus (mean = 3.213, sd = 0.058). We proceed to fit a Gaussian linear model in each region, with the results summarised in Table 3.

For the paraterminal body, sex significantly improves the null model ($F = 8.10, p = 0.01$), and the model is further improved by grouping the mice according to the specific inoculum used ($F = 2.13, p = 0.02$), but not by an interaction term ($F = 1.08, p = 0.40$). A model consisting only of TSE strain is not significantly better than the null model ($F = 0.03, p = 0.86$), suggesting that there is no consistent difference between BSE and scrapie. The preferred model combines groups of mice sharing covariate information (same TSE strain, source animal (cattle or sheep) and stage of infection). The ability to distinguish between scrapie and BSE may be being limited by the significant variation between BSE-challenged mice whose inoculum is of a different origin.

In the thalamus, neither sex nor inoculum significantly improve the null model ($F = 0.94, p = 0.33$ and $F = 0.34, p = 0.99$ respectively). The difference between BSE and scrapie-challenged mice is also non-significant ($F = 0.20, p = 0.66$). In summary, there is no evidence of significant variation in scaled vacuole sizes above and beyond differences between individual mice.

We consider finally the tectum. The effect of specific inoculum is significant ($F = 2.00, p = 0.03$), but the model fit is not compromised by combining the three scrapie groups whilst maintaining the distinction between BSE mice

Table 3. Comparisons and standard F statistics for Gaussian log-linear models of scaled vacuole sizes from the paraterminal body, the thalamus and the tectum of the midbrain. Inoculum indicates specific material used. Inoculum origin indicates combination of covariates. Model (a) is all mice grouped by inoculum origin, model (b) is BSE grouped by inoculum origin, scrapie grouped by inoculum, model (d) is BSE grouped by inoculum, scrapie all combined

Model	Deviance	df	Model comparison	$(q, n - d_c)$	F
Paraterminal body					
Null	6.615	76			
Sex	5.970	75	Null vs Sex	(1, 75)	8.10
Sex + inoculum	4.012	61	Sex vs Sex + inoculum	(14, 61)	2.13
Sex * inoculum	3.036	47	(Sex +) vs (Sex *) inoculum	(14, 47)	1.08
Sex + (a)	4.710	70	Sex + (a) vs Sex + inoculum	(9, 61)	1.18
Sex + TSE strain	5.967	74	Sex + TSE strain vs Sex + (a)	(4, 70)	4.67
TSE strain	6.613	75	Null vs TSE strain	(1, 75)	0.03
Thalamus					
Null	4.169	81			
Sex	4.120	80	Null vs Sex	(1, 80)	0.94
Inoculum	3.890	67	Null vs Inoculum	(14, 67)	0.34
TSE strain	4.159	80	Null vs TSE strain	(1, 80)	0.20
Tectum of midbrain					
Null	4.758	82			
Sex	4.727	81	Null vs Sex	(1, 81)	0.54
Inoculum	3.370	68	Null vs Inoculum	(14, 68)	2.00
(a)	4.278	77	(a) vs Inoculum	(9, 68)	2.04
(b)	4.259	75	(b) vs Inoculum	(7, 68)	2.56
(d)	3.390	70	(d) vs Inoculum	(2, 68)	0.20
TSE strain	4.686	81	Null vs TSE strain	(1, 81)	1.24

challenged using different inocula ($F = 0.20, p = 0.82$). Mouse sex is again not significant in comparison with the null model ($F = 0.54, p = 0.46$). A model consisting only of TSE strain does not show a significant improvement on the null model ($F = 1.24, p = 0.27$). We therefore conclude that there is no significant evidence of a difference between BSE and scrapie, but there are strong differences between BSE mice challenged using different inocula which are not merely a result of differing covariate information.

Conclusions

The preferred models for each region differ considerably. For the paraterminal body, the chosen model divides the mice into groups according to shared covariate information, and sex is also included as a factor. In the thalamus, we are forced to accept the null model as there is no significant evidence of variation beyond individual differences. For the tectum, the preferred model divides the mice into groups according to the specific inocula used. Sex is not included in the model for the tectum data.

Scaled vacuole sizes in the tectum and paraterminal body do not appear to help us distinguish BSE from scrapie. However, the method of modelling sizes is clearly able to distinguish between groups of mice challenged using different inocula. With a larger sample size, and using pure strains of scrapie, it is possible that differences between TSE strains could be detected. Again, the precise interpretation of the results differs between regions. In the paraterminal body, for example, scaled vacuole sizes appear to be strongly associated with the covariates but do not significantly differ between groups of mice which share the same covariate information but have been challenged using different source material. In the tectum, there is again no significant difference between mice challenged with different scrapie inocula but there is significant variation between groups of mice challenged with different BSE inocula, irrespective of whether these groups share covariate information. Scaled vacuole size differs by sex in the paraterminal body but not in the thalamus or tectum.

4.3 Spatial Distribution of Vacuoles

TSEs cause progressive degeneration of neurological tissue, through the formation of vacuoles, and a spatial analysis of the resulting point patterns of vacuole centres may be a useful complement to the analysis of counts and sizes.

In each of the three tissue regions, the estimated K-functions indicate substantial variation between the different groups of mice, with vacuoles tending to be aggregated in some groups and regularly spaced in others (Figs. 5–7).

Paraterminal body

The group-mean inhomogeneous K-function estimates for the paraterminal body are shown in Fig. 5. Monte Carlo testing shows a significant difference between strains (Fig. 5(c), $p = 0.03$), suggesting that within the paraterminal body at least, the locations of vacuoles may have a role to play in strain typing. The average K-function suggests that the arrangement of vacuoles within the paraterminal body of scrapie mice is aggregated over the range of spatial interaction studied, in comparison with that for BSE affected mice, for whom the arrangement appears much closer to spatially random. Monte Carlo significance tests provide no evidence of significant differences in spatial arrangement between the groups of replicates (Fig. 5(a), $p = 0.11$) or the combination of groups sharing covariates (Fig. 5(b), $p = 0.10$).

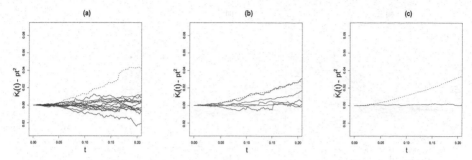

Fig. 5. Group means $\bar{K}_i(t) - \pi t^2$ in paraterminal body region. (*BSE groups are shown as solid lines and scrapie groups as dashed lines*). (**a**) shows groups of mice combined by inoculum (replicated experiments) (**b**) shows groups combined according to shared covariate information, and (**c**) shows BSE groups and scrapie groups combined

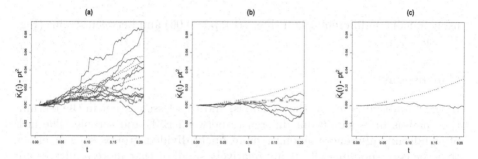

Fig. 6. Group means $\bar{K}_i(t) - \pi t^2$ in thalamus region. (*BSE groups are shown as solid lines and scrapie groups as dashed lines*). (**a**) shows groups of mice combined by inoculum, (**b**) shows groups combined according to shared covariate information, and (**c**) shows BSE groups and scrapie groups combined

Thalamus

Figure 6 shows the group mean inhomogeneous K-function estimates for the thalamus. As with the paraterminal body, a significant difference is found between BSE and scrapie (Fig. 6(c), $p = 0.05$). Over the range of spatial interaction studied, a similar pattern of aggregation in scrapie-affected thalamus and a pattern close to complete spatial randomness in BSE-affected thalamus is observed. No significant differences are found between groups as defined by specific inoculum (Fig. 6(a), $p = 0.10$) or combination of covariates (Fig. 6(b), $p = 0.14$).

Tectum of Midbrain

Figure 7 indicates wide variation in the K-function estimates for the tectum. There are marginally significant differences between groups of mice challenged

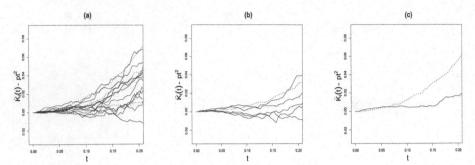

Fig. 7. Group means $\bar{K}_i(t) - \pi t^2$ in tectum of midbrain region. (*BSE groups are shown as solid lines and scrapie groups as dashed lines*). (**a**) shows groups of mice combined by inoculum, (**b**) shows groups combined according to shared covariate information, and (**c**) shows BSE groups and scrapie groups combined

using inocula of different origin (Fig. 7(b), $p = 0.06$) and between strains (Fig. 7(c), $p = 0.07$).

Conclusions

In the parateminal body, the thalamus, and to a lesser extent in the tectum, the analysis finds significant differences between BSE and scrapie. The lack of significant differences amongst more finely divided groups suggest either that the corresponding effects are relatively small or that more replicates are needed to achieve the required sensitivity. The nature of the difference in spatial patterns between mice challenged with BSE and scrapie is strikingly similar for the three brain tissue regions (Figs. 5c, 6c and 7c).

5 Discussion

It is encouraging that there is complementary information to be found in each of the analyses of counts, scaled sizes and spatial distribution. This suggests that an assessment based on one aspect of the vacuolation alone, as in lesion profiling, is at best incomplete, and may be neglecting useful information.

In particular, because the K-function is invariant to scale-changes in intensity and we allow proportional, rather than identical, intensity surfaces for different mice within each group, the spatial analysis is strongly complementary to the count analysis. Our current analyses of vacuole size is confined to an analysis of average vacuole size within each individual pattern. A comprehensive analysis would need to recognise that size is a quantitative mark attached to each vacuole location, i.e. a marked point process.

The results from each of the three analyses are markedly different, but in combination could provide a useful tool for discriminating between TSE strains and helping to identify the strain of TSE affecting a given group of animals. The mice in this study were affected with known TSE strains and we have to date made no attempt to identify unknown strains using these techniques. However, our results suggest that there are clear pathological differences, at least between BSE and scrapie, and these differences may ultimately be used to develop recognisable "fingerprints", indicating a particular disease.

As an example, the vacuoles in all of the studied brain regions of a group of scrapie-affected mice are likely to be aggregated, but mild aggregation in the tectum of midbrain only may be indicative of BSE. The scrapie-affected mice may also show greater differences in vacuole counts across the three regions, and in particular a much lower count in the tectum than in the other two regions. A similar count in all three regions may be suggestive of BSE. Finally, the average scaled vacuole size may provide useful information about the strength of the particular inoculum used. For a given mouse, this will assist the interpretation of a simple vacuole count.

The differences between the three regions are informative in terms of counts, scaled sizes and spatial distribution. As an obvious extension to this work, we propose to investigate whether similar analyses in each of the nine regions as defined in the lesion profiling system might be used to highlight further differences between strains. The groups and numbers of replicates studied here are small, and it is encouraging that the differences between groups and/or strains are sufficiently strong to be detected using our proposed approaches.

The methods outlined have thus far been applied to two TSE strains, namely BSE and a mixed-strain scrapie. Multiple passaging of the scrapie infection can lead to the identification of known strains, which will produce more consistent, distinguishable patterns. Since the material available for this study was only single-passaged, we are unable to say whether the mice in different replicated experiments have been challenged with the same strain of scrapie, as this would only become clear upon further passage. Some of the differences between scrapie groups could, therefore, be attributable to differences in strain between the inocula used. It is our intention to apply these methods to known strains, allowing us to discover whether the methods can help distinguish between different scrapie strains as well as between generic scrapie and BSE.

References

[1] A.J. Baddeley, J. Møller and R.P. Waagepetersen. Non- and semi-parametric estimation of interaction in inhomogeneous point patterns. *Statistica Neerlandica*, 54:329–350, 2000.

[2] M.E. Bruce. Scrapie strain variation and mutation. *British Medical Bulletin*, 49(4):822–838, 1993.

[3] M.E. Bruce, R.G. Will, J.W. Ironside, I. McConnell, D. Drummond, A. Suttie, L. McCardle, A. Chree, J. Hope, C. Birkett, S. Cousens, H. Fraser and C.J. Bostock. Transmissions to mice indicate that new-variant cjd is caused by the bse agent. *Nature*, 389:498–501, 1997.

[4] P.J. Diggle, N. Lange and F. Benes. Analysis of variance for replicated spatial point patterns in clinical neuroanatomy. *Journal of the American Statistical Association*, 86:618–625, 1991.

[5] P.J. Diggle, J. Mateu and H.E. Clough. A comparison between parametric and non-parametric approaches to the analysis of replicated spatial point patterns. *Advances in Applied Probability*, 32:331–343, 2000.

[6] H. Fraser and A.G. Dickinson. The sequential development of brain lesions of scrapie in three strains of mice. *Journal of Comparative Pathology*, 78(3):301–311, 1968.

[7] D.A. Harris. Mad cow disease and related spongiform encephalopathies. *Current topics in Microbiology and Immunology*, 284, 2004.

[8] B.D. Ripley. Modelling spatial patterns. *Journal of the Royal Statistical Society B*, 39:172–212, 1977.

[9] D.B. Rubin. Inference and missing data. *Biometrika*, 63:581–592, 1976.

[10] P. Yarm. *The pathological protein: Mad cow, chronic wasting and other deadly prion diseases*. Copernicus books, 2003.

Modelling the Bivariate Spatial Distribution of Amacrine Cells

Peter J. Diggle[1], Stephen J. Eglen[2] and John B. Troy[3]

[1] Lancaster University and Johns Hopkins University School of Public Health, UK, p.diggle@lancaster.ac.uk
[2] University of Cambridge, UK, S.J.Eglen@damtp.cam.ac.uk
[3] Northwestern University, USA, j-troy@northwestern.edu

Summary. We are interested in studying the spatial dependency between the positions of on and off cholinergic amacrine cells because we hope this will tell us something about how the two cell types emerge during development.

Our goal in is to demonstrate how recently developed Monte Carlo methods for conducting likelihood-based analysis of realistic point process models can lead to sharper inferences about the bivariate structure of the data. In particular, we will formulate and fit a bivariate pairwise interaction model for the amacrines data, and will argue that likelihood-based inference within this model is both statistically more efficient and scientifically more relevant than *ad hoc* testing of benchmark hypotheses such as independence.

Key words: Bivariate pairwise interaction processes, Cholinergic amacrine cells, Likelihood-based inference, Monte-Carlo methods

1 Introduction

1.1 Biological Background

Humans and many vertebrates have a very highly specialised visual system that allows us to perceive the world. Our capacity to see begins at the back of the eye, where a neural structure called the retina converts light into electrical activity. The retina is a three-dimensional structure, composed of several types of cell (Fig. 1). The light is first converted into neural activity by the photoreceptors, which then pass their signals through several types of interneuron. Eventually the activity reaches the retinal ganglion cells, which then send the signals to the brain.

There are many different types of neuron in the retina; with a few exceptions, each type of neuron is arranged in a regular fashion so that the visual world is systematically sampled, without leaving any "holes" in visual space. In this contribution, we will focus on the spatial positioning of two types of retinal neuron, known as the cholinergic amacrine cells [12, 23]. These interneurons modulate the pattern of visual information as it passes through the retina, and are thought to play an important role in the detection of motion in particular directions [11]. There are two types of cholinergic amacrine cell, depending on the depth within the retina at which the cell body is found. Cells found within the inner nuclear layer are termed "off" cells here, whilst cells found in the ganglion cell layer are termed "on" cells.

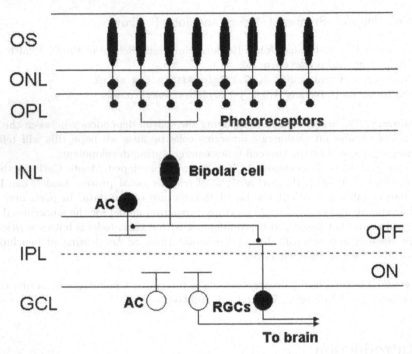

Fig. 1. Cross-section through the different layers of the retina. Layers are named to left, for reference. (*OS: outer segments; ONL: outer nuclear layer; OPL: outer plexiform layer; INL: inner nuclear layer; IPL: inner plexiform layer; GCL: ganglion cell layer*). Light enters the eye through the front (at bottom) and travels through the retina where it is converted to electrical activity by the photoreceptors. Two main cell types can be classified into "on" (*open circles*) or "off" (*filled circles*) depending on whether the cell is excited by an increase or a decrease in illumination. Cholinergic amacrine cells (AC) are found at two different layers, whereas retinal ganglion cells (RGCs) are normally all located within the GCL. RGCs are the only cells that send their information along the optic nerve to the brain. Many cell types have been omitted from this diagram for simplicity

We are interested in studying the spatial dependency between the positions of on and off cells because we hope this will tell us something about how the two cell types emerge during development: do the two cell types emerge from a single undifferentiated population, or do they develop independently of each other? Also, in more general terms, this question that we ask here about the cholinergic amacrine cells could be asked of other cell types. In the special case when the two types of retinal neuron are in different layers, existing approaches [6] may be suitable to test for independence. However, these techniques are *a priori* invalid when both cell types occur in the same layer, because in these circumstances the physical space required by each cell formally precludes statistical independence of the two component arrays.

The data that we shall analyse are shown in Fig. 2. This shows a single, bivariate spatial point pattern taken from the retina of a rabbit, in which the two types of point correspond to the positions of the centres of 152 "on" and 142 "off" amacrine cells; these data are from [25] and were kindly made available to us by Prof. Abbie Hughes. For a general discussion of the biological background to these data, see [17].

The pattern has been recorded within a rectangular section of the retina, of dimension 1060 by 662 μm. Visually, both on and off cells exhibit patterns which are more regular than would be the case for completely random patterns, i.e. realisations of homogeneous Poisson point processes. In particular, there is a pronounced inhibitory effect, meaning that no two on cells, and no two off cells, can be located arbitrarily close together. The inhibitory effect is much less pronounced between cells of opposite type. For example, the minimum observed distance between any two on cells is 21.4μm, between any two off cells is 15.8μm, and between any pair of on and off cells is 5μm. We shall use 1μm as the unit of distance throughout.

Previous analyses of the data have been reported by [6], where nonparametric methods led to the conclusion that the two component patterns were approximately independent, and by [8] and [7] who used the data to illustrate the fitting of univariate models by *ad hoc* and likelihood-based methods, respectively. Our goal in the current contribution is to demonstrate how recently developed Monte Carlo methods for conducting likelihood-based analysis of realistic point process models can lead to sharper inferences about the bivariate structure of the data. In particular, we will formulate and fit a bivariate pairwise interaction model for the amacrines data, and will argue that likelihood-based inference within this model is both statistically more efficient and scientifically more relevant than *ad hoc* testing of benchmark hypotheses such as independence.

One major limitation of the analysis reported here is that the data are unreplicated, i.e. they consist of a single point pattern. The literature on the statistical analysis of replicated spatial point pattern data is surprisingly sparse. [9] and [2] consider methods based on pooled estimates of non-parametric functional summary statistics such as the K-function [19, 20]. [10] compare parametric and non-parametric approaches to testing for differences between replicated patterns in two or more experimental groups. We are assembling a collection of replicated patterns of retinal cells and intend to analyse these using parametric, likelihood-based methods of the kind described in the cur-

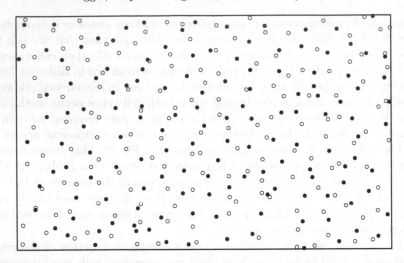

Fig. 2. The cholinergic amacrine data (*on and off cells are shown as open and closed circles, respectively*). The rectangular region on which the cells are observed has dimension 1060 by 662 μm. Cell bodies are drawn to scale (10 μm diameter). Cells of opposite polarity (on vs off) can partially overlap, since they are located in different layers, but cells of like polarity never overlap

rent contribution. We will report separately on the analyses of these data in due course.

2 Pairwise Interaction Point Processes

2.1 Univariate Pairwise Interaction Point Processes

Markov point processes were introduced by [21]. [24] discusses their construction, properties and uses as statistical models for spatial point patterns.

Pairwise interaction point processes are perhaps the most widely used sub-class of Markov point processes, In particular, they provide a flexible, parsimonious class of models for point patterns which display varying degrees of spatial regularity, as exhibited by our data.

Let $X = \{x_i : i = 1, ..., n\}$ be an observed spatial point pattern on a planar region A, hence each $x_i \in A$ and all points in A are observed. According to the scientific purpose of the analysis, it may be more natural to treat n as fixed or random. Here, our focus is on the nature of the interactions amongst the individual cells which determine their overall spatial pattern, given their number. We therefore treat n as fixed. We call the points of the process *events* to distinguish them from arbitrary points $x \in A$. In a pairwise interaction point process, the likelihood ratio for X with respect to a homogeneous Poisson process of unit intensity takes the form

$$c\beta^n \prod_{i=2}^{n} \prod_{j=1}^{i-1} h(||x_i - x_j||). \tag{1}$$

In (1), $||\cdot||$ denotes Euclidean distance, $h(r) : r \geq 0$ is the pairwise inter-action function, β reflects the intensity of the process and c is a normalising constant whose analytic form is typically intractable.

The essence of the model (1) lies in the interaction function, $h(\cdot)$. When $h(r) = 1$ for all r, the process is a homogeneous Poisson process of intensity β. When $h(r) = 0$ for $0 \leq r \leq \delta$, no two events can occur less than a distance δ apart and the process is said to display *strict inhibition*. Values of $h(r)$ intermediate between zero and one correspond to non-strict forms of inhibition in which close pairs of events are relatively unlikely, but not ruled out completely. The smallest distance ρ such that $h(r) = 1$ for all $r > \rho$ is called the *range* of the process. Models with $h(r) > 1$ for certain ranges of r are potentially invalid because the likelihood ratio (1) may not be integrable over A; an early example is the [22] model for clustering, subsequently shown by [?] to be invalid. In theory, models with $h(r) = 0$ for $r \leq \delta$ and $h(r) > 1$ for $\delta < r < \rho$ could be used to model aggregated spatial patterns, but in practice such models are not very useful because they correspond to very extreme forms of spatial aggregation, in a sense made precise by [14].

In (1), conditioning on the observed number of events in A leads to a joint probability density function for X, proportional to

$$f(X) = \prod_{i=2}^{n} \prod_{j=1}^{i-1} h(||x_i - x_j||). \tag{2}$$

In the general inhibitory case, i.e. when $h(r) \leq 1$ for all r, and when n is large, the distinction between processes with a fixed or random number of events in A is relatively unimportant for most purposes (but see below for an example to the contrary). In what follows, we shall consider only the case of fixed n. Hence, we do not attempt to make inferences about the intensity of the process, but only about the form of the interaction function $h(r) = h(r; \theta)$. The log-likelihood for θ is then given by

$$\log L(\theta) = \log f(X; \theta) + \log c(\theta) \tag{3}$$

where $c(\theta)$ is the normalising constant for (2). Figure 3 shows a realisation of a process with interaction function $h(r) = 1 - \exp(-r/\phi)$ for each of $\phi = 0.01, 0.05, 0.10, 0.15$ and, in each case, $n = 100$ events on the unit square. The progressive development of spatial regularity as the value of ϕ increases is clear. The simulations were generated on a toroidal region which was then unwrapped to form the unit square A; this counteracts a tendency for events to be artificially concentrated near the edge of A when the model is strongly inhibitory, i.e. in the present context, when ϕ is large. Note also that the distinction between fixed and random n becomes important as the strength of the interaction function increases with ϕ. Because of the need to place exactly 100 events in A, then as ϕ increases the patterns generated by the model will, with high probability, assume approximately a close-packed lattice configuration and further increases in ϕ will have no discernible effect.

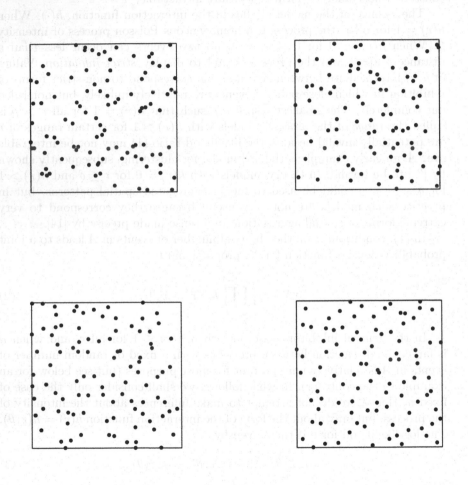

Fig. 3. Simulated realisations of pairwise interaction point processes each with 100 events on the unit square and interaction function $h(r) = 1 - \exp(-r/\phi)$. The values of ϕ are 0.01 (*top-left*), 0.05 (*top-right*), 0.1 (*bottom-left*) and 0.15 (*bottom-right*)

2.2 Bivariate Pairwise Interaction Point Processes

A bivariate spatial pattern consists of two sets of locations corresponding to two distinguishable types of event, which in our application are the on and off cells.

Let $X_1 = \{x_{1i} : i = 1, ..., n_1\}$ and $X_2 = \{x_{2i} : i = 1, ..., n_2\}$ represent a bivariate spatial point pattern of events in a region A. A bivariate pairwise interaction model is specified by three interaction functions, $h_{11}(\cdot)$, $h_{22}(\cdot)$ and $h_{12}(\cdot)$, which operate between pairs of events of type 1, pairs of events of type 2, and pairs of events of opposite type, respectively. Then, if we condition on the observed numbers of events, n_1 and n_2, the probability density of (X_1, X_2) is proportional to

$$f(X_1, X_2) = P_{11} P_{22} P_{12}, \tag{4}$$

where

$$P_{11} = \prod_{i=2}^{n_1} \prod_{j=1}^{i-1} h_{11}(\|x_{1i} - x_{1j}\|), \tag{5}$$

$$P_{22} = \prod_{i=2}^{n_2} \prod_{j=1}^{i-1} h_{22}(\|x_{2i} - x_{2j}\|), \tag{6}$$

and

$$P_{12} = \prod_{i=1}^{n_1} \prod_{j=1}^{n_2} h_{12}(\|x_{1i} - x_{2j}\|). \tag{7}$$

Equation (4) is a natural bivariate counterpart of (2). An important feature of the bivariate model is that its marginal properties depend on all three interaction functions. To illustrate this, we use the family of simple inhibitory interaction functions,

$$h_{ij}(r) = \begin{cases} 0 : r < \delta_{ij} \\ 1 : r \geq \delta_{ij} \end{cases} \tag{8}$$

and specify $\delta_{11} = \delta_{22} = 0.025$. If we also specify $\delta_{12} = 0$, then the two component processes are independent copies of a univariate simple inhibition process. The left-hand panel of Fig. 4 shows a realisation of this bivariate process. The two univariate components each display spatial regularity because of the inhibition effect but, because the two components are independent, arbitrarily close pairs of opposite type can and do occur. If we now introduce a strongly inhibitory interaction between events of opposite type by specifying $\delta_{12} = 0.1$, the effect is very different, as shown in the right-hand panel of Fig. 4. The cross-inhibitory effect between events of opposite type leads to component patterns which are marginally spatially aggregated, albeit with a clearly discernible local inhibitory effect, and jointly spatially segregated.

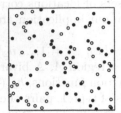
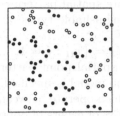

Fig. 4. Simulated realisations of bivariate pairwise interaction point processes each with 50 events of either type on the unit square and simple inhibitory interaction functions. In both panels, the minimum permissible distance between any two events of the same type is 0.025. In the left-hand panel, the two component patterns are independent. In the right-hand panel, the minimum permissible distance between any two events of opposite types is 0.1

3 Monte Carlo Likelihood Inference

The generally agreed "gold standard" for statistical estimation and hypothesis testing is to use likelihood-based methods; specifically, within a classical inferential framework, estimates should be maximum likelihood estimates and tests should be likelihood ratio tests.

The difficulty with applying this gold standard to our model is that the normalising constant for the joint probability density of (X_1, X_2), and hence the likelihood function for θ, is intractable. [16] provided an ingenious solution to this problem, which allows us to use simulations of the process at any fixed value θ_0 to compute an approximation to the likelihood ratio with respect to θ_0 for any value of θ. In the present context of pairwise interaction point processes, the argument runs as follows – we describe only the univariate case explicitly, but the extension to bivariate processes is obvious.

Let $c(\theta)$ be the normalising constant associated with the model (2), hence

$$c(\theta)^{-1} = \int f(X; \theta) dX$$

Now, for any fixed θ_0, write

$$c(\theta)^{-1} = \int f(X; \theta) \times \frac{c(\theta_0)}{c(\theta_0)} \times \frac{f(X; \theta_0)}{f(X; \theta_0)} dX, \qquad (9)$$

define $q(X; \theta, \theta_0) = f(X; \theta)/f(X; \theta_0)$ and re-arrange the right-hand-side of (9) to give

$$c(\theta)^{-1} = c(\theta_0)^{-1} \mathrm{E}_{\theta_0}[q(X;\theta,\theta_0)].$$

Hence, the normalised joint density for X can be expressed as

$$g(X;\theta) = c(\theta_0)f(X;\theta)/\mathrm{E}_{\theta_0}[q(X;\theta,\theta_0)].$$

Since θ_0 is a constant, it follows that the maximum likelihood estimator $\hat{\theta}$ maximises

$$L_{\theta_0}(\theta) = \log f(X;\theta) - \log \mathrm{E}_{\theta_0}[q(X;\theta,\theta_0)]. \tag{10}$$

The Monte Carlo method replaces the expectation on the right-hand-side of (10) by a Monte Carlo estimate, computed from s replicate simulations. Hence the Monte Carlo maximum likelihood estimate maximises

$$L_{\theta_0,s}(\theta) = \log f(X;\theta) - \log s^{-1} \sum_{j=1}^{s}[q(X_j;\theta,\theta_0)], \tag{11}$$

where the $X_j : j = 1, ..., s$ are simulated realisations with $\theta = \theta_0$.

Whilst the computations needed to secure a sufficiently accurate approximation can be time-consuming, the implication of Geyer and Thompson's work is that in principle there is no obstacle to using likelihood-based inference rather than the more *ad hoc* methods which are traditionally used to analyse spatial point pattern data. For a more detailed account, see [15].

Note that (11) defines a whole family of estimation criteria according to the choices made for θ_0 and s, and that for given s the extent of the stochastic variation introduced by the Monte Carlo simulation depends crucially on the choice of θ_0. In practice, the method works best when θ_0 is close to $\hat{\theta}$. Our approach has been to conduct a sequence of numerical optimisations of (11), updating θ_0 to the current maximising value after each stage until no further material change occurs, and increasing s until the Monte Carlo component of variance is negligible compared with the inherent uncertainty in $\hat{\theta}$ as measured by the Hessian matrix.

Whilst we favour Monte Carlo likelihood-based methods for formal parametric inference, in our opinion more *ad hoc* methods still have a useful role to play in the overall analysis. We use them to provide good initial values of θ_0 for the Monte Carlo likelihood calculations, and as checks on the goodness-of-fit of the final models produced by the likelihood-based analysis.

4 Analysis of the Amacrines Data

4.1 Exploratory Analysis

A standard tool for exploratory analysis of spatial point pattern data is the K-function, introduced by [19, 20] and, in the bivariate case, by [18]. In its basic form, the K-function describes the second-order properties of a *stationary* spatial point process. [1] extend its definition to include processes with spatially varying intensities. For the current application, we shall assume a spatially constant intensity.

Figure 5 shows the estimates $\hat{K}_{ij}(r) - \pi r^2$ for the amacrines data. The subscripts i and j refer to the type of event, types 1 and 2 corresponding to on and off cells, respectively. We favour plotting estimates $\hat{K}_{ij}(r) - \pi r^2$, rather than the $\hat{K}_{ij}(r)$ themselves because πr^2 is the natural benchmark relative to which we can assess both departure from complete spatial randomness in the component patterns, and departure from independence between the two components of the bivariate pattern. Note firstly that $\hat{K}_{11}(r)$ and $\hat{K}_{22}(r)$ are close together, suggesting that they may be generated by the same underlying process. Also, both estimates follow the parabola $-\pi r^2$ at small distances, i.e. $\hat{K}_{11}(r) = \hat{K}_{22}(r) = 0$, confirming the visual impression of a strict inhibitory effect within each of the component patterns. In contrast, $\hat{K}_{12}(r) - \pi r^2$ fluctuates around zero at small r. This behaviour, coupled with the fact that the sampling variance of $\hat{K}_{12}(r)$ increases with r, is consistent with the component processes being approximately independent. Note also that the magnitude of the difference between $\hat{K}_{11}(r)$ and $\hat{K}_{22}(r)$ derives from the combination of sampling variation in the estimates and the difference, if any, between the two underlying theoretical functions; it therefore provides an informal upper bound for the sampling variation, and on this basis we can conclude that the much larger difference between the $\hat{K}_{jj}(r)$ and $\hat{K}_{12}(r)$ is incompatible with random labelling.

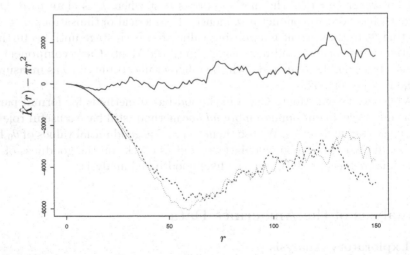

Fig. 5. Estimates of the K-functions for the on and off cells. Each plotted function is $\hat{K}(r) - \pi r^2$. The dashed line corresponds to $\hat{K}_{11}(r)$ (on cells), the dotted line to $\hat{K}_{22}(r)$ (off cells) and the solid line to $\hat{K}_{12}(r)$. The parabola $-\pi r^2$ is also shown as a solid line

4.2 Structural Hypotheses for the Amacrines Data

The exploratory analysis suggests that, purely from a statistical perspective, an inhibitory, bivariate pairwise interaction process with independent components and a common underlying model for the two components may provide a reasonable fit to the data. For many retinal cells, the hypothesis of *statistical independence* is strictly implausible because their cell bodies lie in the same cellular layer and two cells cannot occupy the same space. A more appropriate benchmark hypothesis, which we shall call *functional independence* is that the only form of interaction between type 1 and type 2 events is a simple inhibitory effect due to the physical size of the cells, i.e. an interaction function $h_{12}(r)$ of the form given by (8), with the value of δ_{12} no greater than the typical size of an individual cell.

A second hypothesis which is of some biological interest is *common components*, by which we mean that the data are generated by a bivariate model with $h_{11}(r) = h_{22}(r)$. Our analysis will therefore include formal tests of statistical independence, structural independence and common components.

4.3 Non-parametric Estimation

We use the method of maximum pseudo-likelihood [3, 4] to obtain non-parametric estimates of the interaction functions $h_{11}(r)$ and $h_{22}(r)$. Formally, this is achieved by fitting a deliberately over-parameterised model in which the interaction function is assumed to be piecewise constant, with the heights of the pieces as its parameters.

Figure 6 shows the results. The estimates of the two interaction functions are quite similar, adding weight to the evidence for a common components model.

Figure 6 also suggests what approximate shape a more parsimonious parametric model for the interaction functions would need to accommodate. We shall use functions $h_{ij}(\cdot)$ within the parametric family $h(r, \theta)$ where $\theta = (\delta, \phi, \alpha)$ and

$$h(r; \theta) = \begin{cases} 0 : r \leq \delta \\ 1 - \exp[-\{(r - \delta)/\phi\}^{\alpha}] : r > \delta \end{cases} \tag{12}$$

This allows a wide range of inhibitory interactions within and between types by varying the corresponding parameter vectors θ_{11}, θ_{22} and θ_{12} so as to define the corresponding interaction functions $h_{ij}(r) = h(r; \theta_{ij})$.

Because a large value for the parameter α allows $h(r)$ to take values close to zero even for relatively large values of r, we might expect the parameters of (12) to be poorly identified. Our response to this, following the discussion in Sect. 4.2, is to treat the values of δ_{11} and δ_{22} as fixed constants with a common value 10, corresponding to the approximate physical size of the cells [5, 13]. Of course, the model is at best an approximation to nature, and we

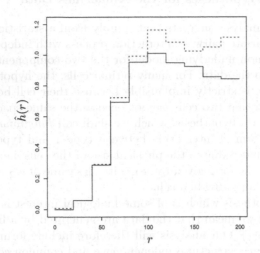

Fig. 6. Non-parametric maximum pseudo-likelihood estimates of the pairwise interaction functions for on cells (*solid line*) and for off cells (*dashed line*)

should not over-interpret this precise value; rather, it represents a plausible lower limit on the physical size of the cells. As we shall see, the model can still capture an effective inhibition distance between cells which is substantially greater than 10.

It is harder to argue for an *a priori* fixed value of δ_{12} because of the vertical displacement between the mature on and off cells. The on cells lie somewhat deeper than the off cells and a pair of cells of opposite type could in principle be almost co-located in the planar projection of the data. We shall therefore treat δ_{12} as a parameter to be estimated; as discussed earlier, inference concerning δ_{12} is of some biological interest in its own right.

4.4 Univariate Parametric Analysis

Under the working assumption of statistical independence, we can analyse the two patterns separately and investigate whether a common set of parameters provides a good fit to both. This analysis is also useful as a prelude to a bivariate analysis, whether or not the independence hypothesis is sustainable.

To obtain initial values for numerical optimisation of the Monte Carlo log-likelihood, we fitted the parametric form of $h(r; \theta)$ to each non-parametric estimate of $h(r)$ shown in Fig. 6 by ordinary least squares. We then obtained Monte Carlo maximum likelihood estimates of ϕ and α separately for each of the two patterns, progressively increasing the number of Monte Carlo samples from 10 to 1000, until the estimates stabilised.

To test whether a common set of parameters fitted both patterns, we repeated the optimisation process, but now maximising a pooled Monte Carlo log-likelihood with common parameter values for the two component patterns. The resulting log-likelihood ratio test statistic was $D = 1.36$ on two degrees of freedom, corresponding to $p = 0.244$. We therefore accepted the common components hypothesis, which gave us the parameter estimates shown in Table 1. Approximate standard errors, and the correlation between $\hat{\phi}$ and $\hat{\alpha}$, were derived from the estimated Hessian matrix of the pooled Monte Carlo log-likelihood at its maximum. All optimisations used the built-in optim() function within R; for details, see http://www.r-project.org. Figure 7 compares the fitted, common parametric form of $h(r)$ with the two non-parametric estimates. The fit appears to be satisfactory, but we postpone a formal goodness-of-fit assessment until we have fitted a bivariate model.

Table 1. Monte Carlo maximum likelihood estimates, standard errors and correlation, assuming independence between on and off amacrine cell patterns and common parameter values

Parameter	Estimate	Std Error	Correlation
ϕ	49.08	2.51	
α	2.92	0.25	-0.06

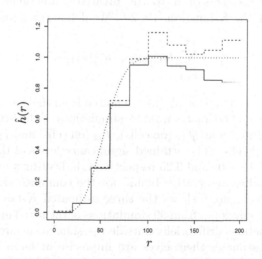

Fig. 7. Non-parametric maximum pseudo-likelihood estimates of the pairwise interaction functions for on cells (*solid line*) and for off cells (*dashed line*), together with parametric fit assuming common parameter values for both types of cell (*dotted line*)

4.5 Bivariate Analysis

The first stage in the bivariate analysis is a simple likelihood ratio rest of *statistical independence* against *functional independence*. To do this, we first begin by estimating δ_{12}, obtaining the maximum likelihood estimate $\hat{\delta}_{12} = 4.9$. We then construct a likelihood ratio test of any fixed value of δ_{12} against $\delta_{12} = 4.9$. The set of values not rejected at the 5% level defines a Monte Carlo 95% confidence interval for δ_{12}. Note that all values of δ_{12} greater than 5, the smallest observed distance between a pair of cells of opposite type, are automatically excluded according to the likelihood criterion, because all such values are incompatible with the data. The resulting 95% confidence interval is $2.3 < \delta_{12} < 5.0$. In particular, this interval excludes zero, implying that statistical independence is rejected at the conventional 5% level; more precisely, the attained significance level of the likelihood ratio test of statistical independence against functional independence is $p = 0.021$ (test statistic $D = 5.30$, $P(\chi_1^2 > 5.30) = 0.021$).

We next investigate whether there is any further degree of dependence between the on and off cells by introducing additional parameters ϕ_{12} and α_{12}, holding the remaining parameters fixed at $\phi_{11} = \phi_{22} = 49.08$, $\alpha_{11} = \alpha_{22} = 2.92$, $\delta_{11} = \delta_{22} = 10$ and $\delta_{12} = 4.9$. The likelihood ratio test statistic to compare functional independence against the general bivariate model is $D = 0.30$ on 2 degrees of freedom, corresponding to $p = 0.861$. Hence, functional independence is not rejected.

To assess the goodness-of-fit to the bivariate, functional independence model we first use the K-function [19, 20]. We define three test statistics

$$T_{ij} = \sum_{r=1}^{150} [\{\hat{K}_{ij}(r) - \bar{K}_{ij}(r)\}/r]^2 \tag{13}$$

where $\hat{K}_{ij}(r)$ is the estimate of $K_{ij}(r)$ calculated from the data and $\bar{K}_{ij}(r)$ the corresponding mean of estimates from 99 simulations of the fitted model. The three statistics of interest are T_{11} (on cells), T_{22} (off cells) and T_{12} (dependence between on and off cells). The attained significance levels of the three Monte Carlo tests were 0.11, 0.05 and 0.25 respectively, indicating a reasonable overall fit; an admittedly conservative bound for the combined significance level is $0.05 \times 3 = 0.15$. Figure 8 shows the three estimated K-functions together with the pointwise envelopes from 99 simulations of the fitted model. Although the estimated functions drift briefly outside the simulation envelopes at large values of r, the estimates themselves are imprecise at large values of r, as indicated by the widths of the simulation envelopes. This also explains why we have chosen to discount progressively the influence of estimates $\hat{K}_{ij}(r)$ at large values of r in our construction of the test statistics (13).

The K-functions assess the goodness-of-fit in terms of second-moment properties. For a complementary goodness-of-fit assessment we now consider nearest neighbour properties. Let $G_{ij}(r)$ denote the distribution function of

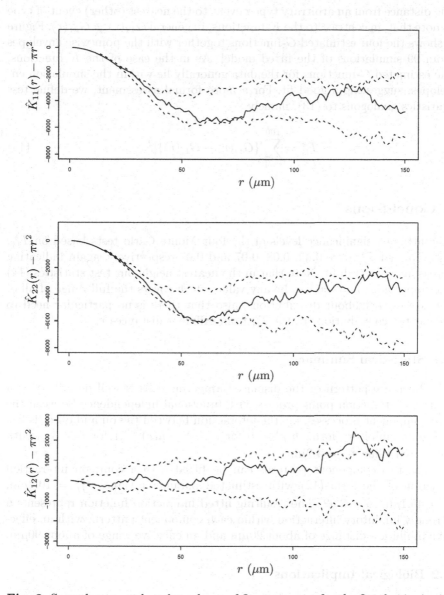

Fig. 8. Second-moment-based goodness-of-fit assessment for the fitted pairwise interaction process with common parameter value for on and off cells. The solid lines are the estimates $\hat{K}_{ij}(r) - \pi r^2$ for the data, the dashed lines are the pointwise envelopes from 99 simulations of the fitted model

the distance from an arbitrary type i event to the nearest (other) event of type j; note that, in contrast to the K-functions, in general $G_{12}(r) \neq G_{21}(r)$. Figure 9 shows the four estimated G-functions together with the pointwise envelopes from 99 simulations of the fitted model. As in the case of the K-functions, the estimated G-functions for the data generally lie within the simulation envelopes, suggesting a good fit. For a more formal assessment, we define test statistics analogous to (13), namely

$$T_{ij}^* = \sum_{r=1}^{100} [\{\hat{G}_{ij}(r) - \bar{G}_{ij}(r)\}]^2. \tag{14}$$

5 Conclusions

The attained significance levels of the four Monte Carlo tests based on T_{11}^*, T_{22}^*, T_{12}^* and T_{21}^* were 0.45, 0.08, 0.07 and 0.07 respectively, again indicating a reasonable overall fit. Note that in the nearest neighbour test statistics (14) the upper limit $r = 100$ can be any value which covers the full range of all of the nearest neighbour distributions, also that there is no particular need to weight the contributions to the T_{ij}^* from different distances r.

5.1 Statistical Summary

The bivariate pattern of the displaced amacrine cells is well described by a pairwise interaction point process with functional independence between the two component processes, i.e. the interaction between the on and off cells has a simple inhibitory form, $h(r) = 0$ for $r < \delta_{12}$, $h(r) = 1$ for $r \geq \delta_{12}$, with estimated value $\hat{\delta}_{12} = 4.9\mu$m.

The two component patterns can be fitted with a common interaction function of the form (12) with estimated parameter values $\hat{\delta}_{jj} = 10.0\mu$m, $\hat{\phi} = 49.1\mu$m, $\hat{\alpha} = 2.92$. The resulting fitted interaction function represents a strongly inhibitory interaction within each component pattern, with an effective inhibition distance of about 20μm and an effective range of about 90μm.

5.2 Biological Implications

The results from the bivariate analysis indicate that there is a small spatial dependency between the positioning of the on and off cells, since one of our conclusions is that δ_{12} is non-zero. This may appear to conflict with earlier assumptions of independence between the two types [6]. However, one advantage of the likelihood-based analysis over the earlier approach is that we can create 95% confidence intervals (here $2.3 \leq \delta_{12} < 5.0$). Hence, the interaction distance between the on and off cells is around 5 μm at most, which is smaller than the typical cell diameter ($\sim 10\mu$m). Therefore, any dependency in the

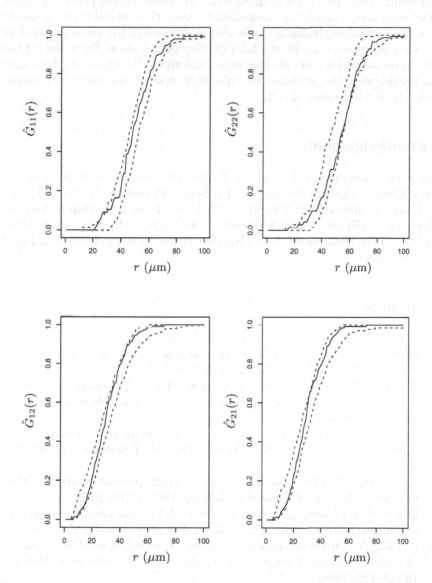

Fig. 9. Nearest-neighbour-based goodness-of-fit assessment for the fitted pairwise interaction process with common parameter value for on and off cells. The solid lines are the estimates $\hat{G}_{ij}(r)$ for the data, the dashed lines are the pointwise envelopes from 99 simulations of the fitted model

positioning of the two types is quite weak. However, this dependency might reflect some early positioning constraints between the two cell types, as would occur if, for example, immature cells were positioned in the same layer before migrating to separate layers at a later developmental stage. By repeating the analysis on many data sets of cholinergic amacrine cells, we aim to determine how consistent is the evidence for this weak dependence between patterns formed by the two types of cell.

Acknowledgements

This work was supported by the UK Engineering and Physical Sciences Research Council through the award of a Senior Fellowship to Peter J. Diggle (Grant number GR/S48059/01), a Wellcome Trust fellowship to Stephen Eglen, and a NIH grant (R01 EY06669) to John Troy. Thanks to Prof. Abbie Hughes and Dr. Elzbieta Wieniawa Narkiewicz for the dataset shown in Fig. 2 and discussions.

References

[1] A.J. Baddeley, J. Møller and R.P. Waagepetersen. Non- and semi-parametric estimation of interaction in inhomogeneous point patterns. *Statistica Neerlandica*, 54(3):329–350, 2000.

[2] A.J. Baddeley, R.A. Moyeed, C.V. Howard and A. Boyde. Analysis of a three-dimensional point pattern with replication. *Applied Statistics*, 42:641–668, 1993.

[3] A.J. Baddeley and R. Turner. Practical maximum pseudolikelihood for spatial point processes. *New Zealand Journal of Statistics*, 42:283–322, 2000.

[4] J.E. Besag, R. Milne and S. Zachary. Point process limits of lattice processes. *Journal of Applied Probability*, 19:210–216, 1982.

[5] C. Brandon. Cholinergic neurons in the rabbit retina: dendritic branching and ultrastructural connectivity. *Brain Research*, 426:119–130, 1987.

[6] P.J. Diggle. Displaced amacrine cells in the retina of a rabbit : analysis of a bivariate spatial point pattern. *Journal of Neuroscience Methods*, 18:115–125, 1986.

[7] P.J. Diggle. *Statistical Analysis of Spatial Point Patterns, 2nd ed.* Oxford University Press, 2003.

[8] P.J. Diggle and R.J. Gratton. Monte carlo methods of inference for implicit statistical models (with discussion). *Journal of the Royal Statistical Society B*, 46:193–227, 1984.

[9] P.J. Diggle, N. Lange and F. Benes. Analysis of variance for replicated spatial point patterns in clinical neuroanatomy. *Journal of the American Statistical Association*, 86:618–625, 1991.

[10] P.J. Diggle, J. Mateu and H.E. Clough. A comparison between parametric and non-parametric approaches to the analysis of replicated spatial point patterns. *Advances in Applied Probability*, 32:331–343, 2000.

[11] T. Euler, P.B. Detwiler and W. Denk. Directionally selective calcium signals in dendrites of starburst amacrine cells. *Nature*, 418:845–852, 2002.

[12] E.V. Famiglietti. "starburst" amacrine cells and cholinergic neurons: mirror-symmetric on and off amacrine cells of rabbit retina. *Brain Research*, 261:138–144, 1983.

[13] E.V. Famiglietti. Starburst amacrine cells: morphological constancy and systematic variation in the anisotropic field of rabbit retinal neurons. *Journal of Neuroscience*, 5:562–577, 1985.

[14] D.J. Gates and M. Westcott. Clustering estimates in spatial point distributions with unstable potentials. *Annals of the Institute of Statistical Mathematics*, 38 A:55–67, 1986.

[15] C.J. Geyer. Likelihood inference for spatial point processes. In O.E. Barndorff-Nielsen, W.S. Kendall and M.N.M. van Lieshout, editors, *Stochastic Geometry: likelihood and computation*, pages 141–172. Chapman and Hall/CRC, 1999.

[16] C.J. Geyer and E.A. Thompson. Constrained monte carlo maximum likelihood for dependent data (with discussion). *Journal of the Royal Statistical Society B*, 54:657–699, 1992.

[17] A. Hughes. New perspectives in retinal organisation. *Progress in Retinal Research*, 4:243–314, 1985.

[18] H.W. Lotwick and B.W. Silverman. Methods for analysing spatial processes of several types of points. *Journal of the Royal Statistical Society*, 44:1982, 406–413.

[19] B.D Ripley. The second-order analysis of stationary point processes. *Journal of Applied Probability*, 13:255–266, 1976.

[20] B.D. Ripley. Modelling spatial patterns. *Journal of the Royal Statistical Society B*, 39:172–212, 1977.

[21] B.D. Ripley and F.P. Kelly. Markov point processes. *Journal of the London Mathematical Society*, 15:188–192, 1977.

[22] D. Strauss. A model for clustering. *Biometrika*, 63:467–475, 1975.

[23] M. Tauchi and R.H. Masland. The shape and arrangement of the cholinergic neurons in the rabbit retina. *Proceedings of the Royal Society, London, B*, 23:101–119, 1984.

[24] M.N.M. van Lieshout. *Markov point processes and their applications*. Imperial College Press, 2000.

[25] E. Wienawa-Narkiewicz. *Light and Electron Microscopic Studies of Retinal Organisation*. PhD thesis, Australian National University, Canberra, 1983.

Analysis of Spatial Point Patterns in Microscopic and Macroscopic Biological Image Data

Frank Fleischer[1], Michael Beil[2], Marian Kazda[3] and Volker Schmidt[4]

[1] Department of Applied Information Processing and Department of Stochastics, University of Ulm, D-89069 Ulm, Germany, frank.fleischer@mathematik.uni-ulm.de
[2] Department of Internal Medicine I, University Hospital Ulm, D-89070 Ulm, Germany
[3] Department of Systematic Botany and Ecology, University of Ulm, D-89069 Ulm, Germany
[4] Department of Stochastics, University of Ulm, D-89069 Ulm, Germany

Summary. Point process characteristics like for example Ripley's K-function, the L-function or Baddeley's J-function are especially useful for cases of data with significant differences with respect to intensities. We will discuss two examples in the fields of cell biology and ecology were these methods can be applied. They have been chosen, because they demonstrate the wide range of applications for the described techniques and because both examples have specific interest- ing properties. While the point patterns regarded in the first application are three dimensional, the second application reveals planar point patterns having a vertically inhomogeneous structure.

Key words: Baddeley's J-function, Centromeric Heterochromatin Structures, CSR, Planar Sections of Root Systems in Tree Stands, Ripley's K-function

1 Introduction

The analysis of spatial point patterns by means of estimated point process characteristics like for example Ripley's K-function [36], the L-function or Baddeley's J-function [7, 46] has proven to be a very useful tool in Stochastic Geometry during the last years (see e.g. [11, 37, 42] and [43]). They offer the possibility to get not only qualitative knowledge about the observed spatial structures of such point patterns, but to quantify them for specific regions of point-pair distances. Other advantages of these methods compared to alternative techniques of spatial analysis like e.g. Voronoi tessellations [26, 30] or minimum spanning trees [14, 15] is their independence of underlying point process intensities, in other words of the average number of points per unit square. Therefore they are especially useful for cases of data with significant differences with respect to intensities. We will discuss two examples in the fields of cell biology and ecology were these methods can be applied. They have been chosen, because they demonstrate the wide range of applications for the described techniques and because both examples have specific interesting properties. While the point patterns regarded in the first application are three dimensional, the second application reveals planar point patterns having a vertically inhomogeneous structure. For a more extensive description of the studied cases, the reader is referred to [4, 5], and [17, 40, 41], respectively.

1.1 Analysis of Centromeric Heterochromatin Structures

The first example deals with the structure of the cell nucleus, notably the distribution of centromeres, during differentiation (maturation) of myeloid cells. These are the precursors of white blood cells and are normally found in the bone marrow. During differentiation, myeloid cells acquire specialised functions by activating a strictly defined set of genes and producing new proteins. In addition, other genes whose function are not needed in maturated cells become silenced. The mechanisms regulating the activity of genes during differentiation remain to be defined in detail.

The architecture of the cell nucleus during the interphase, i.e. the time between consecutive cell divisions, is determined by the packaging of the DNA molecule at various levels of organisation (chromatin structure). The open state of DNA is referred to as euchromatin, whereas heterochromatin is the condensed form of DNA. The production of gene transcripts (mRNA) requires the molecules of the transcriptional machinery to access the DNA molecule. Thus, the regulation of DNA packaging represents an important process for controlling gene activity, i.e. the synthesis of mRNA [9]. In general, transcriptional activity appears to be impeded by a restrictive (compacted) packaging of DNA [34]. This way, the hetrochromatin compartment is an important regulator of gene transcription and, hence, influences the biological function of cells.

There has been a great interest in investigating the processes governing the organisation of chromatin. Whereas the regulation at the level of nucleosomes, e.g. through biochemical modifications of histones, is now becoming elucidated, the long-range remodelling of large portions of the DNA molecule (higher-order chromatin structure) proceeds by yet unknown rules. The condensed form of DNA, i.e. heterochromatin, is generally associated with the telomeres and centromeres of chromosomes [35]. These regions can also induce transcriptional repression of nearby genes [34]. Consequently, the distribution of these regions should be changed during cellular differentiation that is associated with a marked alteration of the profile of activated genes. In fact, previous studies described a progressive clustering of interphase centromeres during cellular differentiation of lymphocytes and Purkinje neurons [1, 31].

However, the overall structural characteristics of the centromeric heterochromatin compartment, e.g. with respect to spatial randomness, remain to be determined.

Leukaemias are malignant neoplasias of white blood cells. They represent an interesting biological model to study cellular differentiation since they can develop at every level of myeloid differentiation. A particular type, acute promyelocytic leukaemia (APL), is characterised by a unique chromosomal translocation fusing the PML gene on chromosome 15 with the gene of the retinoic acid receptor alpha on chromosome 17 [10]. Due to the function of the resulting fusion protein, cellular differentiation is arrested at the level of promyelocytes. However, pharmacological doses of all-trans retinoic acid (ATRA) can induce further differentiation of these cells along the neutrophil pathway [16].

In a recent study, the three-dimensional (3D) structure of the centromeric heterochromatin was studied in the NB4 cell line which was established from a patient with APL [24]. The 3D positions of centromeres served as a surrogate marker for the structure of centromeric heterochromatin. Due to the diffraction-limited resolution of optical microscopy, the notion "chromocenter" was used to define clusters of centromeres with a distance below the limit of optical resolution. During differentiation of NB4 cells as induced by ATRA, a progressive clustering of centromeres was implied from a decreased number of detectable chromocenters. The 3D distribution of chromocenters was evaluated by analysing the minimal spanning tree (MST) constructed from the 3D coordinates of the chromocenters. The results obtained by this method suggested that a large-scale remodelling of higher order chromatin structure occurs during differentiation of NB4 cells.

1.2 Planar Sections of Root Systems in Tree Stands

In our second example we examine the spatial distribution of tree root patterns in pure stands of Norway Spruce (*Picea Abies*) and European Beech (*Fagus sylvatica* (L.) Karst.). While there exists knowledge about the vertical root distribution which can usually be described by one-dimensional depth functions [21, 33], early studies assumed that rooting zones are completely and almost homogeneously exploited by roots [23]. In recent studies it is however shown that fine roots concentrate in distinct soil patches and that they proliferate into zones of nutrient enrichment and water availability [6, 38]. Hence a horizontal heterogeneity of the spatial distribution of fine roots might be expected. Trench soil profile walls can be used for the assessment of two-dimensional root distributions, regarding the roots on the wall as points of different diameter. A new method provided (x,y)-coordinates of each root greater than 2 mm [40]. Using this method, small roots with a diameter between 2 mm and 5 mm were examined in 19 pits on altogether 72 m^2 of soil profiles on monospecific stands of European Beech and of Norway Spruce.

1.3 Detection of Structural Differences

So to summarize, our aims in both studied applications were quite similar from a mathematical point of view. First, as it is generally the case in such a spatial analysis, to compare our data sets with the *null* model of the homogeneous Poisson process, otherwise described as 'complete spatial randomness' (CSR) [11]. If such a hypothesis can be rejected, a goal is to detect structural differences between the two regarded groups in the data, namely nuclei from non-differentiated and differentiated NB4 cells in the first application and beech roots and spruce roots in the second one. For the cell nuclei example we were especially interested in an explanation for the decreasing number of chromocenters during differentiation, while for the tree roots a main aim was to quantify the degree of root aggregation, i.e. the degree of intensity of the exploitation of the soil resources by each tree species. Another question of interest is to provide a suitable and not too complicated mathematical model for underlying generating point processes. Finally, of course it is a necessity to obtain an interpretation of the results from the biological standpoint.

2 Image Data

As it has been mentioned in Sect. 1, there were two different data sets considered, three dimensional point patterns in cell nuclei of a NB4 cell line and two dimensional point patterns in profile walls of European Beech and Norway Spruce.

2.1 NB4 Cell Nuclei

The procedures for cell culture of NB4 cells, specimen preparation, immunofluorescence confocal microscopy and image analysis are described in detail in [4]. In the following two paragraphs the applied techniques are summarised.

Sample Preparation and Image Acquisition

Differentiation of NB4 cells was induced by incubating cells with 5 $\mu mol/l$ ATRA (Sigma, St.Louis, MO) for 4 days. Visualization of centromeres was based on immunofluorescence staining of centromere-associated proteins with CREST serum (Euroimmun Corp., Gross Groenau, Germany). Nuclear DNA was stained with YoPro-3 (Molecular Probes). Two channel acquisition of 3D images was performed by confocal scanning laser microscopy (voxel size: 98 nm in lateral and 168 nm in axial direction).

Image Segmentation

Segmentation of chromocenters as stained by CREST serum was performed in two steps. First, objects at each confocal section were segmented by edge detection followed by a conglomerate cutting procedure. In a second step, 3D chromocenters were reconstructed by analyzing series of 2D profiles. The centre of gravity was used to define the 3D coordinates for each chromocenter. The final analysis included 28 cell nuclei from untreated controls with 68 chromocenters on average and 27 cell nuclei from ATRA-differentiated NB4 cells with 57 chromocenters on average (see Fig. 1).

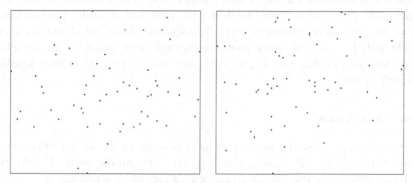

Fig. 1. Projections of the three dimensional chromocenter location patterns of an undifferentiated NB4 cell (*left*) and a differentiated NB4 cell (*right*) onto the xy-plane

2.2 Profile Walls

For details of site description, pit excavation and root mapping and already attained results, see [40]. Our investigations are based upon this article and thus only a short summary of the most important facts is given.

Site Description

Data collection took place near Wilhelmsburg, Austria (exact location at $48°05'51''N$, $15°39'48''E$) in adjoining pure stands of *Fagus sylvatica* and of planted *Picea abies*. One experimental plot of about 0.5 *ha* was selected within each stand. The sites were similar in aspect (NNE), inclination (10%) and altitude (480 *m*). The characteristics of the spruce and beech stands, e.g. the age (55 and 65 years), the dominant tree height (27 *m* and 28 *m*) and the stand density (57.3 and 46.6 trees /ha), also were similar to each other. The soils with only thin organic layer (about 4 *cm*) can be classified as Stagnic Cambisols developed from Flysch sediments. Annual rainfall in Wilhelmsburg averages 843 mm with a mean summer precipitation from May to September of 433 mm. The mean annual temperature is 8.4° C, and the mean summer temperature is 15.7° C.

Pit Excavation and Root Mapping

In every stand 10 soil pits with a size of 2×1 m were excavated. Thus up to 20 profile walls could be obtained in each stand. In most cases $13 - 19$ trees were within a radius of 10 m around the pit centre. The minimum distance from the pit centre to the nearest tree ranged from 0.5 m to 2.8 m. On each wall all coarse roots were identified and divided into living and dead. All living small roots ($2 - 5$ mm) were marked with pins and digitally photographed. These pictures were evaluated and a coordinate plane was drawn over each profile wall W, so that every root corresponds to a point x_n in the plane. Thus, for each profile wall W, a point pattern $\{x_n\} \subset W$ of root locations was determined.

Data Description

Root mapping was performed on 20 profile walls of *Fagus sylvatica* and on 16 profile walls of *Picea abies* (see Fig. 3). The profile walls B with area $\nu_d(B) = 200$ $cm \times 100$ cm are regarded as sampling windows of stochastic point processes in $\mathbb{R}^2$.

Fig. 2. Data collection

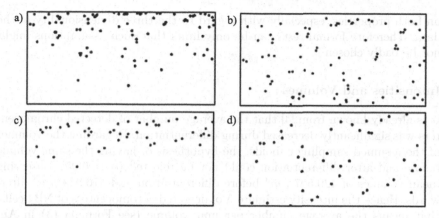

Fig. 3. (a) an original sample of roots for *Picea abies*; (b) the transformed sample for *Picea abies*; (c) an original sample of roots for *Fagus sylvatica*; (d) the transformed sample for *Fagus sylvatica*

3 Statistical Methods and Results

Data analysis for all data groups was performed using the GeoStoch library system. GeoStoch is a Java-based open-library system developed by the Department of Applied Information Processing and the Department of Stochastics of the University of Ulm which can be used for stochastic-geometric modelling and spatial statistical analysis of image data ([27, 28], http://www.geostoch.de).

For both study cases it is important to notice that, considering estimated point process characteristics, means for each group were regarded. This is due to the fact that variability inside a single group (European Beech and Norway Spruce or NB4 cell nuclei, respectively) was large compared to the differences between the two groups for each studied case (tree roots and cell nuclei). For functions, these means were taken in a pointwise sense.

3.1 3D Point Patterns of Chromocenters in NB4 Cells

The real sampling regions for the cell nuclei are not known, therefore assumed sampling regions were constructed as follows: For all three coordinates the smallest and largest values appearing in a sample were determined and denoted as x_{min}, x_{max}, y_{min}, y_{max}, z_{min} and z_{max} respectively. Then the 8 vertices of the assumed sampling cuboid were given by all possible combinations of the three coordinate pairs $\{x_{min}, x_{max}\}$, $\{y_{min}, y_{max}\}$ and $\{z_{min}, z_{max}\}$. Although we performed statistical tests on stationarity and isotropy of the regarded point fields which showed results in favor of such assumptions, we would like to consider stationarity and isotropy as prior assumptions that are not under investigation. This is due to the fact that the numbers of points per sample do not seem to be large enough to provide reliable information on both properties, especially with regard to the three dimensionality of the data. Therefore formal tests can be only hints that such assumptions might not be badly chosen.

Intensities and Volumes

It is already known from [4] that the average number of detected chromocenters was significantly decreased during differentiation. Regarding the volumes of the assumed sampling cuboids, the hypothesis of having the same volume before and after differentiation could not be rejected ($\alpha = 0.05$), observing mean volumes of 429.507 μm^3 before differentiation and 470.929 μm^3 afterwards. Hence the intensity estimate $\hat{\lambda}$ of detected chromocenters of NB4 cells, that means the average number per unit volume (see Formula (3) in Appendix A, is significantly decreased as well.

Averaged Estimated Pair Correlation Function

Figure 4 shows estimations $\hat{g}(r)$ of the pair correlation function $g(r)$ for $c = 0.06$, where c determines the bandwidth of the Epanechnikov kernel used in the definition of $\hat{g}(r)$; see Formula (11) in Appendix A. It is clearly visible that the frequency of point-pair distances for a distance between 350 nm and

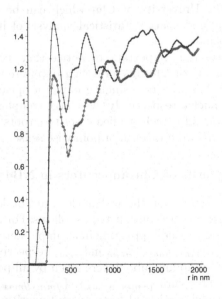

Fig. 4. Averaged estimated pair correlation functions $\hat{g}(r)$ using Epanechnikov kernel and parameter $c = 0.06$. The group of undifferentiated NB4 cells is denoted by $+$, while the group of differentiated NB4 cells is denoted by o

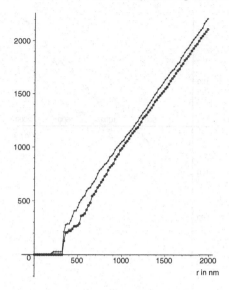

Fig. 5. Averaged estimated functions $\widehat{L}(r)$, where $+$ denotes the group of undifferentiated NB4 cells and o denotes the group of NB4 differentiated cells

800 nm is higher before than after differentiation of NB4 cells. Also a hardcore distance r_0 of about 350 nm can be recognised which is determined by the diffraction-limited spatial resolution of the microscopic imaging method. This means that all point pairs have a distance bigger than r_0. Note that the smaller hardcore values for larger values of c are due to the increased bandwidths of the Epanechnikov kernel in these cases. The results for the estimated pair correlation functions do not depend on the fact that the two groups have different numbers of detectable chromocenters.

Averaged Estimated L-function

We consider the estimator $\widehat{L}(r)$ for $L(r)$ given in Formula (17) of Appendix A. Figure 5 shows the estimated averaged L-function $\widehat{L}(r)$ while Fig. 6 shows $\widehat{L}(r) - r$ where the theoretical value r for Poisson point processes has been subtracted; see Formula (16) in Appendix A.

A similar scenario as for the pair correlation function is observed. Especially for small point-pair distances between 350 nm and 500 nm, there is a higher percentage of point pairs before than after ATRA-induced differentiation of NB4 cells. While for the group of undifferentiated cells the graph $\widehat{L}(r) - r$ has a mostly positive slope in this region, which is an indicator for attraction, the group of differentiated NB4 cells shows a negative slope which is a sign for rejection. The same hardcore distance $r_0 \approx 350$ nm is visible.

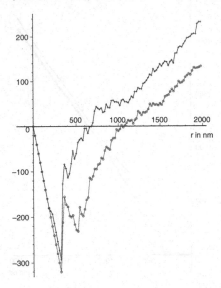

Fig. 6. Averaged estimated functions $\widehat{L}(r) - r$, where + denotes the group of NB4 undifferentiated cells and o denotes the group of NB4 differentiated cells

Again the results do not depend on the different numbers of detectable chromocenters.

Performing a Wilcoxon-Mann-Whitney test for the two group samples for fixed radii shows a significant difference in the values of L-functions before and after differentiation for all radii between 350 nm and 1300 nm, especially for the region between 500 nm and 700 nm ($\alpha = 0.05$).

Averaged Estimated Nearest-Neighbor Distance Distribution and Averaged Estimated J-function

The structural conclusions obtained from the results for the estimated point field characteristics $\widehat{D_H}(r)$ and $\widehat{J}(r)$ were very similar compared to the averaged estimated pair correlation function $\widehat{g}(r)$ and the averaged estimated L-function $\widehat{L}(r)$, where $\widehat{D_H}(r)$ and $\widehat{J}(r)$ are given by Formulae (18) and (22) in Appendix A. Therefore the averaged estimates $\widehat{D_H}(r)$ and $\widehat{J}(r)$ are not displayed here. Again a hardcore distance of 350 nm can be recognised and the two different groups show strong differences in their behavior especially for a range between 350 nm and about 800 nm.

3.2 2D Point Patterns in Planar Sections of Root Systems

From [40] it was already known that the depth densities of the roots of Norwegian Spruce and European Beech can be approximated by exponential and gamma distributions respectively. The data has been homogenised with respect to the vertical axis in order to allow the assumptions of stationarity and isotropy for models of generating point processes. A suitable homogenization can be based on the well-known fact that each random variable Y with a continuous distribution function F_Y can be transformed to a uniformly distributed random variable U on the interval $[0, 1]$ by

$$U = F_Y(Y). \tag{1}$$

Therefore, by denoting the original depths, the total depth of the sampling window and the transformed depths as h_{orig}, h_{tot} and h_{tran} respectively, we get

$$h_{tran} = \frac{F^*(h_{orig})}{F^*(h_{tot})} h_{tot}, \tag{2}$$

where $F^*(x)$ symbolizes the suitable distribution function, i.e. the exponential distribution function in the case of Norway Spruce and the gamma distribution function in the case of European Beech. The total depth was given as $h_{tot} = 100\ cm$. For each sampling window parameters of the distribution functions $F^*(x)$ are estimated individually using maximum-likelihood estimators. Notice that in the following, first only vertically homogenised data is regarded (see Fig. 3), that means considering the vertical coordinate a uniform distribution on $[0, h_{tot}]$ can be assumed. Later on, an inverse transformation is applied to obtain inference for the original data.

Intensities

The average number of points for the samples of *Picea abies* is significantly higher than for the samples of *Fagus sylvatica* ($\alpha = 0.05$). Since sampling windows have the same sizes, the same result is obtained regarding the estimated intensities per cm^2 ($\widehat{\lambda}^{spruce} = 0.00403$ vs. $\widehat{\lambda}^{beech} = 0.00262$). Notice that the following results for the considered point process characteristics are independent of this fact since the functions are scaled with respect to the intensities.

Isotropy and Complete Spatial Randomness

Isotropy was tested by determining the directional distribution of the angles of point pairs to the axes in a quadratic sampling window and testing them for uniform distribution. The hypothesis of isotropy could not be rejected ($\alpha = 0.05$), hence in the following isotropy is assumed. The quadrat count method [43] was used to test on complete spatial randomness. Here, using a 4×4 grid, the hypothesis that the given point patterns are extracts of realisations of homogeneous Poisson processes was rejected ($\alpha = 0.05$).

Averaged Estimated J-function

In Fig. 7 the averaged estimated J-functions $\widehat{J}(r)$ for both groups are displayed. There is a clear indication for attraction between point pairs of a distance less than 12 cm, since both functions are below 1 in this region and have a negative slope. A second observation is that the graph of *Picea abies* lies beneath the graph of *Fagus sylvatica*, which means that the point pairs of spruces are more attracted to each other than the point pairs of beeches for such distances. Notice that for radii larger than 20 cm the estimator becomes numerically unstable, and therefore should not be taken into further consideration. Also one should keep in mind that the J-function is a cumulative quantity.

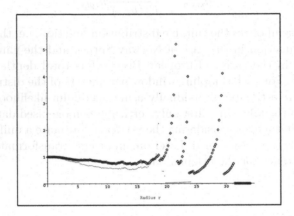

Fig. 7. Averaged estimated J-functions for *Picea abies* ($\cdot$) and *Fagus sylvatica* ($\Diamond$)

Averaged Estimated L-function

In Fig. 8 the graphs for the averaged estimated values of $\widehat{L}(r) - r$ are shown. Since in the Poisson case $L(r) \equiv r$ a positive slope means that there is an attraction, while a negative slope indicates repulsion. Again there are signs of attraction for small point-pair distances, less than 9.5 cm and less than 13.5 cm respectively, and the attraction seems to be stronger for *Picea abies* compared to *Fagus sylvatica* since the slope of $\widehat{L}(r) - r$ is bigger. The negative values for very small distances might indicate a slight hardcore effect between the points.

Averaged Estimated Pair Correlation Function

Further indication for an attraction between point pairs of distances less than 14 cm is provided by the averaged estimated pair correlation functions $\widehat{g}(r)$

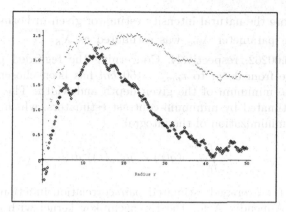

Fig. 8. Averaged estimated functions $\widehat{L}(r) - r$ for *Picea abies* ($\cdot$) and *Fagus sylvatica* ($\Diamond$)

displayed in Fig. 9. Again a stronger attraction is observed for the spruces since the function runs above the function for beeches for r less than 9 cm. Both functions are above 1 for $r < 14$ cm.

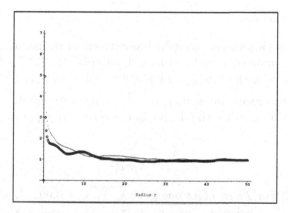

Fig. 9. Averaged estimated pair correlation functions $\widehat{g}(r)$ for *Picea abies* ($\cdot$) and *Fagus sylvatica* ($\Diamond$), estimated using Epanechnikov kernel with parameter $c = 0.15$

3.3 Model Fitting for the Root Data

Homogeneous Matérn-Cluster Model

Regarding the results of the estimated point process characteristics described before and because of its simplicity, Matérn-cluster processes are chosen as a model for the underlying point processes; see Appendix B for a definition.

Using once more the natural intensity estimator given in Formula (3) of Appendix A, the parameter λ_{mc} was estimated as $\widehat{\lambda_{mc}}^{spruce} = 0.00403$ and $\widehat{\lambda_{mc}}^{beech} = 0.00262$, respectively. Concerning the regarded point-pair distances a range from $0\ cm$ to $r_{max} = 50\ cm$ has been chosen, where r_{max} equals half the minimum of the given depth and width. The parameters R and λ_e are estimated by minimum-contrast estimators, which are computed by numerical minimization of the integral

$$\int_0^{r_{max}} (\widehat{g}(r) - g_{theo}(r))^2 dr, \tag{3}$$

where $\widehat{g}(r)$ is the averaged estimated pair correlation function given in Formula (11) of Appendix A for the Epanechnikov kernel with $c = 0.15$, and $g_{theo}(r)$ is the theoretical value for the pair correlation function of the Matérn-cluster process with parameters λ_e, R and λ_{mc}. Since the parameter λ_{mc} has already been estimated, the minimization of the integral in (3) yields an estimation for the pair of parameters R and λ_e. The obtained estimates are $\widehat{R}^{spruce} = 4.9\ cm$ and $\widehat{\lambda_e}^{spruce} = 0.00690$ for spruce roots, while for beech roots $\widehat{R}^{beech} = 7.4\ cm$ and $\widehat{\lambda_e}^{beech} = 0.00603$ are obtained.

Model Conclusions

The given point patterns are modelled as extracts of realisations of stationary Matérn-cluster processes with estimated intensities $\widehat{\lambda}_{mc}^{spruce} = 0.00403$ and $\widehat{\lambda}_{mc}^{beech} = 0.00262$, with cluster radii $\widehat{R}^{spruce} = 4.9\ cm$ and $\widehat{R}^{beech} = 7.4\ cm$, and with parent-process intensities $\widehat{\lambda_e}^{spruce} = 0.00690$ and $\widehat{\lambda_e}^{beech} = 0.00603$. In order to get an idea for the degree of clustering, the quantity

$$\widehat{\lambda}_t = \frac{\widehat{\lambda_{mc}}}{\widehat{\lambda_e}\pi R^2} \tag{4}$$

was evaluated. For *Picea abies* one gets $\widehat{\lambda}_t^{spruce} = 0.00774$, while for *Fagus sylvatica* $\widehat{\lambda}_t^{beech} = 0.00253$ is obtained. From the estimated parameters one can conclude that there is stronger clustering within a smaller cluster radius for spruce roots, while for the beech roots the clustering is weaker, but the cluster radius is slightly larger.

Inhomogeneous Matérn-Cluster Model

For the original data, which shows a vertical distribution property, the Matérn-cluster model fitted for the homogeneous case has to be retransformed, where the inverse transformation

$$h_{orig} = (F^*)^{-1}\left(\frac{F^*(h_{tot})}{h_{tot}}h_{tran}\right) \tag{5}$$

of the depth is considered, with $(F^*)^{-1}(y)$ representing the generalised inverse function of $F^*(x)$. Then, in the retransformed model, the parent process is given by an inhomogeneous Poisson process with intensity function

$$\lambda_e(x, y) = \lambda_e(y) = \lambda_e \frac{f^*(y)}{F^*(h_{tot})} h_{tot}, \tag{6}$$

where x and y represent the horizontal and vertical coordinate, $f^*(x)$ is the density function of the suitable distribution function $F^*(x)$ (exponential distribution for spruces and gamma distribution for beeches), λ_e is the intensity of the parent process of the homogeneous model and h_{tot} represents the total depth of the sampling window. The cluster regions are no longer circles, but the images of these circles under the mapping given in (5). They can be written as

$$\{(x, y) : (x - x_p)^2 + (F^*(y) - F^*(y_p))^2 (\frac{h_{tot}}{F^*(h_{tot})})^2 \leq R^2\}, \tag{7}$$

where the corresponding parent point is denoted as (x_p, y_p). Since the mean total number of points in the given window as well as the mean total number of points in a cluster stay the same compared to the homogeneous model, the intensity function for the inhomogeneous Matérn cluster point process is given as

$$\lambda_{mc}(x, y) = \lambda_{mc}(y) = \lambda_{mc} \frac{f^*(y)}{F^*(h_{tot})} h_{tot}, \tag{8}$$

where λ_{mc} is the corresponding intensity of the homogeneous model. Fig. 11 shows a realisation of the inhomogeneous Matérn-cluster model, which corresponds to the homogeneous realisation displayed in Fig. 10 using an exponential depth distribution. Note that only those simulated data shown in the upper part of Fig. 11 should be used for interpretation purposes. In the lower part of Fig. 11 the influence of transformation and retransformation of data clearly dominates the original spatial structure of those (sparse) root data with larger vertical depths.

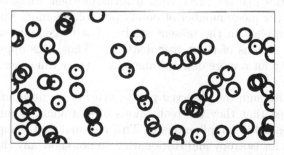

Fig. 10. Realization of the homogeneous Matérn-cluster model fitted in the case of *Picea abies*

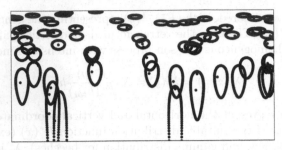

Fig. 11. Realization of the retransformed inhomogeneous Matérn-cluster model

4 Discussion

In both studied cases clear differences between two biologically distinct groups can be recognised using estimated point process characteristics. Apart from rejecting in all cases the null hypotheses of having homogeneous Poisson process as generating processes for the observed point patterns, it has been possible in the first example to detect a distance region where the number of chromocenters differ strongly between the non-differentiated and the differentiated state of NB4 cell nuclei. In the second example differences in the clustering behavior of the fine roots for European Beech compared to Norway Spruce have become visible. Apart from that a simple point process model has been fitted to the tree root data.

4.1 NB4 Cell Nuclei

The centromeric regions of chromosomes represent an important part of the heterochromatin compartment in interphase nuclei. A previous study was focused on the quantitative description of three dimensional distribution patterns of centromeric hetrochromatin in NB4 cells using features of the MST [4]. From a mathematical point of view, this approach has several disadvantages. Quantities like the MST edge lengths or their variance are strongly dependent on the mean number of points per volume unit. Apart from that, the methods applied in the present study allow to get inference about different specific regions of point pair distances. Thus, this approach provides the opportunity for a more detailed analysis of three dimensional centromere distributions.

Notice that, although the observed point patterns are finite and bounded, it can be assumed that they are realisations of stationary point processes restricted to a bounded sampling region. This notional step is supported by the fact that tests for isotropy and stationarity do not show any significant rejections and that the volumes of the assumed sampling regions before and after differentiation are of comparable sizes. The method of assuming unbounded stationary point processes as sources for observed realisations restricted to a bounded sampling region is a quite common practice since very often data is given in finite sampling regions and behave in a rather different non-stationary way outside of these regions [12, 39].

Due to the assumption that the observed point samples in the bounded sampling regions are extracts of unbounded realisations of stationary point processes it is necessary, although having only bounded sampling regions, to perform edge-corrections in order to insure compatibility with the applied methods. We want to emphasize that this procedure has its statistical justification in the facts that tests for isotropy and stationarity do not show any significant rejections and that the sampling regions have similar volumes.

Other types of estimators apart from spatial Horvitz-Thompson style estimators, e.g. of Kaplan-Meier type [3] and other techniques of edge corrections might also be applicable.

Clustering of chromosomal regions in interphase cell nuclei is supposed to be an important mechanism regulating the functional architecture of chromatin. In our previous study, we observed a progressive clustering of centromeric heterochromatin after differentiation of NB4 cells with ATRA [4]. These clusters (chromocenters) represent groups of centromeres with a distance below the limit of spatial resolution of optical microscopy. In the present study, we have analysed the distance of these chromocenters and found a higher frequency of distances between 350 nm and 800 nm for undifferentiated cells in comparison to ATRA-differentiated NB4 cells (Figs. 4 and 6). These new data imply the existence of heterochromatin regions with a range of 350 nm to 800 nm containing functionally related centromeric zones. The centromeres in these regions cluster during ATRA-induced differentiation of NB4 cells as demonstrated by the decreased number of detectable chromocenters, i.e. groups of centromeres within a sphere with a diameter of less than 350 nm. The existence of heterochromatin regions containing centromeres of specific chromosomes would imply that the restructuring of these chromosome territories has to proceed in a coordinated nonrandom way during the differentiation-induced "collapse" of these heterochromatin zones. This model is in accordance with a topological model for gene regulation based on the structural remodelling of chromosome territories during modulation of transcription [9, 32].

Another important result of the present study is the finding that the 3D distribution of chromocenters is not completely random in undifferentiated as well as in ATRA-differentiated NB4 cells. These findings, thus, rejects a previous hypothesis which was based on the comparison of centromere distributions in NB4 cells with simulated completely random patterns using MST features [4]. The result of the present study is in accordance with other studies, which suggested that interphase centromeres are not arranged in a completely random way [18, 22]. Importantly, investigations of interphase chromosome positions indicate that a strictly maintained structure of chromatin appears to be necessary for a normal function of cells even in tumours [8].

4.2 Planar Sections of Root Systems

The point process characteristics using the transformed data described the two dimensional distribution of small roots in pure stands of *Picea abies* and *Fagus sylvatica*. The results for such homogenised data using the averaged estimated pair correlation function (Fig. 9) show that attraction can be observed for point pairs with distances less than approximately 14 *cm*. It means that roots of both species tend to cluster in areas up to this diameter. As roots react to nutrient enriched soil patches by enhanced growth and greater biomass in these areas [13, 29], this attraction of roots within this diameter could also be a direct link to a local occurrence of soil resources. On the other hand, for very small distances of less than 0.5 *cm* there is a hardcore property in the homogeneous case, which can possibly be explained by the thickness of the regarded roots. Also, as the small roots are associated to the uptake-oriented fine roots, concentration of the small roots in clusters of smaller diameters (i.e. less than 0.5 *cm*) is not reasonable. In this context it is important to notice that these structural differences are independent of the observed significant difference in the average number of detected points (roots) for the samples of *Picea abies* and *Fagus sylvatica*.

The homogenised point patterns were modelled as Matérn-cluster processes with estimated parameters described in Sect. 3.2. The Matérn-cluster model chosen has some serious advantages. First the model is of a certain simplicity and theoretical values for point process characteristics are known. Even more important is that the sample data is fitted well by this model. The estimated point process characteristics using the Matérn-cluster processes further differentiated between the species. The results show for spruces a stronger clustering in a smaller range of attraction ($\widehat{R}_{spruce} = 4.9\ cm$), while the clustering is weaker for beeches, but the range of attraction ($\widehat{R}_{beech} = 7.4\ cm$) seems to be slightly larger. This finding is in accordance with another investigation [41] calculating influence areas for each root. Their results indicated that the root system of spruce requires more roots to achieve a similar degree of space acquisition and thus beech exploits patchily distributed soil resources at lower root numbers. In summary there is a combination of two effects, the depth distribution already described in [40] and the cluster effects analysed in the present paper. Structural differences between spruce and beech indicated in [41] have been mathematically described. Stronger clustering in the case of spruce than in the case of beech can be seen also regarding the characteristics mentioned above as well as by a comparison of the estimated parameters for the homogeneous model $\widehat{\lambda}_t^{\,spruce} = 0.00774$ vs. $\widehat{\lambda}_t^{\,beech} = 0.00253$, keeping in mind that the estimated parent intensities are almost equal ($\widehat{\lambda}_p^{\,spruce} = 0.00690$ vs. $\widehat{\lambda}_p^{\,beech} = 0.00603$). The previous GIS-based investigation of root distribution [41] was not able to quantify the differences of clustering between the two species so precisely as the applied modelling by point processes.

Finally a non-homogeneous Matérn-cluster model has been constructed by a retransformation of the homogeneous model, thereby reflecting the observed depth distribution of the tree roots. The visualisation of the retransformed data suggests a depth-dependent size and shape of root clusters. Close to the soil surface, roots form clusters along the horizontal axis. This shape agrees also with the horizontally distributed root points in the original samples (c.f. Fig. 3). Horizontally distributed roots as well as the shape of generated clusters may reflect the attractive soil patches in the nutrient-rich topsoil layers. Deeper, the real size of clusters is larger and more circular. However, because the transformation and retransformation of root data at low intensities in the deep parts of the soil profile makes the results unstable, the lower third of Fig. 11 is not really useful for interpretation of spatial structures. The investigated small roots were also described regarding water and nutrient uptake [25] and mediates to the most active fine roots ($< 2\ mm$). Thus, clusters of small roots reflect the presence of nutrient patches or zones of better water availability [20, 33, 38]. As the number of small roots and their clustering was independent of the distance to the surrounding trees and of their diameter [41], the root clusters are suggested as an inherent property of below-ground space acquisition.

Acknowledgements

Data collection of the profile walls was financially supported by the Austrian Science Foundation within the Special Research Program "Restoration of Forest Ecosystems", F008-08.

Appendix A: Point Process Characteristics and their Estimators

In the following let $\mathbf{x} = \{\mathbf{x}_n\}$ be a random point process in $\mathbb{R}^d$, where $d \in \{2, 3, \ldots\}$ and let $N(B) = \#\{n : \mathbf{x}_n \in B\}$ denote the number of points $\mathbf{x}_n$ of $\mathbf{x}$ located in a sampling window B

Intensity Measure

The intensity measure Λ is defined as

$$\Lambda(B) = EN(B) \tag{1}$$

for a given set B. Hence $\Lambda(B)$ is the mean number of points in B. In the homogeneous case it suffices to regard an intensity λ since then

$$\Lambda(B) = \lambda \nu_d(B) \tag{2}$$

where $\nu_d(B)$ denotes the volume of B. A natural estimator for λ is given by

$$\widehat{\lambda} = \frac{N(B)}{\nu_d(B)}. \tag{3}$$

However, for the estimation of the nearest-neighbour distance distribution a different estimator

$$\widehat{\lambda}_H = \sum_{\mathbf{x}_n \in B} \frac{1_{B \ominus b(o,s(\mathbf{x}_n))}(\mathbf{x}_n)}{\nu_d(B \ominus b(o, s(\mathbf{x}_n)))} \tag{4}$$

is recommended [45], where $s(\mathbf{x}_n)$ denotes the distance of $\mathbf{x}_n$ to its nearest neighbour and $b(x, r)$ is the ball with radius r and midpoint x.

Notice that, following the recommendation in [44], λ^2 has been estimated by

$$\widehat{\lambda^2} = \frac{N(B)(N(B) - 1)}{(\nu_d(B))^2}, \tag{5}$$

since even in the Poisson case $(\widehat{\lambda})^2$ is not an unbiased estimator for λ^2.

Moment Measure and Product Density

Let B_1 and B_2 be two sets. The second factorial moment measure $\alpha^{(2)}$ of $\mathbf{x}$ is defined by

$$\alpha^{(2)}(B_1 \times B_2) = \mathrm{E}(\sum_{\substack{\mathbf{x}_1, \mathbf{x}_2 \in N \\ \mathbf{x}_1 \neq \mathbf{x}_2}} 1_{B_1}(\mathbf{x}_1) 1_{B_2}(\mathbf{x}_2)). \tag{6}$$

Often $\alpha^{(2)}$ can be expressed using a density function $\varrho^{(2)}$ as follows

$$\alpha^{(2)}(B_1 \times B_2) = \int_{B_1} \int_{B_2} \varrho^{(2)}(x_i, x_j) dx_i dx_j. \tag{7}$$

The density function $\varrho^{(2)}$ is called the second product density. If one takes two balls C_1 and C_2 with infinitesimal volumes dV_1 and dV_2 and midpoints x_1 and x_2 respectively, the probability for having in each ball at least one point of $\mathbf{x}$ is approximately equal to $\varrho^{(2)}(x_1, x_2) dV_1 dV_2$. In the homogeneous and isotropic case $\varrho^{(2)}(x_1, x_2)$ can be replaced by $\varrho^{(2)}(r)$, where $r = ||x_1 - x_2||$. As an estimator

$$\widehat{\varrho^{(2)}}(r) = \frac{1}{d b_d r^{d-1}} \sum_{\substack{\mathbf{x}_i, \mathbf{x}_j \in B \\ i \neq j}} \frac{k_h(r - ||\mathbf{x}_i - \mathbf{x}_j||)}{\nu_d(B_i \cap B_j)} \tag{8}$$

has been used [44], where $k_h(x)$ denotes the Epanechnikov kernel

$$k_h(x) = \frac{3}{4h}(1 - \frac{x^2}{h^2}) 1_{(-h,h)}(x), \tag{9}$$

$B_{\mathbf{x}_j} = \{x + \mathbf{x}_j : x \in B\}$ is the set B translated by the point $\mathbf{x}_j$, and the sum in (8) extends over all pairs of points $\mathbf{x}_i, \mathbf{x}_j \in B$ with $i \neq j$. The bandwidth h has been chosen as $h = c\widehat{\lambda}^{-1/d}$ with a fixed parameter c.

Pair Correlation Function

The product density $\varrho^{(2)}(r)$ is used to obtain the pair correlation function $g(r)$ as

$$g(r) = \frac{\varrho^{(2)}(r)}{\lambda^2}. \tag{10}$$

The pair correlation function at a certain value r can be regarded as the frequency of point pairs with distance r, where $g(r) = 1$ is a base value. The pair correlation function can be estimated by the usage of estimators for $\varrho^{(2)}(r)$ and λ^2 respectively. In particular, we consider the estimator

$$\widehat{g}(r) = \frac{\widehat{\varrho^{(2)}(r)}}{\widehat{\lambda^2}}, \tag{11}$$

where $\widehat{\lambda^2}$ and $\widehat{\varrho^{(2)}}$ are given by (5) and (8), respectively. Note that $g(r) \geq 0$ for all distances r. In the Poisson case $g_{Poi}(r) \equiv 1$, therefore $g(r) > 1$ indicates that there are more point pairs having distance r than in the Poisson case, while $g(r) < 1$ indicates that there are less point pairs of such a distance.

K-function

Ripley's K-function [36] is defined such that $\lambda K(r)$ is the expected number of points of the stationary point process $\mathbf{x} = \{\mathbf{x}_n\}$ within a ball $b(\mathbf{x}_n, r)$ centred at a randomly chosen point $\mathbf{x}_n$ which itself is not counted. Formally

$$\lambda K(r) = E \sum_{\mathbf{x}_n \in B} \frac{N(b(\mathbf{x}_n, r)) - 1}{\lambda \nu_d(B)}. \tag{12}$$

The K-function has been estimated by

$$\widehat{K}(r) = \frac{\kappa(r)}{\widehat{\lambda^2}}, \tag{13}$$

where

$$\kappa(r) = \sum_{\substack{\mathbf{x}_i, \mathbf{x}_j \in B \\ i \neq j}} \frac{1_{b(o,r)}(\mathbf{x}_j - \mathbf{x}_i)}{|B_{\mathbf{x}_j} \cap B_{\mathbf{x}_i}|}, \tag{14}$$

For Poisson processes it is easy to see that $K_{Poi}(r) = b_d r^d$.

L-function

Often it is more convenient to scale the $K(r)$ in order to get a function equal to r for the Poisson case. Hence $L(r)$ is defined as

$$L(r) = \sqrt[d]{\frac{K(r)}{b_d}}, \tag{15}$$

where b_d denotes the volume of the d-dimensional unit sphere. Thus, in the Poisson case, we have

$$L(r) - r = 0. \tag{16}$$

A natural estimator for $L(r)$ is given by

$$\widehat{L}(r) = \sqrt[d]{\frac{\widehat{K}(r)}{b_d}}. \tag{17}$$

Nearest-Neighbor Distance Distribution

The nearest-neighbour distance distribution D is the distribution function of the distance from a randomly chosen point $\mathbf{x}_n$ of the given stationary point process $\mathbf{x}$ to its nearest neighbour. Hence $D(r)$ is the probability that a randomly chosen point $\mathbf{x}_n$ of $\mathbf{x}$ has a neighbour with a distance less than or equal to r. According to [45] we used the Hanisch estimator $\widehat{D}_H(r) = D_H(r)/\widehat{\lambda}_H$ [2, 19] with

$$\widehat{D_H}(r) = \sum_{\mathbf{x}_n \in B} \frac{1_{B \ominus b(o,s(\mathbf{x}_n))}(\mathbf{x}_n) 1_{(0,r]}(s(\mathbf{x}_n))}{\nu_d((B \ominus b(o,s(\mathbf{x}_n))))}. \tag{18}$$

A useful property of the nearest-neighbour distance distribution is that in the case of stationary Poisson processes we have

$$D_{Poi}(r) = 1 - \exp\left(-\lambda b_d r^d\right). \tag{19}$$

Therefore one can conclude that $D(r) < D_{Poi}(r)$ indicates rejection between points, on the other hand $D(r) > D_{Poi}(r)$ indicates attraction, keeping in mind that the nearest-neighbour distance distribution function is a cumulated quantity.

Spherical Contact Distribution Function

The spherical contact distribution function $H_s(r)$ is the distribution function of the distance from an arbitrary point, chosen independently of the point process $\mathbf{x}$, to the nearest point belonging to $\mathbf{x}$. Notice that the value $H_s(r)$ can be interpreted as the probability that at least one point $\mathbf{x}_n$ of $\mathbf{x}$ is in the sphere of radius r centred at the origin. As an estimator for $H_s(r)$,

$$\widehat{H}_s(r) = \frac{\nu_d((B \ominus b(0,r)) \bigcup_{\mathbf{x}_n \in B} b(\mathbf{x}_n,r))}{\nu_d(B \ominus b(0,r))} \tag{20}$$

is used.

J-function

Based on $H_s(r)$ and on $D(r)$, Baddeley's J-function is defined by

$$J(r) = \frac{1 - H_s(r)}{1 - D(r)}. \tag{21}$$

where

$$\widehat{J}(r) = \frac{1 - \widehat{H}_s(r)}{1 - \widehat{D}(r)}, \tag{22}$$

is a natural estimator for $J(r)$. In the case of Poisson point processes $J_{Poi}(r) \equiv 1$ and therefore if $J(r) > 1$ one can conclude that there is repulsion between point pairs of distance r. On the other hand if $J(r) < 1$ there is attraction between point pairs compared to the case of complete spatial randomness.

Appendix B: Matérn-Cluster Model

The Matérn-cluster point process $\mathbf{x}_{mc}$ is based on a Poisson process with intensity λ_e whose points are called parent points. Around each parent point a sphere with radius R is taken in which the points of the Matérn-cluster process are scattered uniformly where the number of points in such a sphere is Poisson distributed with parameter $R^d b_d \lambda_t$. Notice that λ_t is the mean number of points per unit area generated by a single parent point in a sphere of radius R. Since the parent points themselves are not part of the Matérn-cluster process, its intensity is given as

$$\lambda_{mc} = R^d b_d \lambda_t \lambda_e. \tag{23}$$

Thus, the Matérn-cluster point process $\mathbf{x}_{mc}$ is uniquely determined by three of the four parameters λ_e, λ_t, R and λ_{mc} Obviously, for small distances, points of the Matérn-cluster process are attracted to each other, in other words there is a bigger expected number of points of $\mathbf{x}_{mc}$ in a sphere around an arbitrarily chosen point of $\mathbf{x}_{mc}$ than for Poisson processes of comparable intensity $\lambda_{Poi} = \lambda_{mc}$. For $\mathbf{x}_{mc}$, closed formulae for the point process characteristics described in Appendix A are known [43].

References

[1] I. Alcobia, A.S. Quina, H. Neves, N. Clode and L. Parreira. The spatial organization of centromeric heterochromatin during normal human lymphopoiesis: evidence for ontogenically determined spatial patterns. *Experimental Cell Research*, 290:358–369, 2003.

[2] A.J. Baddeley. Spatial sampling and censoring. In W. S. Kendall, M.N.M. van Lieshout and O.E. Barndorff-Nielsen, editors, *Current Trends in Stochastic Geometry and its Applications*. Chapman and Hall, 1998.

[3] A.J. Baddeley and R.D. Gill. Kaplan-Meier estimators of distance distribution for spatial point processes. *The Annals of Statistics*, 25(1):263–292, 1997.

[4] M. Beil, D. Durschmied, S. Paschke, B. Schreiner, U. Nolte, A. Bruel and T. Irinopoulou. Cytometry. *Spatial distribution patterns of interphase centromeres during retinoic acid-induced differentiation of promyelocytic leukemia cells*, 47:217–225, 2002.

[5] M. Beil, F. Fleischer, S. Paschke and V. Schmidt. Statistical analysis of 3d centromeric heterochromatin structure in interphase nuclei. *Journal of Microscopy*, 217:60–68, 2005.

[6] M.M. Caldwell, J.H. Manwaring and S.L. Durham. Species interaction at the level of fineroots in the field: influence of soil nutrient heterogeneity and plant size. *Oecologia*, 106:440–447, 1996.

[7] S.N. Chiu and D. Stoyan. Estimators of distance distributions for spatial patterns. *Statistica Neerlandica*, 52(2):239–246, 1998.

[8] M. Cremer, K. Kupper, B. Wagler, L. Wizelman, J. von Hase, Y. Weiland, L. Kreja, J. Diebold, M. R. Speicher and T. Cremer. Inheritance of gene density-related higher order chromatin arrangements in normal and tumor cell nuclei. *The Journal of Cell Biology*, 162:809–820, 2003.

[9] T. Cremer and C. Cremer. Chromosome territories, nuclear architecture and gene regulation in mammalian cells. *Nature Reviews Genetics*, 2:292–301, 2001.

[10] H. de The, C. Chomienne, M. Lanotte, L. Degos and A. Dejean. The t(15;17) translocation of acute promyelocytic leukaemia fuses the retinoic acid receptor alpha gene to a novel transcribed locus. *Nature*, 347:558–561, 1990.

[11] P.J. Diggle. *Statistical Analysis of Spatial Point Patterns, 2nd ed.* Oxford University Press, 2003.

[12] P.J. Diggle, J. Mateu and H.E. Clough. A comparison between parametric and non-parametric approaches to the analysis of replicated spatial point patterns. *Advances in Applied Probability*, 32:331–343, 2000.

[13] M.C. Drew. Comparison of the effects of a localized supply of phosphate, nitrate, ammonium and potassium on the growth of the seminal root system, and the shoot, of barley. *New Phytologist*, 75:479–490, 1975.

[14] C. Dussert, G. Rasigni, J. Palmari, M. Rasigni, A. Llebaria and F. Marty. Minimal spanning tree analysis of biological structures. *Journal of Theoretical Biology*, 125:317–323, 1987.

[15] C. Dussert, G. Rasigni, M. Rasigni, J. Palmari and A. Llebaria. Minimal spanning tree: a new approach for studying order and disorder. *Physical Revue B*, 34:3528–3531, 1986.

[16] P. Fenaux, C. Chomienne and L. Degos. Acute promyelocytic leukemia: biology and treatment. *Seminars in Oncology*, 24:92–102, 1997.

[17] F. Fleischer, S. Eckel, I. Schmid, V. Schmidt and M. Kazda. Statistical analysis of the spatial distribution of tree roots in pure stands of *fagus sylvatica* and *picea abies*. Preprint, 2005.

[18] T. Haaf and M. Schmid. Chromosome topology in mammalian interphase nuclei. *Experimental Cell Research*, 192:325–332, 1991.

[19] K.-H. Hanisch. Some remarks on estimators of the distribution function of nearest-neighbor distance in stationary spatial point patterns. *Statistics*, 15:409–412, 1984.

[20] R.B. Jackson and M.M. Caldwell. Geostatistical patterns of soil heterogeneity around individual perennial plants. *Journal of Ecology*, 81:683–692, 1993.

[21] R.B. Jackson and M.M. Caldwell. Integrating resource heterogeneity and plant plasticity: modeling nitrate and phosphate uptake in a patchy soil environment. *Journal of Ecology*, 84:891–903, 1996.

[22] J. Janevski, P.C. Park and U. De Boni. Organization of centromeric domains in hepatocyte nuclei: rearrangement associated with de novo activation of the vitellogenin gene family in xenopus laevis. *Experimental Cell Research*, 217:227–239, 1995.

[23] G. Krauss, K. Müller, G. Gärtner, F. Härtel, H. Schanz and H. Blanckmeister. Standortsgemässe durchführung der abkehr von der fichtenwirtschaft im nordwestsächsischen niederland. *Tharandter Forstl. Jahrbuch*, 90:481–715, 1939.

[24] M. Lanotte, V. Martin-Thouvenin, S. Najman, P. Balerini, F. Valensi and R. Berger. Nb4, a maturation inducible cell line with t(15;17) marker isolated from a human acute promyelocytic leukemia (m3). *Blood*, 77:1080–1086, 1991.

[25] J. Lindenmair, E. Matzner, A. Göttlein, A.J. Kuhn and W. H. Schröder. Ion exchange and water uptake of coarse roots of mature norway spruce trees. In W.J. Horst et al., editor, *Plant Nutrition – Food Security and Sustainability of Agro-Ecosystems*. Kluwer Academic Publishers, 2001.

[26] R. Marcelpoil and Y. Usson. Methods for the study of cellular sociology: voronoi diagrams and parametrization of the spatial relationships. *Journal of Theoretical Biology*, 154:359–369, 1992.

[27] J. Mayer. *On quality improvement of scientific software: Theory, methods, and application in the GeoStoch development*. PhD thesis, University of Ulm, 2003.

[28] J. Mayer, V. Schmidt and F. Schweiggert. A unified simulation framework for spatial stochastic models. *Simulation Modelling Practice and Theory*, 12:307–326, 2004.

[29] E.C. Morris. Effect of localized placement of nutrients on root-thinning in self-thinning populations. *Annals of Botany*, 78:353–364, 1996.

[30] A. Okabe, B. Boots, K. Sugihara and S.N. Chiu. *Spatial Tessellations*. John Wiley & Sons, 2000.

[31] P.C. Park and U. de Boni. Spatial rearrangement and enhanced clustering of kinetochores in interphase nuclei of dorsal root ganglion neurons

in vitro: association with nucleolar fusion. *Experimental Cell Research*, 203:222–229, 1992.

[32] P.C. Park and U. de Boni. Dynamics of structure-function relationships in interphase nuclei. *Life Sciences*, 64:1703–1718, 1999.

[33] M.M. Parker and D.H. van Lear. Soil heterogeneity and root distribution of mature loblolly pine stands in piedmont soils. *Soil Science Society of America Journal*, 60:1920–1925, 1996.

[34] S. Perrod and S.M. Gasser. Long-range silencing and position effects at telomeres and centromeres: parallels and differences. *Cellular and Molecular Life Sciences*, 60:2303–2318, 2003.

[35] E.J. Richards and S.C.R. Elgin. Epigenetic codes for heterochromatin formation and silencing. *Cell*, 108:489–500, 2002.

[36] B.D. Ripley. The second-order analysis of stationary point processes. *Journal of Applied Probability*, 13:255–266, 1976.

[37] B.D. Ripley. *Spatial Statistics*. John Wiley & Sons, 1981.

[38] R.J. Ryel, M.M. Caldwell and J.H. Manwaring. Temporal dynamics of soil spatial heterogeneity in sagebrush-wheatgrass steppe during a growing season. *Plant Soil*, 184:299–309, 1996.

[39] K. Schladitz, A. Särkkä, I. Pavenstädt, O. Haferkamp and T. Mattfeldt. Statistical analysis of intramembranous particles using fracture specimens. *Journal of Microscopy*, 211:137–153, 2003.

[40] I. Schmid and M. Kazda. Vertical distribution and radial growth of coarse roots in pure and mixed stands of *fagus sylvatica* and *picea abies*. *Canadian Journal of Forest Research*, 31:539–548, 2001.

[41] I. Schmid and M. Kazda. Clustered root distribution in mature stands of *fagus sylvatica* and *picea abies*. *Oecologia*, page (in press), 2005.

[42] D. Stoyan, W.S. Kendall and J. Mecke. *Stochastic Geometry and its Applications*. John Wiley & Sons, 1995.

[43] D. Stoyan and H. Stoyan. *Fractals, Random Shapes and Point Fields*. John Wiley & Sons, 1994.

[44] D. Stoyan and H. Stoyan. Improving ratio estimators of second order point process characteristics. *Scandinavian Journal of Statistics*, 27:641–656, 2000.

[45] D. Stoyan, H. Stoyan, A. Tscheschel and T. Mattfeldt. On the estimation of distance distribution functions for point processes and random sets. *Image Analysis and Stereology*, 20:65–69, 2001.

[46] M.N.M. van Lieshout and A.J. Baddeley. A nonparametric measure of spatial interaction in point patterns. *Statistica Neerlandica*, 50:344–361, 1996.

Spatial Marked Point Patterns for Herd Dispersion in a Savanna Wildlife Herbivore Community in Kenya

Alfred Stein[1] and Nick Georgiadis[2]

[1] International Institute for Geo-Information Science and Earth Observation (ITC), PO Box 6, 7500 AA Enschede, The Netherlands, stein@itc.nl
[2] Mpala Research Centre, PO Box 555, Nanyuki, Kenya, njg@mpala.org

Summary. Quantitative descriptions of animal species' distributions at the ecosystem level are rare. In this study we used marked spatial point pattern analysis to characterize herd spatial distributions of several species comprising a savanna large herbivore community in Laikipia, central Kenya. Points are the herd centres, marks are the herd sizes. Previous research [15] identified possible discrepancies between prey and non-prey species on the basis of the nearest neighbour distance function. In this paper we make a similar distinction and analyse possible consequences. Analysis concentrated on Ripley's K-function on several data subsets. A digitised boundary of the area has been included. The herd patterns of Thomson gazelle and of the plains zebra were modelled with a Strauss marked point process. The pattern of the Thomson gazelle showed a single mode, whereas that of the plains zebra showed multiple modes. This can be well explained by the ecosystem behavior (habitat specialist versus habitat generalist) of the two species.

Key words: Herbivores, Laikipia, Nearest neighbour distances, Savanna, Spatial point pattern

1 Introduction

Herbivores living freely in nature tend to aggregate in groups or herds of various sizes. These herds usually do not randomly distribute and therefore display spatial distribution patterns [15]. An explanation for variation in animal grouping and distribution can be given on the basis of physiological grounds, invoking metabolic requirements, on ecological grounds, invoking habitat preference [8], feeding style, competition, facilitation [1, 11], and food distribution [17], and on climatic grounds [18]. In the past, both a statistical and an ecological study have been devoted to modelling herds of herbivores in space [14, 15]. None of these studies, however, employs a marked point pattern spatial statistics approach – herd size has not been taken into account so far. A marked point pattern analysis explicitly uses the locations of herds and distances between the herds, whereas the size of a herd is likely to be related with ecological conditions.

Advances in Global Positioning System technology (GPS; see [19]), and spatial point pattern analysis [3, 13], permitted us to characterize spatial distributions of the nine most abundant large herbivore species in the Laikipia ecosystem of central Kenya. The data, representing marked point measurements of herds, were collected in 1996 during a total count of wildlife in an area of 7,100 km², using 10 aircraft equipped with GPS receivers [5]. Apart from the characterization, also herd size has been measured by counting the number of animals grouping together. We distinguish and further differentiate those that can be preyed upon by large carnivores (Plains zebra *Equus burchelli* (Gray)), impala *Aepyceros melampus*, Grant's gazelle *Gazella grantii*, eland *Taurotragus oryx* and hartebeest *Alcelaphus buselaphus*), those that are too large to be preyed upon (buffalo *Syncerus caffer*, elephant *Loxodonta africana* and giraffe *Giraffa, camelopardalis*) and those that are too small to be preyed upon (Thomson's gazelle *Gazelli thomsoni* (Gunther)).

Three methods of spatial point pattern analysis were used to characterize the distributions of wild herbivore herds for each species separately, and for all species combined: the marked K-function, the pair correlation function and fitting of the Strauss process to both the marked and the unmarked point process. Possible causes of observed patterns of dispersion within and among species are discussed.

The aim of this study has been to further explore the relations between herds of different animals species and combinations of species using marked points processes. We aimed to combine the sizes of the herds as additional information, having in mind that large herds of herbivores may show different behavior than herds of a smaller size. The study is illustrated with the unique point pattern data set from the Laikipia area.

2 Materials and Methods

2.1 Distribution Data

Data in this study were collected during a total count within Laikipia District over three days in September 1996 ([5]; the region is also described in [6]). The area of 7,100 km² was divided into daily counting blocks of approximately 200 to 300 km², and each block was allocated to one aircraft per day. Ten high winged aircraft were used simultaneously to systematically search each block. Each aircraft flew at heights between 70 and 130 m above ground level, following transects spaced 1 km apart. Whenever an animal or a group of animals was spotted, the aircraft deviated from its flight-line to circle the observed animals until their number was counted. Geographical co-ordinates of positions of the centers of the herds were recorded using a Trimble GPS receiver. Overlaps and double counts at the boundaries of the blocks were identified and subtracted from the total wildlife numbers as a correction for count overlaps. This resulted in a data set of 1828 locations where at least one animal was observed (Fig. 1).

The observation of one or more animals at a given location is termed a herd. We make the basic assumption that each location is equally likely to host a herd. Deviations from randomness may occur due to external factors influencing the pattern. Such deviations are then of interest, both for the animals individually, and in their mutual relationships. Estimates of dispersion used here were affected by subjective variation among observers in their assignment of individuals to a her, but to the same degree.

In total, 55,201 animals of the 9 species were observed in this study, distributed over 3,025 herds. The maximum herd size equals 473 animals, whereas 322 solitary animals were observed, i.e. herds of size 1. Abundance varied widely among the nine species, largest herds occurring for plains zebra (Table 1). Also mean herd size varied widely, with plains zebra having the most herds (1,034) and a median herd size of (18 individuals herd^{-1}). The Thomson gazelle has less herds (211), and the animals tend to aggregate in herds of a smaller size (12 individuals herd^{-1}). Densities, i.e. number of herds per km^2, varies between the species in the area, the highest density occurring for Plains zebra (average intensity = 0.160 herds km^2).

Table 1. Summary statistics of the two selected animal species, preyed and non-preyed species and all species

Process	Species	No.of herds	Total count	Group size mean	median	stdev	Average intensity
X_1	Plains zebra	1,034	31,517	30.48	18	39.7	0.160
X_2	Thomson's gazelle	211	4,255	20.17	12	31.0	0.0326
$\tilde{X}_1$	Preyed Species	2,365	45,576	20.22	10	29.8	0.365
$\tilde{X}_2$	Non-preyed Species	660	9,445	15.30	8	23.1	0.102
$X_\bullet$	All Species	3,025	55,201	19.15	10	28.6	0.479

From a previous study [15] there was evidence that prey preference complementarity was an important factor in the distribution of herds. We found that herds of mid-sized prey species such as the plains zebra, which are more likely to be preferred by the dominant predators in this ecosystem (hyenas, lions and leopards), are expected to display less aggregated (even random) distributions. By contrast, herds of the smallest-sized species (Thomson's gazelle), as well as those of large-sized herbivores, are expected to experience lower predation pressure, and thus to be more aggregated. The Thomson gazelle, though, does not necessarily experience low predation pressure, as they are preyed upon by jackals.

To investigate prey preference, the dataset was split into two subsets: the combined data for large- and small-sized (habitat specialist or 'non-preferred' prey) species, and on the combined data for the mid-sized (habitat generalist or 'preferred' prey) species. If habitat preference complementarity was operating, the result should be a tendency by both groups to shift towards a more regular dispersion pattern. If prey preference complementarity was operating, the result for the non-preferred prey group should be to remain aggregated, while the preferred prey group should become regularly spaced.

2.2 Measures of Dispersion

Spatial processes yielding herds of various sizes are characterised by a simple stochastic model applied to a region A, here the Laikipia area. Herds Y are represented by the coordinates of their centre of gravity and the animal species and are marked by the observed number of animals. Marks, contained in a set M. A marked point process is hence denoted by (Y, M). As a result, A is summarised by a marked point pattern, consisting of the presence of at least one herbivore.

To describe the spatial point pattern generated by the distribution of the herbivores, we let (Y, M) be a nine-variate point process in A with jointly stationary components. The process consisting of all marked points regardless of type is denoted by $(X_\bullet, M) = \cup_{i=1}^{9}(X_i, M_i)$, where (X_i, M_i) denotes the marked point process for the ith herbivore. Similarly, the set (Y, M) can be decomposed as well into sets $X_\bullet = \cup_{i=1}^{2}\tilde{X}_i$ with $\tilde{X}_1$ corresponding to the preyed species and $\tilde{X}_2$ to the non-preyed species, respectively. The density of the processes is denote by λ, λ_i for each of the nine species and $\tilde{\lambda}$ for the preyed and nonpreyed species. In this paper, statistical inference for Y is based on distances. Although patterns may be non-stationary for several reasons, we take stationary processes as the starting point for our research. Non-homogeneity then appears as a result from the analysis, and could be analysed on the basis of [4].

The marked K-function

We consider the K-function to an arbitrary marked point pattern, in this study the centres of the herds. We consider here small herds and solitary species (herds of size up to 5), medium size herds (herds of sizes 6–25) and large sized herds (herds of sizes > 25 animals) and applied the multitype K-function.

We assume that X can be treated as a realisation of a stationary (spatially homogeneous) random spatial point process in the plane, observed through a bounded window W. The window W is in this study given by the edge of the Laikipia area. For edge correction we applied two correction procedures:

- the border method or *reduced sample* estimator [12]. This is the least efficient (statistically) and the fastest to compute.
- the translation correction method [9].

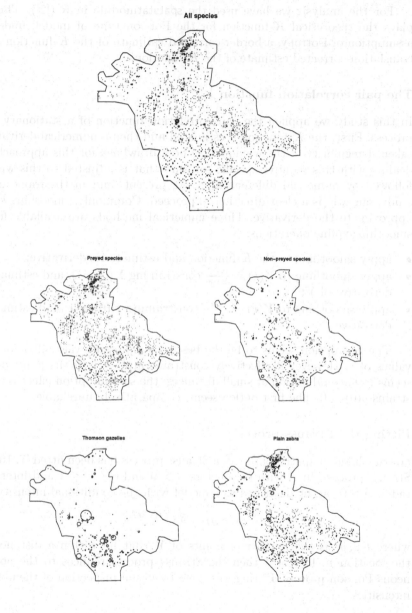

Fig. 1. Marked spatial point patterns of herds for the nine herbivores (*top*), the preyed and non-preyed species (center) and the Thomson gazelle and plains zebra (*bottom*) in the study area

For the analysis we have used the spatstatmodule in R ([2]). This displays the theoretical K-function for the Poisson type of model, under the assumption of isotropy, a border-corrected estimate of the K-function and a translation-corrected estimate of the K-function.

The pair correlation function

In this study we applied the pair correlation function of a stationary point process. First, the K-function is estimated and then a numerical derivative is taken. Irregularity of the window may be a drawback for this approach, but dealing with this requires specific software that is adjusted to this window, followed by numerical differentiation. At present, and in the frame of this study, our way is a clear direction to proceed. Commonly, smoothing splines approximate the derivative. Three numerical methods are available for the smoothing spline operations:

- apply smoothing to the K-function and estimate its derivative;
- apply smoothing to $Y(r) = \frac{K(r)}{2\pi \cdot r}$ constraining $Y(0) = 0$, and estimate the derivative of Y;
- apply smoothing to $Y(r) = \frac{K(r)}{\pi \cdot r^2}$ constraining $Y(0) = 1$, and estimate its derivative.

The last option seems to be the best at suppressing variability for small values of r. However it effectively constrains $g(0) = 1$. If the point pattern seems to have inhibition at small distances, the second option effectively constrains $g(0) = 0$. The first option seems comparatively unreliable.

Fitting the Strauss process

For the different spatial patterns, a Strauss process has been fitted [7, 16]. The Strauss process on A with parameters $\beta > 0$ and $0 \le \gamma \le 1$ and interaction radius $\delta > 0$ can be described as a model with the conditional intensity

$$\lambda(u, y) = \beta \cdot \gamma^{t(u,y)} \tag{1}$$

where $t(u, y)$ is the number of points of Y that lie within a distance δ of the location u. If $\gamma = 1$ then the Strauss process reduces to the homogeneous Poisson process. Fitting was done by visual inspection of the observed intensities.

3 Results

3.1 Spatial Point Patterns

Spatial point patterns for the plains zebra and for the Thomson gazelle are displayed in Fig. 1, where the herd sizes are displayed as the size of the circles. Clearly, the Thomson gazelle clusters in the South Western part of the area, the Plains zebra is more regularly distributed with both a high density and a highly aggregated spatial point pattern, showing evidence of clustering throughout the area. The Thomson gazelle shows clustering in the South-Western part. Both species appear to exhibit some aggregation. At a somewhat higher level, the preyed animals do not show any clustering, and the non-preyed species again show some clustering within the south-western sub-area. Finally, the plot of the combined spatial pattern is fairly dense with no apparent spatial pattern. A section with almost no herds occurs in the northern part of the area.

Figure 2 shows the three versions of the marked $K(r)$ functions, for the set of all herds (top), the preyed and the non-preyed species and for the individual Thomson gazelle and Plains zebra, respectively. The two corrected marked K-functions are relatively close together, but differ markedly from the theoretical one, as might be expected. In particular, the marked K-function for the Thomson gazelle shows a noisy behavior, but none of the latter two seems to deviate much from a quadratic behavior.

Pair correlation functions (fig. 3) were estimated for both the Thomson gazelle and the Plains zebra individually. Both pair correlation functions show an alternating sequence of values. Non of the two pictures shows a hard core, therefore the two species have herds that may be close to each other. The maximum for both species occurs at approximately 2 km, showing that this distance is preferred for herds of both species. A peak at 1 km for the Thomson gazelle is probably caused by the relatively low number of data. A much higher peak occurs for the Thomson gazelle (up to 5) as compared to the Plains zebra (up to 2.2). Therefore, the distribution of Plain's zebra herds is somewhat closer to the Poisson process than that of the Thomson gazelle. Further, the tail of the pair correlation function is thinner for the Plains zebra than for the Thomson gazelle, at least for distances larger than 10 km. This indicates that randomness occurs for distances beyond 10 km for the Plains zebra, whereas that for the Thomson gazelle only appear for distances of 25 km and more. Herds of the Plains zebra do not show any regularity beyond distances of 10 km, whereas the Thomson gazelle shows regularity for distances up to 25 km.

A similar picture emerges when considering Strauss models. We first fitted those processes to the unmarked patterns (fig. 4) and applied distance parameters of 25 km for the Thomson gazelle and 15 km for the Plains zebra. These choices of parameters well represent the key factors for the distribution: a single mode distribution for the Thomson gazelle, with a peak somewhere in the South Western part of the Laikipia area, and a multi-modal distribution for the Plains zebra, representing their large abundance throughout the area.

We finally repeated fitting of the Strauss process to the marked process as well. Marked Strauss processes rarely occur in the literature [10]. We redistributed the marks according to herd size. Marks = 1 were assigned to herds of size up to 5, marks = 2 to herds of sizes 6–25, and marks = 3 to herds of sizes > 25 animals. As such, small, medium and large herds are distinguished

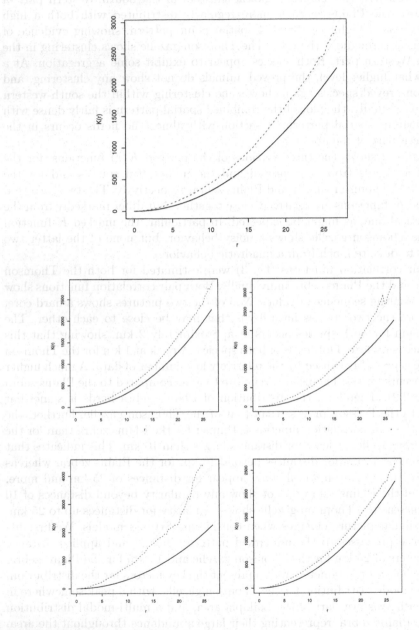

Fig. 2. Estimated marked $K(r)$ functions for herds of the nine herbivores (*top*), the preyed and non-preyed species (*center*) and the Thomson gazelle and plains zebra (*bottom*). The solid line is the theoretical $K(r)$, the dashed one is the $K(r)$ corrected for boundary of the area, the dotted one is the translation corrected $K(r)$

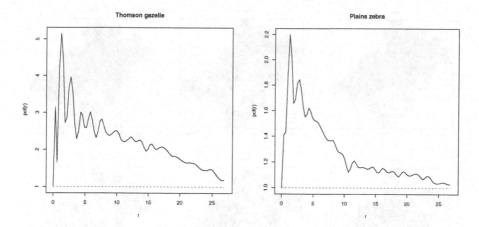

Fig. 3. Estimated pair correlation functions for the Thomson gazelle (*left*) and plains zebra (*right*)

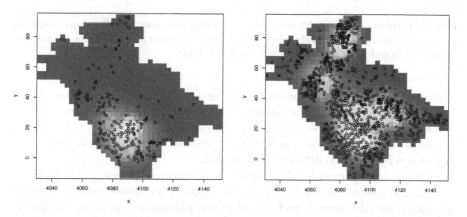

Fig. 4. Estimated unmarked Strauss process for the Thomson gazelle (*left*), using an $r = 25$ km parameter and the plains zebra (*right*), using an $r = 15$ km parameter.

for the two species (fig. 5. The results thus obtained do not violate earlier results.

4 Discussion

In this study we have used standard statistical software to analyse the data [2]. As a consequence, methods for stationary point patterns have been used, whereas some species, e.g. the Thomson gazelle, showed some clear non-stationarity. Currently, however, non-stationary analysis tools are not readily

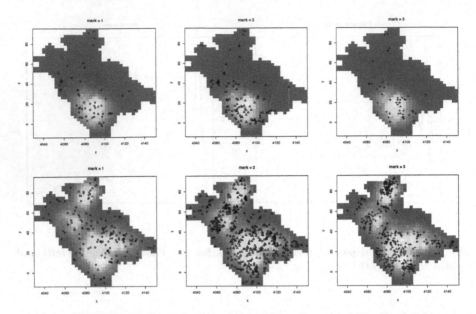

Fig. 5. Estimated marked Strauss process for the Thomson gazelle (*top*), using an r = 25 km parameter and the plains zebra (*bottom*), using an r = 15 km parameter. Marks are equal to 1 (*left*), 2 (*middle*) and 3 (*right*)

available. Non-standard software may serve as an extension in a future study. Also, a different method of estimating the pair correlation function, such as differentiating the $K(r)-$ function, might give different results. Our method is one possible method at hand, and it remains to be seen whether other methods might lead to different conclusions.

Spatial aggregation is a frequently encountered dispersion pattern in ecosystems, due to prevalence of potent aggregating forces such as habitat specificity, social structure and organization, philopatry, predator avoidance, and limited dispersal. Herbivore species in this study are subject to all these forces. Also the sizes of the herds in the study area are influenced not only by natural forces affecting herd dispersion, but also by 'unnatural' factors, such as displacement of wildlife by humans, cultivation, and livestock. In ecosystems such as this, where wildlife are displaced from some areas by humans and livestock, all species are likely to violate the assumption of random dispersion patterns typically made when sample counting. Factors that have a potentially organizing influence on herds within species, such as territoriality (Thomson's gazelle), or intra-specific competition, may have been operating. By contrast, plains zebra harems are known to associate and disassociate on a daily and seasonal basis (Rubenstein. *pers. comm.*), but this evidently does not result in significant aggregation at the landscape level. At least in this woodland-dominated habitat, 'exogenous' forces such as patchiness of preferred habitat are more likely account for aggregations of Thomson's gazelle herds, which prefer open, short grassland.

Factors causing herds of individual species to be aggregated or randomly dispersed in relation to their sizes either 1) complement each other when combined across space, or 2) are organised by factors that exert their influence on the entire community, or both. As an example of the former, which we refer to as 'habitat preference complementarity', we expect habitat generalists not only to be randomly dispersed, but also to be more abundant than habitat specialists. Also we expect specialists to be clustered in larger herds, then generalists. Plains zebra showed random distributions, but without overall association between rank of relative abundance and dispersion pattern ($P >$ 0.05). By contrast, habitat specialists are expected to be aggregated within preferred patches, to display lower herd densities in transitional habitats, and to be absent from unsuitable habitats. When all species are pooled, the net effect is for herds to become regularly spaced across the landscape.

As an example of the latter, which we refer to as 'prey preference complementarity', predators are hypothesised to have a disaggregating effect on dispersion of preferred prey herds, which, when prey species are pooled, is manifested as an organizing effect by predators on the dispersion of preferred prey. Herds of preferred prey species, which could be aggregated in the absence of predation, react to predator functional responses by moving apart, becoming less aggregated, and alleviating pressure exerted by predator functional responses. Since predator functional responses are cued to multiple prey species, the net effect on combined prey herds is to cause a more regular pattern of dispersion.

We observe Thomson's gazelle as a small-sized species requiring open habitats with low biomass, and plains zebra as mid-sized species distributed across a variety of savanna habitats featuring grasslands associated with a range of tree densities. Because extreme habitat types, featuring either high or low vegetation biomass, are likely to be rarer and more patchy than intermediate habitat types, herbivore species preferring extreme habitat types are likely to display more aggregated distributions than are species preferring intermediate habitat types.

5 Conclusions

Combination of GPS technology with spatially explicit statistical techniques, in particular the marked K-function and the point correlation function, yield novel ways of characterizing dispersion patterns of wild herbivore herds and corresponding herd sizes. In particular, we found an interesting difference between the pair correlation function for the plains zebra with a correlation length of approximately 10 km and that of the Thomson gazelle with a correlation length of approximately 25 km.

Also, the herd patterns of Thomson gazelle was modelled with a Strauss marked point process, showing a single mode, whereas the herd pattern of the plains zebra showed multiple modes. This can be well explained by the ecosystem behavior (habitat specialist versus habitat generalist) of the two species.

References

[1] R. Arsenault and N. Owen-Smith. Facilitation versus competition in grazing herbivore assemblages. *Oikos*, 97:313–318, 2002.

[2] A.J. Baddeley and R. Turner. *Introduction to SPATSTAT, spatstat version 1.5-4*. http://www.maths.uwa.edu.au/~adrian/spatstat/current/Intro.pdf.

[3] P.J. Diggle. *Statistical Analysis of Spatial Point Patterns, 2nd ed.* Oxford University Press, 2003.

[4] P.J. Diggle and A.G. Chetwynd. Second-order analysis of spatial clustering for inhomogeneous populations. *Biometrics*, 47:1155–1163, 1991.

[5] N. Georgiadis. Numbers and distribution of large herbivores in laikipia district: sample counts for february and september 1997. Laikipia Wildlife Forum, Nanyuki, Kenya, 1997.

[6] N. Georgiadis, M. Hack and K. Turpin. The influence of rainfall on zebra population dynamics: implications for management. *Journal of Applied Ecology*, 40:125–136, 2003.

[7] F.P. Kelly and B.D. Ripley. On strauss's model for clustering. *Biometrika*, 63:357–360, 1976.

[8] H. Lamprey. Ecological separation of the large mammal species in the tarangire game reserve, tanganyika. *East African Wildlife Journal*, 2:1–46, 1963.

[9] J. Ohser. On estimators for the reduced second moment measure of point processes. *Mathematische Operationsforschung und Statistik series Statistics*, 14(1):63–71, 1983.

[10] N.V. Petersen. A spatio-temporal model for fmri data. In *Proceedings of the Eleventh European Young Statisticians Meeting, Marly-le-Roi*, pages 196–201, 1999.

[11] H.H.T. Prins and H. Olff. Species richness of african grazers assemblages: Towards a functional explanation. In D.M. Newbery, H.H.T. Prins and H. Brown, editors, *Dynamics of Tropical Communities*, pages 449–490. Blackwell Science, 1998.

[12] B.D Ripley. *Statistical Inference for Spatial Processes*. Cambridge University Press, Cambridge, 1988.

[13] B.D. Ripley. *Spatial Statistics – 2nd edition*. Wiley, 2004.

[14] A. Stein, N. Georgiados and W. Khaemba. Spatial statistics to quantify patterns of herd dispersion in a savanna herbivore community. In A.J. Baddeley, P. Gregori, J. Mateu, R. Soica and D. Stoyan, editors, *Spatial*

point process modelling and its applications, pages 239–260. Publicacions de la Universitat Jaume I, 2004.

[15] A. Stein, N. Georgiados and W. Khaemba. Patterns of herd dispersion in savanna herbivore communities. In F. van Langeveld and H.H.T. Prins, editors, *Spatial and temporal dynamics of foraging*. Kluwer, (in press).

[16] D. Strauss. A model for clustering. *Biometrika*, 63:467–475, 1975.

[17] M.M. Voeten. *Living with wildlife: Coexistence of wildlife and livestock in an East African Savannah system*. PhD thesis, Wageningen University, 1999.

[18] P.A. Walker. Modelling wildlife distributions using a geographic information system: Kangaroos in relation to climate. *Journal of Biogeography*, 17:279–289, 1990.

[19] W. Wint. Drylands: Sustainable use of rangelands into the twenty-first century. In V.R. Squires and A.E. Sidahmed, editors, *Resource assessment and environmental monitoring using low level aerial surveys*, pages 277–301. IFAD publication, 1998.

Diagnostic Analysis of Space-Time Branching Processes for Earthquakes

Jiancang Zhuang[1], Yosihiko Ogata[1] and David Vere-Jones[2]

[1] Institute of Statistical Mathematics, Japan, zhuangjc@ism.ac.jp
[2] Victoria University of Wellington, New Zealand, dvj@mcs.vuw.ac.nz

Summary. It is natural to use a branching process to describe occurrence patterns of earthquakes, which are apparently clustered in both space and time. The clustering features of earthquakes are important for seismological studies.

Based on some empirical laws in seismicity studies, several point-process models have been proposed in literature, classifying seismicity into two components, background seismicity and clustering seismicity, where each earthquake event, no matter it is a background event or generated by another event, produces (triggers) its own offspring (aftershocks) according to some branching rules. There are further ideas on probability separation of background seismicity from the clustering seismicity assuming a constant background occurrence rate throughout the whole studied region and other authors proposed a stochastic declustering method and made the probability based declustering method practical.

In this paper, we show some useful graphical diagnostic methods for improving model formulation.

Key words: Branching processes, Patterns of earthquakes, Point process models, Stochastic declustering

1 Introduction

It is natural to use a branching process to describe occurrence patterns of earthquakes, which are apparently clustered in both space and time. The clustering features of earthquakes are important for seismological studies. For the purpose of long-term earthquake prediction, such as zoning and earthquake hazard potential estimation, people try to remove the temporary clustering to estimate background seismicity; on the other hand, for short-term or real-time prediction, we need a good understanding to earthquake clusters. Thus, separating background seismicity from earthquake clusters is believed to be of central importance.

Based on some empirical laws in seismicity studies, several point-process models have been proposed in [3, 4, 6, 9, 11, 12, 15]. In general, all of those models classify seismicity into two components, background seismicity and clustering seismicity, where each earthquake event, no matter it is a background event or generated by another event, produces (triggers) its own offspring (aftershocks) according to some branching rules.

The ideas on probability separation of background seismicity from the clustering seismicity first appears in Kagan and Knopoff [7]. They assume a constant background occurrence rate throughout the whole studied region, which should be location dependent as we can see in this article. Zhuang et al. [22, 23] propose a stochastic declustering method and made the probability-based declustering method practical. The core of the stochastic declustering method is an iterative approach to simultaneously estimate the background intensity, assumed to be a function of spatial locations but constant in time, and the parameters associated with clustering structures. Making use of these estimates and the thinning operation, one can obtain the probabilities for each event being a background event or a triggered event. These probabilities are the key to realising stochastic versions of the clustering family trees in the catalogue, and, of course, also to separating the background events from the earthquake clusters.

Because these probabilities are estimated through a particular model, the closeness between the model and the reality is the essentially important factor that influences the output. The closer the model to the real data, the more reliable the output. Of course, some model selection procedures can be used to choose the best model among many models fitted to the same set of data. But these procedures usually give us only a number indicating the overall goodness-of-fit for the model, and rarely tell whether there are some good points in a model even if its overall fit is not the best. Moreover, it is also difficult to find clues about how to improve the formulation of the clustering models through model selection procedures. In this paper, we are going to show some useful graphical diagnostic methods for improving model formulation.

2 The Space-time ETAS Model

Model formulation

In empirical studies on seismicity, the Omori law ([14, 18]; and see, [19], for a review) has been used to describe the decay of aftershock frequencies with time, i.e.,

$$n(t) = \frac{K}{(t+c)^p},\tag{1}$$

where $n(t)$ is the occurrence rate of events at the time t after the occurrence of the mainshock, and K, c and p are constants. Another commonly accepted empirical law is the Gutenberg-Ritcher law, which describes the relationship between the magnitudes and occurrence frequencies of earthquakes, taking the form

$$\log_{10} N(\geq M) = a - bM, \tag{2}$$

i.e., the number of earthquake events in a catalog decreases exponentially when we increase the magnitude threshold.

The above empirical laws have been considered as the foundation for statistical modelling in seismicity. Several point process models have been proposed by [3, 4, 6, 9, 11, 12, 15]. Those models classify seismicity into two components, background seismicity and clustering seismicity, where each earthquake event, no matter it is a background event or generated by another event, produces (triggers) its own offspring (aftershocks) according to some branching rules. All these models can be formulated in the form of the conditional intensity function (see, e.g., [5, Chap. 7]), i.e., at time t, location (x, y) and magnitude M, the conditional intensity function is defined by

$$\lambda(t, x, y, M)dt\, dx\, dy\, dM = \mathbf{E}[\mathbf{N}(dt\, dx\, dy\, dM)|\mathcal{H}_t], \tag{3}$$

where $\mathcal{H}_t$ is the observational history up to time t, but not including t. In this study, we base our analysis on the formulation of the space-time epidemic type aftershock sequence (ETAS) model (see [12]),

$$\lambda(t, x, y, M) = \lambda(t, x, y)J(M) \tag{4}$$

$$\lambda(t, x, y) = \mu(x, y) + \sum_{i:\, t_i < t} \kappa(M_i)g(t - t_i)f(x - x_i, y - y_i | M_i), \tag{5}$$

where

1. $\mu(x, y)$ is the background intensity, a function of spatial locations but constant in time;
2. $\kappa(M)$ is the expected number of events triggered from an event of magnitude M, given by

$$\kappa(M) = A\exp[\alpha(M - M_C)], \quad M \geq M_c; \tag{6}$$

where A and α are constant, and M_C is the magnitude threshold (see [20]);

3. $g(t)$ is the p.d.f of the occurrence times of the triggered events, taking the form

$$g(t) = \frac{p - 1}{c}\left(1 + \frac{t}{c}\right)^{-p}, \quad t > 0; \tag{7}$$

i.e., the p.d.f form of (1);

4. $f(x, y|M)$ is the p.d.f of the locations of the triggered events, which is formulated as

$$f(x, y; M) = \frac{1}{2\pi De^{\alpha(M - M_C)}}\exp\left[-\frac{x^2 + y^2}{2De^{\alpha(M - M_C)}}\right] \tag{8}$$

for the short-range Gaussian decay (light tail), or

$$f(x, y; M) = \frac{q-1}{\pi D e^{\alpha(M-M_C)}} \left(1 + \frac{x^2 + y^2}{D e^{\alpha(M-M_C)}} \right)^{-q} ; \qquad (9)$$

for the long-range inverse power decay (heavy tail); and

5. $J(M)$ is the probability density of magnitudes for all the events, taking the form of the Gentenburg-Richter law, i.e.,

$$J(M) = \beta e^{-\beta(M-M_c)}, \text{ for } M \geq M_c \qquad (10)$$

where β is linked with the Gutenberg-Richter's b-value in (2) by $\beta = b \log 10$ and M_c is the magnitude threshold considered.

In the text below, we call an ETAS model Model I or Model II, if it is equipped with (8) or (9), respectively. The expected number of offspring that an event can trigger, $\kappa(M)$ in (6), is also called its *triggering ability*.

In (6), (8) and (9), the spatial scaling factor for the direct aftershock region is proportional to the triggering ability of the ancestor. This judgment is from [20]. In the Sect. 6, we will use the stochastic reconstruction method to verify whether it is a good choice.

Maximum likelihood estimates

Given a set of observed earthquake data, say $\{(t_i, x_i, y_i, M_i) : i = 1, 2, \ldots, N\}$, if the background rate $\mu(x, y) = \nu u(x, y)$ where $u(x, y)$ is known, the parameters in (5) can be estimated by maximizing the log-likelihood (cf. [5], Chap. 7)

$$\log L(\boldsymbol{\theta}) = \sum_{j:(t_j, x_j, y_j) \in S \times [T_1, T_2]} \log \lambda(t_j, x_j, y_j) - \iint_S \int_{T_1}^{T_2} \lambda(t, x, y) \mathrm{d}t \, \mathrm{d}x \, \mathrm{d}y,$$

$$(11)$$

where the parameter vector is, respectively, $\boldsymbol{\theta} = (\nu, A, \alpha, c, p, D)$ for Model I, and $\boldsymbol{\theta} = (\nu, A, \alpha, c, p, D, q)$ for Model II, and j runs over all the events in the study region S and time period $[T_1, T_2]$. Because the events occurring outside of the study region, or before the study time period, may also trigger seismicity inside the study region and time period, particularly the large ones, we include these events in the observation history $\mathcal{H}_t$ and call them *complemental events*. Events inside the study space-time zone conversely, are called *target events*.

The thinning method and stochastic declustering

The technical key point of the stochastic declustering method is the thinning operation to a point process (i.e., random deletion of points, c.f. [8, 10]). Observe (5), the relative contribution of a previous ith event to the total seismicity rate at the occurrence time and location of the jth event, (t_j, x_j, y_j), is

$$\rho_{ij} = \begin{cases} \zeta_i(t_j, x_j, y_j)/\lambda(t_j, x_j, y_j), & \text{when } j > i, \\ 0, & \text{otherwise,} \end{cases} \qquad (12)$$

where

$$\zeta_i(t, x, y) = \kappa(M_i)g(t - t_i)f(x - x_i, y - y_i; M_i), \qquad (13)$$

represents the rate triggered by the ith event. That is to say, for each $j = 1, 2, \ldots, N$, if we select the jth event with probability ρ_{ij}, we can realise a subprocess that consists of the direct offspring of the ith event. In this way, ρ_{ij} can be naturally regarded as the probability that the jth event is a direct offspring of the ith event. Similarly, the probability that the event j is a background event is

$$\varphi_j = \frac{\mu(x_j, y_j)}{\lambda(t_j, x_j, y_j)}. \qquad (14)$$

and the probability that the jth event is triggered is given by

$$\rho_j = 1 - \varphi_j = \sum_i \rho_{ij}, \qquad (15)$$

In other words, if we select each event j with probabilities φ_j, we can then form a new processes, the background subprocess with a rate function $\mu(x, y)$, and its complement, the clustering subprocess.

Variable kernel estimates of seismicity rates

The total spatial seismicity rate can be estimated by using variable kernel estimates

$$\hat{m}(x, y) = \frac{1}{T} \sum_j Z_{h_j}(x - x_j, y - y_j), \qquad (16)$$

where T is the length of the time period of the process, subscript j runs over all the event in the process and Z is the Gaussian density function. The variable bandwidth h_j (the standard deviation of the Gaussian density) is determined by

$$h_j = \max\{\epsilon, \inf(r : N[B(x_i, y_i; r)] > n_p)\}, \qquad (17)$$

where ϵ is the allowed minimum bandwidth, $B(x, y; r)$ is the disk of radius r centred at (x, y), and n_p is a positive integer, i.e., h_j is the distance to its n_pth closest neighbour.

Once the thinning probabilities $\{\varphi_j\}$ are obtained, we can estimate the spatial background seismicity rate by using weighted variable kernel estimates [22],

$$\hat{\mu}(x, y) = \frac{1}{T} \sum_j \varphi_j Z_{h_j}(x - x_j, y - y_j), \qquad (18)$$

where T, Z and h_j are defined as in (16).

By simply taking the difference between the total seismicity rate and the background seismicity rate, we can get the estimate of the clustering rate function, i.e., the estimate of the clustering rate is

$$\hat{C}(x,y) = \hat{m}(x,y) - \hat{\mu}(x,y) = \frac{1}{T}\sum_j (1 - \varphi_j)Z_{h_j}(x - x_j, y - y_j) \qquad (19)$$

Estimation algorithm

Background seismicity and the parameters can be determined in the clustering structures simultaneously using an iterative approach [22, 23]. Firstly, we assume some initial background seismicity rate, using the maximum likelihood procedure to obtain the parameters in the branching structure. We then calculate the background probabilities $\{\varphi_j : j = 1, 2, \cdots N\}$ for all of the events using (14). Substituting these φ_j into (18) we get a better estimate of the background seismicity rate, and use this newly estimated background seismicity rate to replace the initial background rate. We repeat these steps many times until the results converge.

3 Data and Preliminary Results

Two of sets of data are considered in this study. The first data set is the JMA catalogue in a range of longitude $121° - 155°E$, latitude $21° \sim 48°N$, depth $0\sim100$ km, time 1926/Janary/1$\sim$1999/December/31 and magnitude $\geq M_J 4.2$. The second data set is the Taiwan CWB (Center Weather Bureau) catalogue in a range of longitude $120° \sim 123°E$, latitude $21° \sim 25.2°N$, depth $0\sim55$ km, time 1900/Janary/1$\sim$2001/December/31, and magnitude ≥ 5.3. There are 19,139 and 892 events in the JMA data set and the Taiwan data set, respectively.

For an earthquake catalogues covering records of a long history, completeness and homogeneity are always problems causing troubles for statistical analysis. To tackle these problems, we choose a target space-time range, in which the seismicity seems to be relatively complete and homogeneous. The incompleteness of the early period and inhomogeneity in the JMA data set can be easily seen from Figs. 1. We choose the target range of longitude $130° - 146°E$, latitude $33° - 42.5°N$, a time period of $10,000 - 26,814$ days after 1926/Jan/1, and same depth and magnitude ranges as the whole data set. With similar reason, we chose the target range as time 1941/Janary/1$\sim$2001/December/31, longitude $120° \sim 122°E$, latitude $22° \sim 25°$ for the Taiwan CWB data (Fig. 2). There are 8283 and 491 events occurring in the target space-time range of the JMA data set and the Taiwan data set, respectively.

We fit both models I and II to both the JMA and the Taiwan data. The results are outlined in Table 1. Model II fits better than Model I for both data sets, indicating that the locations of triggered events decay in a long range

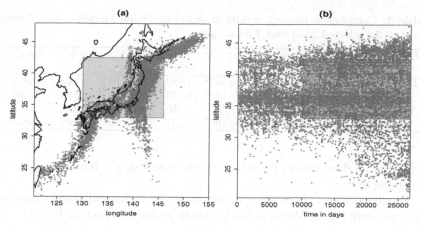

Fig. 1. Seismicity in the Japan region and nearby during 1926–1999 ($M_J \geq 4.2$). (**a**) Epicenter locations; (**b**) Latitudes of epicenter locations against occurrence times. The shaded region represents the study space-time range

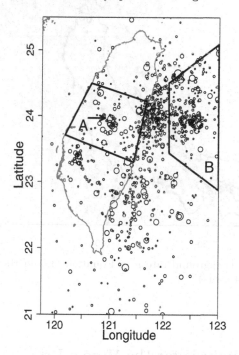

Fig. 2. Epicenter locations of earthquakes in the Taiwan region during 1900–2000 ($M \geq 5.3$) and subdivision. Sizes of the circle indicate magnitudes from 5.3 to 8.2. The arrow indicates the Chi-Chi earthquake (1999/9/21, $M_S 7.3$)

Table 1. Comparison between results obtained by fitting Models I and II respectively to the selected JMA and Taiwan data

Data	Model	A mag^{-1}	α day	c	p deg^2	D	q	$\log L$
JMA	I	0.191	1.365	1.726×10^{-2}	1.089	1.414×10^{-3}	na	-45,658
	II	0.198	1.334	1.903×10^{-2}	1.103	8.663×10^{-4}	1.691	-45,068
TW	I	0.182	1.694	6.692×10^{-3}	1.150	2.868×10^{-3}	na	-1,411.1
	II	0.234	1.504	5.089×10^{-3}	1.141	3.194×10^{-3}	1.839	-1,397.6

rather than a short range. We will show this conclusion in a more explicit way in coming Sect. 6.

For illustration, the spatial variation of the estimated background rate obtained from fitting Model II to the Japan region is shown in Fig. 3.

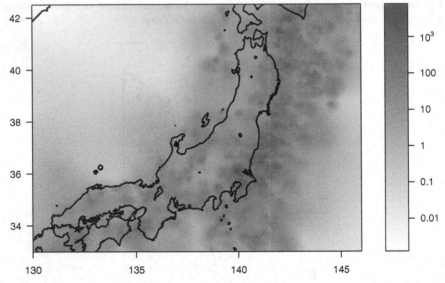

Fig. 3. Estimated background rate $\hat{\mu}$ in Equation (18) in the study region (unit: events/(degree$^2 \times$74 years))

4 Residual Analysis for the Whole Process

The goodness-of-fit of the space-time ETAS model to the earthquake data can be found by using residual analysis. For example, Ogata et al. [13] re-fitted the data with a time-variant version of a space-time ETAS model and then took the ratio between its conditional intensity and the conditional intensity of the stationary ETAS model as the residual process. Schoenberg [16] proposed the thinning residuals, i.e., to keep each event i with probability

$$\frac{\min\{\lambda(t_k, x_k, y_k) : k = 1, \cdots, N\}}{\lambda(t_i, x_i, y_i)}$$

to obtain a homogeneous Poisson process, which is essentially the analogue of the Stoyan and Grabarnik weights in spatial point process [17], as discussed in this section.

Because the process

$$I_{B,t} \equiv \int_0^t \iint_B N(\mathrm{d}x\,\mathrm{d}y\,\mathrm{d}u) - \lambda(u, x, y)\mathrm{d}x\,\mathrm{d}y\,\mathrm{d}u \tag{20}$$

is a zero-mean martingale as a function of t (see, e.g., [2], for justification), where B is an arbitrary regular spatial region,

$$\mathbf{E}\left[\int_0^t \int_B h(u, x, y)\{N(\mathrm{d}x\,\mathrm{d}y\,\mathrm{d}u) - \lambda(u, x, y)\mathrm{d}x\,\mathrm{d}y\,\mathrm{d}u\}\right] = 0, \tag{21}$$

for any predictable function $h(t, x, y)$. If we take $h(t, x, y) = 1/\lambda(t, x, y)$, then

$$\mathbf{E}\left[\sum_{i:(t_i, x_i, y_i) \in B} w_i\right] = |B| \tag{22}$$

where $|\cdot|$ represents the volume or the Lebesgue measure, and we also call $w_i = h(t_i, x_i, y_i)$ the Stoyan-Grabarnik weight [1, 17]. Baddeley et al. [1] called $\sum_{i:(t_i, x_i, y_i) \in B} w_i - |B|$ as the inverse-lambda residual.

To apply the above Stoyan-Grabarnik weights to the Taiwan data, we consider the following functions,

$$s_1(t) = \sum_{i:\, t_i < t} w_i, \tag{23}$$

$$s_2(x) = \sum_{i:\, x_i < x} w_i, \tag{24}$$

$$s_3(y) = \sum_{i:\, y_i < y} w_i. \tag{25}$$

If the target space-time range is a direct product of intervals, s_1, s_2 and s_3 should increase with a constant rate approximately. The results for the Taiwan data are plotted in Fig. 4. We can see that there are some departures of s_1, s_2 and s_3 from their expectation. In Fig. 4(a), the slope of the cumulative wights against the time axis changes at the years around 1940 and 1975, or even around 1994. These changes are mainly caused by the changes of the monitoring systems. In Fig. 4(b), along the latitude axis, the seismicity

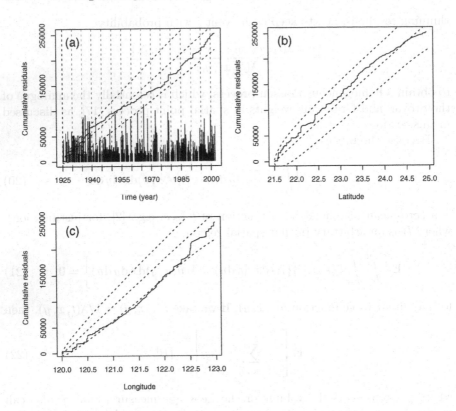

Fig. 4. Results from residual analysis for the Taiwan catalogue. The cumulative weights against times, latitudes and longitudes are plotted in (**a**), (**b**), and (**c**), respectively

pattern changes around $22°N$ and $24°N$. In Fig. 4(c), along the longitude axis, the accumulation rate of the weights changes around $120.5°E$, $121.5°E$ and $122.5°E$, showing differences in seismicity from the west to the east. More detailed explanations on these results and the local tectonic structures in the Taiwan region can be found in [21].

If we set $h(t, x, y) \equiv 1$ or $h(t, x, y) = 1/\sqrt{\lambda(t, x, y)}$, then we form up the raw residual and the Pearson residual, respectively [1]. Both of them can be used for the residual analysis as well as the inverse-lambda residual.

5 Verifying Stationarity of the Background

In the ETAS model, we assume that the background process is stationary. To test this assumption, we set

$$h(t,x,y) = \frac{u(x,y)}{\lambda(t,x,y)}, \tag{26}$$

in (21) and then, according to (14),

$$S(t) \equiv \mathbf{E}\left[\sum_{i:\,(t_i,x_i,y_i)\in[0,t)\times B} \varphi_i\right] = t\iint_B u(x,y)\mathrm{d}x\,\mathrm{d}y \tag{27}$$

where B is a spatial region. If the model fits the seismicity well or the background occurrence rate is constant in time, the function $S(t)$ defined by (27) increases approximately in a constant rate with time t. If the slope of $S(t)$ decreases, we call it quiescence in background seismicity, or simply background quiescence; otherwise, if the slope of $S(t)$ increases, we call it an activation in the background seismicity, or simply background activation.

As given in Fig. 5(a), the background seismicity $S(t)$ in Region A shows a quiet period from 1960-1990, followed by recovery of activity culminating with the Chi-Chi rupture. In Region B, the background seismicity given in Fig. 5(b), shows steady activity after the 1940's. This indicates a conspic-

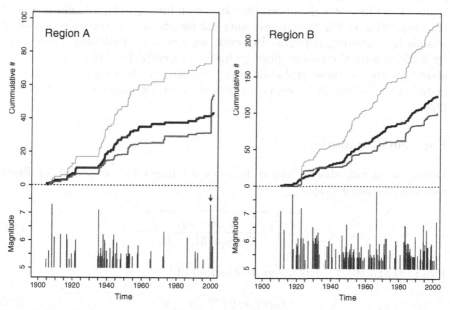

Fig. 5. Cumulative background seismicity $S(t)$ in (27) (*black step lines*), cumulative clustering seismicity $\#(t)-S(t)$ (*dark gray step lines*) and cumulative total seismicity $\#(t)$ (*light gray step lines*) for Regions A and B (see Fig. 2). The magnitudes against the occurrence times of the events are plotted in the lower part of each panel

uously quiet period lasting up to several decades, prior to recovery of the activity, culminating in the1999 Chi-Chi earthquake ($M_S 7.3$), while we find that other major seismic regions remain active stationarily as Region B. [21] interpreted this phenomenon as an effect of the aseismic slip in the Chi-Chi rupture fault, whereby the inland region around the Chi-Chi source becomes a stress-shadow.

6 Verifying Formulation of Branching Structures by Stochastic Reconstruction

As mentioned in Sect. 2, if we select each event j with probabilities ρ_{ij}, ρ_j or φ_j, we can get a new process being the triggered process by the ith event, the clustering process or the background processes, respectively. That is to say, we can separate the whole catalogue into different family trees. This method tackles the difficulties in testing hypotheses associated with earthquake clustering features, which are caused by the complicated overlapping of the background seismicity and different earthquake clusters in both space and time. We can repeat such thinning operations for many times to get different stochastic versions of separations of the earthquake clusters. The non-uniqueness of such realisations illustrates the uncertainty in determining earthquake clusters, and thus repetition can help us to evaluate the significance of some properties of seismicity clustering patterns. However, we can also implement these tests by working with the probabilities φ_j and ρ_{ij} directly. In this section, we will show how to use these probabilities to reconstruct the characteristics associated with earthquake clustering features. More examples can be found in [23].

Location distributions

Define the standardised distance between a triggered event j and its direct ancestor, assumed i, by

$$r_{ij} = \sqrt{\frac{(x_j - x_i)^2 + (y_j - y_i)^2}{D \exp[\alpha(M_i - M_c)]}}. \tag{28}$$

From (8) and (9), r_{ij} has a density function of

$$f_R(r) = 2re^{-r^2}, \ r \geq 0; \tag{29}$$

for Model I and

$$f_R(r) = \frac{2r(q-1)}{(1+r^2)^q}, \quad r \geq 0; \tag{30}$$

for Model II, respectively. The distribution with a density of (29) is called a Rayleigh distribution. On the other hand, $f_R(r)$ can be reconstructed through

$$\hat{f}_R(r) = \frac{\sum_{i,j} \rho_{ij} I(|r_{ij} - r| < \Delta r/2)}{\Delta r \sum_{i,j} \rho_{ij}}, \tag{31}$$

where Δr is a small positive number, and I is the index function such that

$$I(x) = \begin{cases} 1, \text{ if the logical statement } x \text{ holds}, \\ 0, \text{ else.} \end{cases} \tag{32}$$

The comparison between $\hat{f}_R$ and f_R for the two models are shown in Fig. 6. It

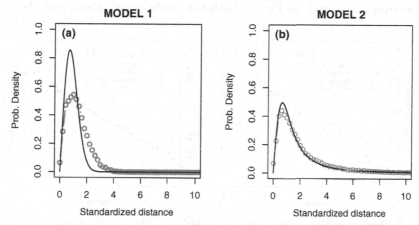

Fig. 6. Reconstruction results for the distribution of the standardised triggering distances $\hat{f}_R(r)$ in (31) (*gray circles*) by using Model I (**a**) and Model II (**b**) for the JMA catalogue. The theoretical curves of $f_R(r)$ in (29) and (30) are plotted in solid lines in (a) and (b), respectively

can be seen that, if Model I is used, the reconstructed probability density of the standardised distances between the ancestors and the direct offspring is quite different from the theoretical one. When Model II is used, the reconstructed probability density is very close to the theoretical one. These results confirm that the aftershocks decay in a long range in space rather than a short range [4, 3, 12]. These results also imply the robustness of the reconstruction method, for we can get a reconstructed probability density function very close to the corresponding function in Model II, even if an improper model like Model I is employed.

Since Model II fits the seismicity much better than Model I, we only consider Model II for reconstruction in the following sections.

Difference in triggering ability between the background events and the triggered events

The triggering abilities of an event sized M from all the events, the background events and the triggered events in a catalogue can be reconstructed by using

$$\hat{\kappa}(M) = \frac{\sum_i \sum_j \rho_{ij} I(|M_i - M| < \Delta M/2)}{\sum_i I(|M_i - M| < \Delta M/2)} \qquad (33)$$

$$\hat{\kappa}_b(M) = \frac{\sum_i \sum_j \varphi_i \rho_{ij} I(|M_i - M| < \Delta M/2)}{\sum_i \varphi_i I(|M_i - M| < \Delta M/2)} \qquad (34)$$

and

$$\hat{\kappa}_t(M) = \frac{\sum_i \sum_j (1 - \varphi_i) \rho_{ij} I(|M_i - M| < \Delta M/2)}{\sum_i (1 - \varphi_i) I(|M_i - M| < \Delta M/2)}, \qquad (35)$$

respectively, as shown in Fig. 7. Both the background events and the trig-

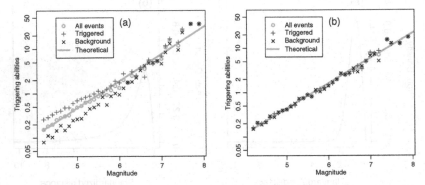

Fig. 7. Reconstruction of the triggering abilities, $\hat{\kappa}_b(M)$ in (34) of the background events and $\hat{\kappa}_t(M)$ in (35) of the triggered events for (**a**) the JMA catalogue and (**b**) the simulated catalogue. For comparison, the empirical functions of the triggering abilities, $\hat{\kappa}(M)$ in (33) for all the events are plotted in gray circles, and the corresponding theoretical functions, $\kappa(M) = A e^{\alpha(M - M_C)}$, are represented by the straight lines

gered events generate offspring approximately according to different exponential laws. For the same ancestor magnitude, a triggered event generates more offspring than a background event. The higher is the magnitude, the smaller is the difference. Applying the same procedures to a synthetic catalogue simulated by using the ETAS model with the same background rate and the same parameters estimated from the JMA catalogue, the results show that there is no differences in triggering abilities between these two types of events, indicating that these differences are not caused by numerical procedures.

In the ETAS model, the background events and triggered events generate offspring in the same way. The reconstruction results show that it is a possible direction to improving the current ETAS models to have different exponential laws for the triggering abilities of these two types of event. One reason for the higher triggering abilities of the triggered events may be because they occur in an environment where the stress field is adjusting to the stress changes caused by their ancestors.

Distributions of offspring locations from different magnitude classes

Because of the historical reasons mentioned in Sect. 2, we model the distribution of locations of the direct offspring from an earthquake as an inverse power one with a scaling factor associated with the ancestor's magnitude, $De^{\alpha(M-M_C)}$. Immediate questions for such a choice are:

(a) Is this scaling factor necessary? Or, can we use a constant D_0 to replace the scaling factor $De^{\alpha(M-M_C)}$ in the model?

(b) Is it necessary to link the scaling factor to the triggering ability, or should we introduce a new parameter γ instead of α for the scaling factor?

To answer the above questions, for a small interval $\mathcal{M}$ of magnitudes, we select the pairs $\{(i,j)\}$ such that $M_i \in \mathcal{M}$ and then estimate the scaling factor $D_{\mathcal{M}}$ for $\mathcal{M}$ in the following way. Given ρ_{ij}, consider the following pseudo log-likelihood function,

$$
\log L = \sum_{M_i \in \mathcal{M}} \sum_j \rho_{ij} \log \left[\frac{(q-1)D^{(q-1)}R_{ij}}{(R_{ij}^2 + D)^q} \right], \tag{36}
$$

where R_{ij} is the distance between the events i and j. To maximize it, let

$$
\frac{\partial \log L}{\partial D} \bigg|_{D=D_{\mathcal{M}}} = 0,
$$

i.e.,

$$
\frac{q-1}{D_{\mathcal{M}}} \sum_{M_i \in \mathcal{M}} \sum_j \rho_{ij} - q \sum_{M_i \in \mathcal{M}} \sum_j \frac{\rho_{ij}}{R_{ij}^2 + D_{\mathcal{M}}} = 0. \tag{37}
$$

Thus, we can construct the following iteration to solve the above equation

$$
D_{\mathcal{M}}^{(n+1)} = \frac{(q-1)\sum_{M_i \in \mathcal{M}} \sum_j \rho_{ij}}{q \sum_{M_i \in \mathcal{M}} \sum_j \dfrac{\rho_{ij}}{R_{ij}^2 + D_{\mathcal{M}}^{(n)}}}. \tag{38}
$$

Figure 8 shows the values of $D_{\mathcal{M}}$ against the magnitude classes. We can see that values of $D_{\mathcal{M}}$ have a slope different from $\kappa(M)$. Thus, it is not suitable use $\kappa(M)$ as the scaling factor, a better choice is to introduce another parameter γ as the coefficient in the exponential part.

7 Conclusion

In this paper, we have outlined the general routines of using the ETAS model to fit the clustering features of earthquake processes and some methods on

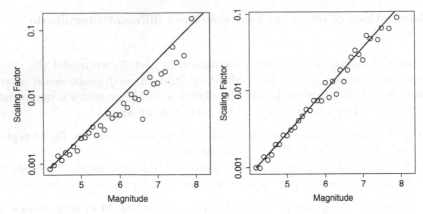

Fig. 8. Re-estimated $D_{\mathcal{M}}$ for the JMA catalogue (*left*) and the simulated catalogue (*right*). Theoretical fitting curves, $De^{\alpha(M-M_C)}$, are represented by the straight lines

how to evaluate the goodness-of-fit. The results show that the ETAS model is a good starting point for modelling the earthquake clusters.

We use the inverse-lambda residuals to test the overall goodness-of-fit, and departures from their expectation may indicate some essential features of the data, such as data inhomogeneity caused by the changes of monitoring abilities or different local tectonic environment. The stationarity of the background process is tested by using the function $S(t)$ defined in (27), which helps us to detect the existence of quiescence or activation in background seismicity prior to some large earthquakes. For the components associated with the branching structure, we make use of the stochastic reconstruction method. We first reconstruct the distribution of the locations of the triggered events relatively to their parents, which is shown to have a heavy tail than a light tail. We have also shown that a background event triggers less children than an triggered events of the same magnitude, and that the two exponential laws, one for the triggering ability and the other for the spatial density of the offspring, should have different exponents. All of these indicate that the methods discussed in this article are powerful in testing the hypotheses associated with earthquake clusters, and in finding clues for improving the model formulation.

References

[1] A.J. Baddeley, R. Turner, J. Möller and M. Hazelton. Residual analysis for spatial point processes. Technical report, School of Mathematics & Statistics, University of Western Australia, 2004.

[2] P. Brémaud. *Point Processess and Queues*. Springer-Verlag, 1980.

[3] R. Console and M. Murru. A simple and testable model for earthquake clustering. *Journal of Geophysical Research*, 106:8699–8711, 2001.

[4] R. Console, M. Murru and A.M. Lombardi. Refining earthquake clustering models. *Journal of Geophysical Research*, 108(B10):doi:10.1029/2002JB002130, 2003.

[5] D.J. Daley and D. Vere-Jones. *An Introduction to the Theory of Point Processes, 2nd edition.* Springer, New York, 2003.

[6] Y.Y. Kagan. Likelihood analysis of earthquake catalogues. *Journal of Geophysical Research*, 106(B7):135–148, 1991.

[7] Y.Y. Kagan and L. Knopoff. Statistical study of the occurrence of shallow earthquakes. *Geophysical Journal of the Royal Astronomical Society*, 1978.

[8] P.A.W. Lewis and E. Shedler. Simulation of non-homogeneous poisson processes by thinning. *Naval Research Logistics Quarterly*, 26:403–413, 1979.

[9] F. Musmeci and D. Vere-Jones. A space-time clustering model for historical earthquakes. *Annals of the Institute of Statistical Mathematics*, 44:1–11, 1992.

[10] Y. Ogata. On lewis' simulation method for point processes. *IEEE translations on information theory*, IT-27:23–31, 1981.

[11] Y. Ogata. Statistical models for earthquake occurrences and residual analysis for point processes. *Journal of American Statistical Association*, 83:9–27, 1988.

[12] Y. Ogata. Space-time point-process models for earthquake occurrences. *Annals of the Institute of Statistical Mathematics*, 50:379–402, 1998.

[13] Y. Ogata, K. Katsura and M. Tanemura. Modelling heterogeneous space-time occurrences of earthquake and its residual analysis. *Journal of the Royal Statistical Society: Applied Statistics*, 52(4):499–509, 2003.

[14] F. Omori. On after-shocks of earthquakes. *Journal of the College of Science, Imperial University of Tokyo*, 7:111–200, 1898.

[15] S.L. Rathbun. Modeling marked spatio-temporal point patterns. *Bulletin of the International Statistical Institute*, 55(2):379–396, 1993.

[16] F.P. Schoenberg. Multi-dimensional residual analysis of point process models for earthquake occurrences. *Journal of the American Statistical Association*, 98:789–795, 2004.

[17] D. Stoyan and P. Grabarnik. Second-order characteristics for stochastic structures connected with gibbs point processes. *Mathematische Nachritchten*, 151:95–100, 1991.

[18] T. Utsu. Aftershock and earthquake statistics (i): Some parameters which characterize an aftershock sequence and their interrelations. *Journal of the Faculty of Science, Hokkaido University*, 3, Ser. VII (Geophysics):129–195, 1969.

[19] T. Utsu, Y. Ogata and R.S. Matsu'ura. The centenary of the omori formula for a decay law of aftershock activity. *Journal of Physical Earth*, 1995:1–33, 1995.

[20] Y. Yamanaka and K. Shimazaki. Scaling relationship between the number of aftershocks and the size of the main shock. *Journal of Physical Earth*, 1990:305–324, 1990.

[21] J. Zhuang, C.-P. Chang, Y. Ogata and Y.-I. Chen. A study on the background and clustering seismicity in the taiwan region by using a point process model. *Journal of Geophysical Research*, 110(B05S18):doi:10.1029/2004JB003157, 2005.

[22] J. Zhuang, Y. Ogata and D. Vere-Jones. Stochastic declustering of space-time earthquake occurrences. *Journal of the American Statistical Association*, 97:369–380, 2002.

[23] J. Zhuang, Y. Ogata and D. Vere-Jones. Analyzing earthquake clustering features by using stochastic reconstruction. *Journal of Geophysical Research*, 109(No. B5):doi:10.1029/2003JB002879, 2004.

Assessing Spatial Point Process Models Using Weighted K-functions: Analysis of California Earthquakes

Alejandro Veen[1] and Frederic Paik Schoenberg[2]

[1] UCLA Department of Statistics, 8125 Math Sciences Building, Box 951554, Los Angeles, CA 90095-1554, USA, veen@stat.ucla.edu

[2] UCLA Department of Statistics, 8125 Math Sciences Building, Box 951554, Los Angeles, CA 90095-1554, USA, frederic@stat.ucla.edu

Summary. We investigate the properties of a weighted analogue of Ripley's K-function which was first introduced by Baddeley, Møller, and Waagepetersen. This statistic, called the *weighted* or *inhomogeneous* K-function, is useful for assessing the fit of point process models. The advantage of this measure of goodness-of-fit is that it can be used in situations where the null hypothesis is not a stationary Poisson model. We note a correspondence between the weighted K-function and thinned residuals, and derive the asymptotic distribution of the weighted K-function for a spatial inhomogeneous Poisson process. We then present an application of the use of the weighted K-function to assess the goodness-of-fit of a class of point process models for the spatial distribution of earthquakes in Southern California.

Key words: Goodness-of-fit of spatial point process models, Inhomogeneity, Spatial distribution of earthquakes, Weighted Ripley's K-function

1 Introduction

Ripley's K-function [21], $K(h)$, is a widely used statistic to detect clustering or inhibition in point process data. It is commonly used as a test, where the null hypothesis is that the point process under consideration is a homogeneous Poisson process and the alternative is that the point process exhibits clustering or inhibitory behavior. Previous authors have described the asymptotic distribution of the K-function for simple point process models including the homogeneous Poisson case (see [11],Rip88 and [27, pp. 28–48]).

The K-function has also been used in conjunction with point process residual analysis techniques in order to assess more general classes of point process models. For instance, a point process may be rescaled (see [16, 17, 24]) or thinned [25] to generate residuals which are approximately homogeneous Poisson, provided the model used to generate the residuals is correct. The

K-function can then be applied to the residual process in order to investigate the homogeneity of the residuals, and the result can be interpreted as a test of the goodness-of-fit of the point process model in question. Hence, residual analysis of a point process model involves two steps, the transformation of the data into residuals and a subsequent test for whether the residuals appear to be well approximated by a homogeneous Poisson process.

Of course, other methods for assessing the homogeneity of a point process exist, including tests for monotonicity [23], uniformity (see [9, 14, 15]), and tests on the second and higher-order properties of the process (see [4, 8, 12]). Likelihood statistics, such as Akaike's Information Criterion (AIC, [1]) and the Bayesian Information Criterion (BIC, [26]) are often used to assess more general classes of models; see e.g. [18] for an application to earthquake occurrence models.

We focus here on Ripley's K-function, in particular on a modified version of the statistic which we call the *weighted K-function*, K_W, and which was first introduced as the *inhomogeneous K-function* in [2]. It may be used to test a quite general class of null hypothesis models for the point process under consideration and it provides a direct test for goodness-of-fit, without having to assume homogeneity or to transform the points using residual analysis, the latter of which often introduces problems of highly irregular boundaries and large sampling variability when the conditional intensity in question is highly variable (see [25]).

This paper is outlined as follows. In Sect. 2, the definitions of the ordinary and weighted K-functions are reviewed, a connection between K_W and thinned residuals is noted, and the asymptotic distribution of K_W is derived under certain conditions. The weighted K-function is then used in Sect. 3 to assess the goodness-of-fit for competing models for the spatial background rate of California earthquakes. Some concluding remarks are given in Sect. 4.

2 The Weighted K-function

In this section, we derive its distributional properties of the weighted K-function, $K_W(h)$, under certain conditions. $K_W(h)$ is a weighted analogue of Ripley's K-function and it is similar to the mean of K-functions applied to a repeatedly thinned point pattern, denoted here as $K_M(h)$, an application of which can be found in [25]. We begin with a review of Ripley's K-function.

2.1 Ripley's K-function and Variants

Consider a Poisson process of intensity λ on a connected subset $\mathcal{A}$ of the plane $\mathbf{R}^2$ with finite area A, and let the N points of the process be labelled $\{p_1, p_2, \ldots, p_N\}$. Ripley's K-function $K(h)$ is typically defined as the average number of further points within h of any given point divided by the overall rate λ, and is most simply estimated via

$$\hat{K}(h) = \frac{1}{\hat{\lambda}N} \sum_r \sum_{s \neq s} \mathbf{1}(|p_r - p_s| \leq h), \tag{1}$$

where $\hat{\lambda} = N/A$ is an estimate of the overall intensity, $\mathbf{1}(\cdot)$ is the indicator function and h is some inter-point distance of interest. As pointed out in [28], one can also estimate $K(h)$ using an estimator for the squared intensity $\tilde{\lambda}^2 = N(N-1)/A^2$:

$$\tilde{K}(h) = \frac{1}{\tilde{\lambda}^2 A} \sum_r \sum_{s \neq s} \mathbf{1}(|p_r - p_s| \leq h). \tag{2}$$

In applications, estimates of K are typically calculated for several different choices of h. For a homogeneous Poisson process, the expectation of $\hat{K}(h)$ is πh^2 (similarly for $\tilde{K}(h)$). Values which are higher than this expectation indicate clustering, while lower values indicate inhibition. However, it should be noted that a point pattern can be clustered at certain scales and inhibitory at others. Note also that two very different point processes may have identical K-functions, as $K(h)$ only takes the first two moments into account. An example of such a situation can be found in [3].

Under the null hypothesis that the point process is homogeneous Poisson with rate λ, $\hat{K}(h)$ is asymptotically normal:

$$\hat{K}(h) \sim N\left(\pi h^2, \frac{2\pi h^2}{\lambda^2 A}\right), \tag{3}$$

as the area of observation A tends to infinity (see p. 642 of [6] or pp. 28–48 of [22]). As is pointed out in [28], it is crucial to use an estimate of λ or λ^2 rather than their true values, even if they are known. Situations where the true intensity is known can arise in simulation studies, where one may feel tempted to plug in the true value for the intensity in (1) or (2). Somewhat surprisingly, however, using the true value for λ or λ^2 will actually inflate the variance of $\hat{K}(h)$ by a factor of $1 + 2\pi h^2 \lambda$ (see [11]).

Several variations on $\hat{K}(h)$ have been proposed. Many deal with corrections for boundary effects, as found in [13, 19, 21]. Variance-stabilizing transformations of estimated K-functions which are more easily interpretable have been proposed (see [5]), such as $\hat{L}(h)$ and $\hat{L}(h) - h$ where

$$\hat{L}(h) = \sqrt{\frac{\hat{K}(h)}{\pi}}. \tag{4}$$

2.2 Definition and Distribution of the Weighted K-function

Suppose that a given planar point process in a connected subset $\mathcal{A}$ of $\mathbf{R}^2$ with finite area A may be specified by its conditional intensity with respect to some filtration on $\mathcal{A}$, for $(x, y) \in \mathcal{A}$ (see [7]). The point process need not be Poisson; in the simple case where the point process is Poisson, however, the conditional intensity and ordinary intensity coincide. Suppose that the conditional intensity of the point process is given by $\lambda(x, y)$.

The *weighted K-function*, used to assess the model $\lambda_0(x, y)$, may be defined as

$$K_W(h) = \frac{1}{\lambda_*^2 A} \sum_r w_r \sum_{s \neq r} w_s \mathbf{1}(|p_r - p_s| \leq h) \tag{5}$$

where $\lambda_* := \inf\{\lambda_0(x, y); (x, y) \in \mathcal{A}\}$ is the infimum of the conditional intensity over the observed region for the model to be assessed and $w_r = \lambda_*/\lambda_0(p_r)$, where $\lambda_0(p_r)$ is the modelled conditional intensity at point p_r.

One can think of the weighted K-function as a combination of Ripley's K-function and the thinning method used for residual analysis in [25]. In [25], $K(h)$ is repeatedly applied to thinned data where the probability of retaining a point is inversely proportional to the conditional intensity at that point. The computation of the weighted K-function $K_W(h)$ uses these same retaining probabilities as weights for the points in order to offset the inhomogeneity of the process. By incorporating all pairs of the observed points, rather than only the ones that happen to be retained after an iteration of random thinning, the statistic $K_W(h)$ eliminates the sampling variability in any finite collection of random thinnings. Indeed, simulations appear to indicate that $K_W(h)$ has approximately the same distribution as $K_M(h)$, the mean of K-functions on a repeatedly thinned point pattern, as the number of random thinnings approaches infinity.

We conjecture that, provided the conditional intensity λ is sufficiently smooth, $K_W(h)$ will be asymptotically normal as the area of observation A approaches infinity. Indeed, for the Poisson case where λ is locally constant on distinct subregions whose areas $A_i^{(n)}$ are large relative to the interpoint distance h_n, we have the following result.

Theorem 1. *Let $N^{(n)}$ be a sequence of inhomogeneous Poisson processes with intensities $\lambda^{(n)}$ and weighted K-functions $K_W^{(n)}$, defined on connected subsets $\mathcal{A}^{(n)} \subset \mathbf{R}^2$ of finite areas $A^{(n)}$. Suppose that for each n, the observed region $\mathcal{A}^{(n)}$ can be broken up into disjoint subregions $\mathcal{A}_1^{(n)}, \mathcal{A}_2^{(n)}, \ldots, \mathcal{A}_{I_n}^{(n)}$ each having area $A_i^{(n)} = A^{(n)}/I_n$, and that the intensity $\lambda_i^{(n)}$ is constant within $\mathcal{A}_i^{(n)}$. Suppose also that for some scalar $\lambda_{\min}$, $0 < \lambda_{\min} \leq \lambda_i^{(n)} < \infty$ for all i, n. In addition, suppose that, as $n \to \infty$, $I_n \to \infty$ and $h_n^2/A_i^{(n)} \to 0$. Further, assume that the boundaries of $\mathcal{A}_i^{(n)}$ are sufficiently regular that the number of pairs of points (p_r, p_s) with $|p_r - p_s| \leq h_n$ such that p_r and p_s are in distinct subregions is small, satisfying $R^{(n)} := \frac{1}{A^{(n)}} \sum_{p_r, p_s} \frac{\mathbf{1}(|p_r - p_s| \leq h_n)\mathbf{1}(i \neq j)}{\lambda_i^{(n)} \lambda_j^{(n)}} \to 0$ in probability as $n \to \infty$, where the sum is over all $p_r \in \mathcal{A}_i^{(n)}, p_s \in \mathcal{A}_j^{(n)}$. Then $K_W^{(n)}(h_n)$ is asymptotically normal as $n \to \infty$:*

$$\frac{K_W^{(n)}(h_n) - \pi h_n^2}{\sqrt{\frac{2\pi h_n^2}{A^{(n)} H\big((\lambda^{(n)})^2\big)}}} \approx N(0,1),$$

where $H\big((\lambda^{(n)})^2\big)$ represents the harmonic mean of the squared intensity within the observed region $\mathcal{A}^{(n)}$.

Proof. We first show that $K_W^{(n)}(h_n)$ can be represented as the arithmetic mean of K-functions computed individually on each of the squares $i = 1, 2, \ldots, I_n$, plus the remainder term $R^{(n)}$ defined above:

$$K_W^{(n)}(h_n) = \frac{1}{\lambda_*^2 A^{(n)}} \sum_r w_r \sum_{s \neq r} w_s \mathbf{1}(|p_r - p_s| \leq h_n) \tag{6}$$

$$= \frac{1}{\lambda_*^2 A^{(n)}} \sum_{i=1}^{I_n} \frac{\lambda_*^2}{(\hat{\lambda}_i^{(n)})^2} \sum_{r_i} \sum_{s_i \neq r_i} \mathbf{1}(|p_{r_i} - p_{s_i}| \leq h_n) + R^{(n)} \tag{7}$$

$$= \frac{1}{I_n} \sum_{i=1}^{I_n} \frac{1}{(\hat{\lambda}_i^{(n)})^2 A_i^{(n)}} \sum_{r_i} \sum_{s_i \neq r_i} \mathbf{1}(|p_{r_i} - p_{s_i}| \leq h_n) + R^{(n)}$$

$$= \frac{1}{I_n} \sum_{i=1}^{I_n} \hat{K}_i^{(n)}(h_n) + R^{(n)} \tag{8}$$

Since the intensity $\lambda_i^{(n)}$ is constant on each square $\mathcal{A}_i^{(n)}$, the weights w_r, w_s assigned to a pair of points in $\mathcal{A}_i^{(n)}$ within distance h_n are each $\lambda_*^2/(\hat{\lambda}_i^{(n)})^2$, which is used in going from (6) to (7). Thus, since $R^{(n)}$ converges to zero in probability by assumption, the distribution of the weighted K-function is equivalent to that of the mean of the I_n ordinary K-functions in (8).

Under the conditions of the theorem, $\hat{K}_i^{(n)}$ is asymptotically normal from [22], and since the point process on $\mathcal{A}_i^{(n)}$ is homogeneous Poisson with rate $\lambda_i^{(n)} \geq \lambda_{\min} > 0$, the variance of $\hat{K}_i^{(n)}$ is bounded above by the variance of a homogeneous Poisson process on $\mathcal{A}_i^{(n)}$ with rate $\lambda_{\min}$. This implies that the collection of random variables $\left\{ \dfrac{\hat{K}_i^{(n)}(h_n) - \pi h_n^2}{I_n \sqrt{Var\big(\hat{K}_i^{(n)}(h_n)\big)}} \right\}$ satisfies the Lindeberg condition (see e.g. [10, p. 98]), and therefore the mean $\frac{1}{I_n} \sum_{i=1}^{I_n} \hat{K}_i^{(n)}(h_n)$ is asymptotically normal. The variance of $K_W^{(n)}(h) = Var\left(\frac{1}{I_n} \sum_{i=1}^{I_n} \hat{K}_i^{(n)}(h)\right) + o(n)$, which can be computed as

$$Var\left(\frac{1}{I_n}\sum_{i=1}^{I_n}\hat{K}_i^{(n)}(h)\right) = \frac{1}{I_n^2}\sum_{i=1}^{I_n}Var\left(\hat{K}_i^{(n)}(h)\right)$$

$$= \frac{1}{I_n^2}\sum_{i=1}^{I_n}\frac{2\pi h^2}{(\lambda_i^{(n)})^2 A_i^{(n)}}$$

$$= \frac{2\pi h^2}{A^{(n)}H\left((\lambda^{(n)})^2\right)}, \tag{9}$$

where (9) follows from the fact that $A_i^{(n)} = A^{(n)}/I_n$.

□

Note that a variance-stabilised version of the weighted K-function can be defined in analogy with (4), namely:

$$L_W(h) = \sqrt{\frac{K_W(h)}{\pi}}. \tag{10}$$

3 Application

The test statistic $K_W(h)$ in (5) is applicable to a very general class of planar point process models. We investigate their application to models for the spatial background rate for the occurrences of Southern California earthquakes.

3.1 Data Set

Data on Southern California earthquakes are compiled by the Southern California Earthquake Center (SCEC). The data include the occurrence times, magnitudes, locations, and often waveforms and moment tensor solutions, based on recordings at an array of hundreds of seismographic stations located throughout Southern California, including over 50 stations in Los Angeles County alone. The catalog is maintained by the Southern California Seismic Network (SCSN), a cooperative project of the California Institute of Technology and the United States Geological Survey. The data are available to the public; information is provided at http://www.data.scec.org.

We focus here on the spatial locations of a subset of the SCEC data occurring between 01/01/1984 and 06/17/2004 in a rectangular area around Los Angeles, California, between longitudes $-122°$ and $-114°$ and latitudes $32°$ and $37°$ (approximately $733\,km \times 556\,km$). The data set consists of earthquakes with magnitude not smaller than 3.0, of which 6,796 occurred within the given 21.5-year period. The epicentral locations of these earthquakes are shown in Fig. 1.

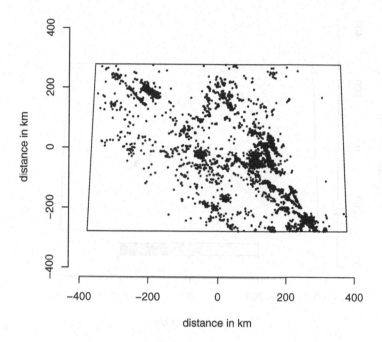

Fig. 1. Earthquakes in Southern California 1984-2004: The data set consists of 6796 earthquakes with magnitude 3.0 or larger

3.2 Analysis

Spatial background rates are commonly estimated by seismologists by smoothing the larger events only. For instance [18] suggests anisotropic kernel smoothing of larger events in order to estimate the spatial background intensity for all earthquakes. In this application, we investigate various spatial background seismicity rate estimates involving kernel smoothings of only the 2030 earthquakes of magnitude 3.5 and higher, by using $K_W(h)$ to assess their fit to the earthquake data set. The local seismicity at location (x, y) may be estimated using a bivariate kernel smoothing $\mu(x, y)$ of the events of magnitude at least 3.5. Figure 2 shows such a kernel smoothing, using an anisotropic bivariate normal kernel with a bandwidth of 8 km and a correlation of -0.611. That is,

$$\mu(x, y) = \sum_{r=1}^{N} f(x - x_r, y - y_r), \qquad (11)$$

where the sum is over all points (x_r, y_r) with magnitude $m_r \geq 3.5$, and f is the bivariate normal density centred at the origin with standard deviation

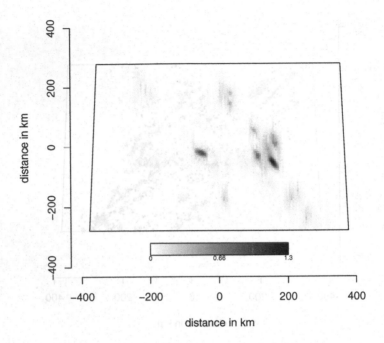

Fig. 2. Kernel smoothing of seismicity in Southern California 1984-2004: An anisotropic bivariate normal kernel with a bandwidth of 8 km ($\rho = -0.611$, $\sigma_x = \sigma_y = 8$ km) is applied to 2030 earthquakes with magnitude not smaller than 3.5

$\sigma_x = \sigma_y = 8$ km and correlation $\rho = -0.611$. This correlation is estimated using the empirical correlation of the values of x_r and y_r, and the bandwidth is selected by inspection. The agreement of Figs. 1 and 2 does not seem grossly unreasonable.

Since such a kernel smoothing uses only the observed seismicity over the last 20 years (a relatively small time period by geological standards), one may wish to allow for the possibility of seismicity in regions where no earthquakes of magnitude 3.5 or higher have recently been observed. One way to do this is by estimating the spatial background intensity via a weighted average of the kernel-smoothed seismicity of magnitude at least 3.5 and a positive constant representing an estimate of the spatial background intensity under the assumption that the process is homogeneous Poisson. Hence we consider the estimate of the form

$$\hat{\lambda}_a(x, y) = a\mu(x, y) + (1 - a)\nu, \tag{12}$$

where $\nu = N/A$ is the estimated conditional intensity for a homogeneous Poisson model and a is some constant with $0 \le a \le 1$.

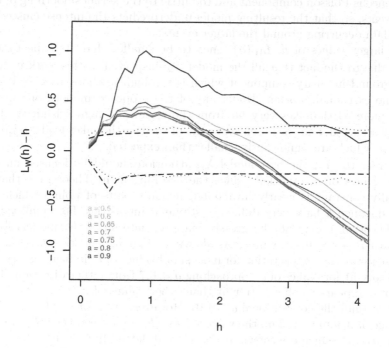

Fig. 3. Weighted L-function for competing models: The difference between the weighted L-function and its expectation h is shown for different values of a in the background intensity model $\hat{\lambda}_a$ as described in (12). The dashed and dotted lines are 95% bounds for $L_W(h) - h$ using model $\hat{\lambda}_{a=0.7}$ based on the theoretical result of Theorem 1 (*dashed*) and simulations (*dotted*)

Instead of plotting the weighted K-function for visual inspection, we will show the difference between $L_W(h)$ as given by (10) and its expectation h, because the latter highlights the departures of the estimate from its hypothetical expectation. Figure 3 shows $L_W(h) - h$ applied to several spatial intensity estimates, each of the form (12), using different values for the parameter a. For the competing estimates $\hat{\lambda}_a$, a takes on the values 0.5, 0.6, 0.65, 0.7, 0.75, 0.8, and 0.9, where a darker line color indicates a higher value of a. The lower values of a give more weight to the homogeneous background rate than higher values of a.

High values of a, such as $a = 0.9$ or greater, fit very poorly to the data, especially for small values of h, as shown in Fig. 3. For such values of a, the intensity estimate gives most of the weight to the kernel smoothing, so that pairs of small earthquakes in areas where there were no earthquakes of magnitude greater than or equal to 3.5 have a very small probability and are hence given enormous weight in the computation of K_W. Similarly, for values of $a = 0.6$ or less, the intensity estimate gives too much weight to the homogeneous Poisson component and too little to the kernel smoothing of the large events, so that the resulting model underpredicts the intense clustering in the data occurring around the larger events.

For larger values of h, $L_W(h)$ tends to be smaller than its expectation. This is due to the fact that all the models $\hat{\lambda}_a$ inspected in this work include a background intensity component which is too high in those areas of Fig. 1 where no earthquakes occur. Under any of the proposed models, one would expect more earthquakes very far from the regions of high seismicity than actually occurred, and the absence of pairs of such earthquakes leads to values of $L_W(h)$ which are significantly smaller than expected.

In order to pick the best model $\hat{\lambda}_a$, attention should be focused on the smaller values of h, especially since the assumption in Theorem 1 that λ be locally constant is clearly invalidated if many pairs of points which are within distance h have very different estimated intensities. For small values of h, Theorem 1 may not be grossly inappropriate since the models for λ are continuous in this example. As shown in Fig. 3, $L_W(h) - h$ seems to decrease towards its expectation for most small values of h, indicating a rather satisfactory fit for values of a approaching $a = 0.7$ from either direction. This value of a appears to offer better fit than other values of a (and certainly is far better than the conventional $a = 1.0$). However, even for $a = 0.7$, for h in the range of 0.3km to 1.3km, the values of $L_W(h) - h$ exceed the 95% bounds for $L_W(h) - h$, which are shown as dashed and dotted lines in Fig. 3.

The dashed lines in Fig. 3 are derived using the result in Theorem 1 for model $\hat{\lambda}_{a=0.7}$. The dotted lines are based on empirical 95% bounds for $L_W(h) - h$ based on 150 simulations of model $\hat{\lambda}_{a=0.7}$. The simulated bounds line up quite well with the theoretical bounds, which indicates that the conditions of the theorem are sufficiently satisfied in our application. In particular, the observed area seems to be sufficiently large, the intensity sufficiently smooth (at least for the values of h used in this work), and boundary effects do not seem to affect the estimation of $K_W(h)$ in any substantial way.

In summary, the data set contains many more small earthquakes in areas far removed from any of the larger events than predicted by a kernel smoothing of the larger events only, and clearly contains much more clustering than would be predicted by a homogeneous Poisson model. However, there is significant short-range clustering of the smaller earthquakes that occur in these locations not covered by the larger events, which explains the positive departure of $L_W(h)$ for small ranges of h. At the same time, the total number of earthquakes occurring in these remote areas is small; that is, the preponderance of these smaller earthquakes are occurring much closer to the large events than one would expect from a homogeneous Poisson process, which explains why $L_W(h)$ is smaller than expected for larger values of h. Although a mixture of a kernel smoothing of the larger events and a homogeneous Poisson background appears to fit much better than either of these individually, no such mixture can thoroughly account for the observed patterns mentioned above.

4 Concluding Remarks

The application of the weighted K-function to spatial background rate estimates for Southern California seismicity shows the power of K_W in testing for goodness-of-fit. The weighted K-function is easily able to detect the major departures from the data for simple kernel or Poisson estimates of the spatial distribution of earthquakes. In addition, even for the optimally-chosen mixture model for the background events, the weighted K-function is able to detect deficiencies and to indicate potential areas for improvement.

K_W has some advantages to alternative goodness-of-fit procedures like thinning or re-scaling, especially in situations where the intensity on the observed region has high variability. For the mixture estimate with a−0.7, for instance, estimates of $\hat{\lambda}_a$ ranged from 0.0049978 to 0.96792. With intensity estimates varying over such a wide range, the application of thinning procedures can by quite problematic. Since the estimated lowest intensity is very low, only very few points will be kept after a random iteration of thinning, which introduces a high degree of sampling variability. Re-scaling procedures, on the other hand, would lead to highly irregular boundaries, which would make it rather difficult to compute any test statistics on the re-scaled process.

In contrast to standard kernel smoothing of the larger events in the catalog, the method of spatial background rate estimation which mixes the kernel estimate with a homogeneous constant rate appears to offer somewhat superior fit to the SCEC dataset. This suggests that spatial background rate estimates in commonly used models for seismic hazard, such as the epidemic-type aftershock sequence (ETAS) model of [18], might possibly be improved in this way as well. Seismologically, the results are consistent with the notion that Southern California earthquakes, though certainly far more likely to occur on known faults, can potentially occur on unknown faults as well, and these faults may be quite uniformly dispersed. The results suggest that a spatial background rate estimate incorporating both of these possibilities could provide improved fit to existing models for seismic hazard. Such a modification may be especially relevant given the occurrences in California of blind (i.e. previously unknown) faults such as the one which ruptured during the Northridge earthquake in 1994, causing at least 33 deaths and 138 injuries as well as extensive public and private property damage [20].

Further study is needed in order to confirm the seismological results suggested herein, for several reasons. First, it remains to be seen whether the features observed here may be reproduced elsewhere or are particular to Southern California. Second, in the estimation of the intensities of the form (12), the bandwidth and choice of kernel were not optimally selected, but chosen rather arbitrarily. Another issue worth mentioning is that the earthquakes of magnitude greater than 3.5 were used both in the fitting and in the testing. This is in keeping with common practice in seismology, though in statistical terms this is certainly non-standard. Also note that the clustering of small earthquakes in areas where the model assigns low intensity, as suggested by the high values of $L_W(h) - h$ for small h in Fig. 3, may or not be causal clustering. That is, these high values of $L_W(h) - h$ may be attributable to clustering of these small earthquakes not accounted for by any mixture model of type (12), or may instead be attributable to inhomogeneity of the process not accounted for by the model. However, the weighted K-function cannot discriminate between these alternatives. It is similarly unclear how robust the estimator $K_W(h)$ is to various departures from our assumptions, and in particular whether the weighted K-function is more or less robust than alternative measures of goodness-of-fit, such as thinned and re-scaled residuals. This is an important subject for future research. In addition, the problem of boundary effects in the estimation of the weighted K-function has not been addressed in this paper. Instead, we have attempted to give a simplified presentation in introducing $K_W(h)$ and its application. It should be noted, however, that exactly the same standard boundary-correction techniques which are used for the ordinary K-function (see Sect. 2.1) can be used for the weighted K-function as well. Fortunately, in our application the fraction of points within distance h of the boundary was so small for all values of h considered as to make such considerations rather negligible.

Acknowledgements

This material is based upon work supported by the National Science Foundation under Grant No. 0306526. We thank Dave Jackson, Yan Kagan, and anonymous referees for helpful comments, and the Southern California Earthquake Center for its generosity in sharing their data.

References

[1] H. Akaike. A new look at statistical model identification. *IEEE Transactions on Automatic Control*, AU–19:716–722, 1974.
[2] A.J. Baddeley, J. Møller and R.P. Waagepetersen. Non- and semi-parametric estimation of interaction in inhomogeneous point patterns. *Statistica Neerlandica*, 54(3):329–350, 2000.
[3] A.J. Baddeley and B.W. Silverman. A cautionary example on the use of second-order methods for analyzing point patterns. *Biometrics*, 40:1089–1093, 1984.

[4] M. Bartlett. The spectral analysis of two-dimensional point processes. *Biometrika*, 51:299–311, 1964.

[5] J.E. Besag. Comment on "modelling spatial patterns" by b.d. ripley. *Journal of the Royal Statistical Society, Series B*, 39:193–195, 1977.

[6] N.A.C. Cressie. *Statistics for spatial data, revised edition.* Wiley, New York, 1993.

[7] D.J. Daley and D. Vere-Jones. *An Introduction to the Theory of Point Processes, 2nd edition.* Springer, New York, 2003.

[8] R. Davies. Testing the hypothesis that a point process is poisson. *Adv. Appl. Probab.*, 9:724–746, 1977.

[9] J. Dijkstra, T. Rietjens and F. Steutel. A simple test for uniformity. *Statistica Neerlandica*, 38:33–44, 1984.

[10] R. Durret. *Probability: Theory and Examples.* Wadsworth, Belmont, CA, 1991.

[11] L. Heinrich. Asymptotic gaussianity of some estimators for reduced factorial moment measures and product densities of stationary poisson cluster processes. *Statistics*, 19:87–106, 1988.

[12] L. Heinrich. Goodness-of-fit tests for the second moment function of a stationary multidimensional poisson process. *Statistics*, 22:245–278, 1991.

[13] Ohser J. and Stoyan D. On the second-order and orientation analysis of planar stationary point processes. *Biometrical Journal*, 23:523–533, 1981.

[14] A. Lawson. On tests for spatial trend in a non-homogeneous poisson process. *Journal of Applied Statistics*, 15:225–234, 1988.

[15] B. Lisek and M. Lisek. A new method for testing whether a point process is poisson. *Statistics*, 16:445–450, 1985.

[16] E. Merzbach and Nualart. D. A characterization of the spatial poisson process and changing time. *Annals of Probability*, 14:1380–1390, 1986.

[17] Y. Ogata. Statistical models for earthquake occurrences and residual analysis for point processes. *Journal of the American Statistical Association*, 83:9–27, 1988.

[18] Y. Ogata. Space-time point-process models for earthquake occurrences. *Annals of the Institute of Statistical Mathematics*, 50:379–402, 1998.

[19] J. Ohser. On estimators for the reduced second-moment measure of point processes. *Mathematische Operationsforschung und Statistik series Statistics*, 14:63–71, 1983.

[20] C. Peek-Asa, J.F. Kraus, L.B. Bourque, D. Vimalachandra, J. Yu and J. Abrams. Fatal and hospitalized injuries resulting from the 1994 northridge earthquake. *International Journal of Epidemiology*, 27 (3):459–465, 1998.

[21] B.D Ripley. The second-order analysis of stationary point processes. *Journal of Applied Probability*, 13:255–266, 1976.

[22] B.D Ripley. *Statistical Inference for Spatial Processes.* Cambridge University Press, Cambridge, 1988.

[23] J.G. Saw. Tests on intensity of a poisson process. *Communications in Statistics*, 4 (8):777–782, 1975.

[24] F.P. Schoenberg. Transforming spatial point processes into poisson processes. *Stochastic Processes and their Applications*, 81:155–164, 1999.

[25] F.P. Schoenberg. Multi-dimensional residual analysis of point process models for earthquake occurrences. *Journal of the American Statistical Association*, 98:789–795, 2003.

[26] G. Schwartz. Estimating the dimension of a model. *Annals of Statistics*, 6:461–464, 1979.

[27] B.W. Silverman. Distances on circles, toruses and spheres. *Journal of Applied Probability*, 15:136–143, 1978.

[28] D. Stoyan and H. Stoyan. Improving ratio estimators of second order point process characteristics. *Scandinavian Journal of Statistics*, 27:641–656, 2000.

Functional Approach to Optimal Experimental Design

V.B. Melas

The book presents a novel approach for studying optimal experimental designs. The functional approach consists of representing support points of the designs by Taylor series. It is thoroughly explained for many linear and nonlinear regression models popular in practice including polynomial, trigonometrical, rational, and exponential models. Using the tables of coefficients of these series included in the book, a reader can construct optimal designs for specific models by hand.

2005. 336 p. (Lecture Notes in Statistics, Vol. 184) Softcover
ISBN 0-387-98741-X

Space, Structure and Randomness
Contributions in Honor of Georges Matheron in the Fields of Geostatistics, Random Sets, and Mathematical Morphology

M. Bilodeau, F. Meyer and M. Schmitt (Editors)

This volume is divided in three sections on random sets, geostatistics and mathematical morphology. They reflect Georges Matheron's professional interests and his search for underlying unity.

2005. 416 p. (Lecture Notes in Statistics, Vol. 183) Softcover
ISBN 0-387-20331-1

Nonparametric Monte Carlo Tests and Their Applications

L. Zhu

A fundamental issue in statistical analysis is testing the fit of a particular probability model to a set of observed data. Monte Carlo approximation to the null distribution of the test provides a convenient and powerful means of testing model fit. *Nonparametric Monte Carlo Tests and Their Applications* proposes a new Monte Carlo-based methodology to construct this type of approximation when the model is semistructured.

2005. 190 p. (Lecture Notes in Statistics, Vol.182) Softcover
ISBN 0-387-25038-7